Professor Sheridan draws you into the biochemical workings and fine-tuned operational controls of this fabulous, adaptive and essential lifelong organ. He finishes with a plea to preserve it and how and why we should.

Mr Gary Brow, General Surgeon, Dublin

This is a most informative and pleasurable book to read, full of details about every aspect of cardiovascular science; presented in a very readable form which leads you effortlessly through the amazing complexity and interconnectivity that has emerged from hundreds of millions of years of evolution. It finishes with one of the clearest discussions of how we can combat the epidemic of heart disease through practical and achievable changes in our diet and lifestyles.

Professor Kim Parker, Mathematician, Imperial College London, UK

This book is a tour de force. The range is awesome, and it is conveyed with an enthusiasm and clarity that make it accessible to a wide range of "lay" readers. Covering not only its evolution and biology but also why "heart" continues to be used as a metaphor since the origins of writing in Mesopotamia, ancient Egypt and China. There is something here to fascinate everyone. Buy it for a family member; buy it for a library; buy it for yourself. It is a delight.

Dr Michael Pike, Neurologist, University of Oxford, UK and Oxford Children's Hospital, UK

Professor Sheridan applies, with great effect, the timeless sequence used to understand complex matters; observation, contemplation, deduction and then, and only then, the masterly explanation. He offers the reader a profound book about the heart presented with a light touch. It will be of interest to many disciplines, well beyond medicine and cardiology. It merits a place on the bookshelf of any enquiring mind imbued with curiosity about this remarkable organ.

Dr Steven Reid, General Medical Practitioner, UK

This book is a critical read for anyone working in this area; whether you have an interest in cardiology, engineering, biology, genetics or the medical sciences you will gain immensely from reading this book and wonder how such an incredible system could ever have come into existence.

Dr Christopher Vincent, Bioengineer, University College London, UK

A very readable and stimulating book, not just for those with an academic or medical background but anyone with an interest in this vital organ; how it evolved and endures to pump about 2 billion times in the average lifetime from its appearance at 18 days in the womb. Be thankful for your heart and enjoy this wonderful book.

Professor John Widdows, Marine Biologist, Plymouth Marine Laboratory, UK

The Heart An Elegant Pump

Its Origins and Partners

The Heart An Elegant Pump

Its Origins and Partners

Desmond J Sheridan
Imperial College London, UK

World Scientific

NEW JERSEY • LONDON • SINGAPORE • BEIJING • SHANGHAI • HONG KONG • TAIPEI • CHENNAI • TOKYO

Published by

World Scientific Publishing Europe Ltd.

57 Shelton Street, Covent Garden, London WC2H 9HE

Head office: 5 Toh Tuck Link, Singapore 596224

USA office: 27 Warren Street, Suite 401-402, Hackensack, NJ 07601

Library of Congress Cataloging-in-Publication Data
Names: Sheridan, Desmond J. (Desmond John), author.
Title: The heart — an elegant pump : its origins and partners /
Desmond J. Sheridan, Imperial College London, UK.
Description: New Jersey : World Scientific, [2023] | Includes bibliographical references and index.
Identifiers: LCCN 2022006741 | ISBN 9781800612501 (hardcover) |
ISBN 9781800612518 (ebook for institutions) | ISBN 9781800612525 (ebook for individuals)
Subjects: LCSH: Heart. | Heart--Physiology. | Heart--Evolution.
Classification: LCC QP111.4 .S54 2023 | DDC 612.1/7--dc23/eng/20220321
LC record available at https://lccn.loc.gov/2022006741

British Library Cataloguing-in-Publication Data
A catalogue record for this book is available from the British Library.

For any available supplementary material, please visit
https://www.worldscientific.com/worldscibooks/10.1142/Q0369#t=suppl

Desk Editors: Soundararajan Raghuraman/Adam Binnie/Shi Ying Koe

Typeset by Stallion Press
Email: enquiries@stallionpress.com

For Jackie
my life's friend

Preface

During the 20th century, diseases of the heart and circulation emerged as the most frequent cause of death in the world. Consequently, heart disease has become the focus of medical and popular attention in recent times. For most of us, however, the heart is an exquisitely efficient and long lasting pump that supports us throughout our lives. This book is about how it does this. How can a heart pump blood for up to 100 years while the best Formula 1 racing engine has a working life measured in hours? Why is the heart so efficient and how is it serviced? We all know that the heart is a pump, but how does it generate the force to do this, and how is it controlled electrically? Unlike many animals, human hearts cannot regenerate when injured, and yet the heart can heal itself in most people following injury such as a heart attack. This book is about how the heart works, why it is so exquisitely matched to the needs of the body and how it can function so well and for so long. These questions are addressed from physiological and molecular perspectives in language that aims to be accessible to all with an interest in biology. The book also features a liberal use of figures to illustrate mechanisms such as the molecular motor that drives the heart and the electrical systems which regulate it.

A constant theme throughout the book is the origins of the heart. Chapter 1 takes a step aside from biology and recognises that for millennia the heart has also been at the centre of human culture and experience. Until the fifth century BCE, it was thought to be the source of reason and emotion and although these functions are now known to originate in the brain, the heart remains deeply embedded in our culture as a metaphor for reason and emotion. This is exemplified by the figure on the front cover showing an image of the heart drawn by the author alongside one by his granddaughter. As a four year old, Alice has no knowledge of anatomy but has already absorbed a lively concept of the heart as a symbol of affection from her cultural environment. In this chapter the use of heart metaphors is traced in modern and ancient languages to show that they have been present throughout human history in all cultures, based on the shared human experience of how emotion is felt.

Modern humans emerged around 200,000 years ago; however, the biological origin of the heart is far older. Mammals have four-chambered hearts pumping two parallel circulations at different pressures. The most critical function of the heart is to deliver oxygen to the organs and tissues of the body and as organisms developed larger and more complex forms their need for energy and therefore oxygen also increased. A wide variety of heart configurations evolved to meet these needs from simple tube hearts in insects, two-chambered hearts in fish and three-chambered hearts in amphibians so that in each species the heart perfectly matches its energy needs. The evolution of all of these is traced back to the oxygenation of the earth's atmosphere which began around 2.5 billion years ago following the onset of oxygenic photosynthesis.

The molecular and genetic origins of the heart are also ancient and can be traced back to the earliest life forms on earth. The molecules of the heart's motor and electrical systems emerged in some of the earliest unicellular organisms and have been conserved over billions of years to function in our muscles. So too, the genes which regulate the early formation of our heart in the womb are found in other animals such as the fruit fly from which our evolutionary path diverged 300 million years ago. The book aims to show that the elegance and efficiency of human hearts are the result of evolutionary steps that allowed our ancestors to flourish and which have been conserved over billions of years to function in our molecular machinery.

A critical feature of the heart's efficiency and longevity is the partnerships in which it functions. The blood vessels which make up the circulation, referred to as its roots and branches exemplify this. Far from being passive tubes, Chapter 7 explains how the structure and function of arteries and veins evolved to exquisitely match to the function of the heart in ways that greatly enhance its performance, reduce its workload and enable its longevity. So too, the composition of blood co-evolved with the heart and circulation; details such as the size and structure of human red corpuscle and the particular molecular conformation of haemoglobin are shown to have evolved to perfectly match them in a way that optimises the uptake and delivery of oxygen.

A broader aim of the book is that by drawing the reader's attention to the astonishing efficiency and elegance of the heart he or she may come to feel that it is worth preserving. Although we are constantly reminded of its presence in response to emotion or exertion, we rarely consider the demands we place on the heart and all too often we act in ways that may harm it. The last chapter is devoted to how we can avoid this by choosing lifestyles that are more in keeping with its evolutionary history and the lifestyle for which it evolved. Since modern humans emerged around 200,000 years ago our genes have not changed in any significant way. However, the lifestyles of people living in high income nations and increasingly in middle and low income nations have changed radically from those of our ancestors. Much of this is harmful and accounts for most cardiovascular deaths worldwide. Reversing this is achievable and up to date evidence in support of healthy dietary modifications and levels of physical activity is discussed alongside the importance of avoidance of tobacco and excess alcohol consumption. Successful implementation of these could prevent around 70% of cardiovascular deaths worldwide.

About the Author

Desmond J Sheridan is Emeritus Professor of Cardiology at Imperial College London. He qualified in medicine from Trinity College Dublin and received his PhD from Newcastle University.

His research interests relate to understanding the consequences of normal and abnormal growth of the heart with a particular focus on disturbances in the electrical properties of heart muscle and the coronary circulation. His recent books have been concerned with the impact of healthcare configurations on clinical research.

Acknowledgements

I am grateful to James Kinnier Wilson and Dr Edward Reynolds who provided invaluable and stimulating advice, comments and discussion in relation to early medical writings.

My thanks also to the library staff at Imperial College, who have always been helpful in sourcing reference materials, and to Imperial Health Charity, for office support.

Contents

Preface		ix
About the Author		xiii
Acknowledgements		xv
Chapter 1	Not Just a Pump: Giving Life to Emotion	1
Chapter 2	Origins of the Heart	49
Chapter 3	Molecular Motors	93
Chapter 4	An Intelligent Pump	135
Chapter 5	A Pump for Life: In Good Times and in Bad	179
Chapter 6	Broken Hearts: Repair and Recovery	229
Chapter 7	Roots and Branches	277
Chapter 8	Intelligent Tubes	323
Chapter 9	Heart and Blood: Co-evolved Intimate Partners	367
Chapter 10	Preserving the Heart	429
Index		467

1

Not Just a Pump: Giving Life to Emotion

The Origin of Heart

This book is devoted to telling the story of the heart as a truly extraordinary pump, one that works efficiently and mostly without our being consciously aware of it throughout our lives. As such, its pages are mostly concerned with biological concepts of how the heart works and how it evolved. This chapter begins by exploring the origins of the word heart and how it came to have such a wide range of metaphorical meanings conveying emotion and feelings. This is highly relevant in a book primarily concerned with the physical heart because some have expressed fears that advances in medical science and cardiac medicine have come to dominate our understanding of the word heart and thereby contributed to undermining those human experiences associated with it. It also matters because emotional experiences such as fear and anxiety are inextricable and inescapable consequences of illness and many may be expressed as heart metaphors. For people who experience them, they can be just as important as the biological manifestations of their ill health. My aim here, then, is to trace the origins of the word heart in a wide range of languages in order to understand how and when it emerged. Two questions arise.

(i) Did metaphorical representations of the heart appear in one or more cultures and then spread to others or did they arise independently as a universal form of expression? The answer to this question will help to clarify whether heart metaphors began as a linguistic/cultural device or if they derived from a shared human experience such as the heart's physical response to emotion.
(ii) Did our early ancestors use the word "heart" to convey the same physical and metaphorical meanings we understand today or has our deeper understanding of cardiovascular biology changed how the word heart is used and understood now?

The word heart has few rivals in terms of the range of meanings it conveys. Its entry in the Oxford English Dictionary exceeds 15,000 words, reflecting an extensive etymology and diversity of uses, especially its role in expressing thoughts and feelings. The English word *heart* is most closely linked to other Northern European languages and like much of modern English can be traced to the arrival of Anglo-Saxon peoples in the seventh century. It shares its origins with the modern word *herz* in German, *hjerte* in Danish, *hart* in Dutch, *hjärta* in Swedish, *hjerte* in Norwegian and *hjarta* in Icelandic. In Old English (before 1100), *heorte* and *hearta* were used and are similar to *herte* and *hirte* in Old Frisian and *hearta* in Old Swedish. Through the Proto-Indo-European root, *kerd*, it is also related to *cor* in Latin, *kardia* in Greek and through them to all the Romance languages, as in *coeur* in French, *corazón* in Spanish, as well as *cuore* in Italian.

For a cardiologist, the word heart most often brings to mind ideas of anatomy, physiology, pathology and therapeutics. However, for all of us the word heart means much more. I have been reminded of this from time to time by patients, some of whom have asked directly if my expertise is limited to the heart as a simple pump or extends to its much wider metaphorical uses and meanings. Patients with heart problems inevitably wish to understand as far as possible the technical nature of their illness; however, many aspects of their experience of it may be more closely related to the metaphorical uses of heart. Feelings of sadness, anxiety and fear can all be expressed using heart metaphors. "Sick at heart" is much more likely to refer to an emotional state than a cardiac diagnosis. For this reason, some patients faced with a clinical encounter, in which "heart" is used in an exclusively technical sense, may feel it lacks empathy for what they are experiencing or to put in another way, "heartless". It is a reminder of how powerful the word is and what it represents for all of us.

Heart Metaphors

For physicians, the anatomical meaning of heart tends to dominate their thoughts for obvious work-related reasons. In a physiological context, the heart is associated with a large body of information recalled on a regular basis in a highly conscious manner. However, in everyday usage, the heart is most often used to express emotions and feelings. We have no difficulty in distinguishing these physical and emotional meanings, relying on the context to indicate and comprehend which is intended. In a similar way, the words "heart" and "cardiac" are freely intermixed in English without any loss of sense. This is especially the case in medicine, reflecting its linkage to classical Greek literature following the Renaissance. We also use many other words with obvious links to the Latin word *cor* for heart. For example, courageous — meaning brave or fearless — would be equally well understood in English as "stout-hearted" or "strong of heart"; and "accord", meaning to agree, also comes from *cor* — literally "to be of one heart". Similarly, "creed", "credulous", "miscreant", "discord", "core", "cordial", and "credit" also derive from *cor* and its root *kerd*. Few words have such a rich variety of meanings derived from a single concept as "heart". It is a remarkable feature of language that we can use all these words in conversations without being conscious of their origin and many meanings. It is also an indication of how central those aspects of human experience, to which heart relates, play in our lives.

This book is primarily intended to discuss and explain how the heart and circulation work and for the most part, it relates to the workings of a pump, the tubes it is connected to and the blood which flows through them. However, the heart has been much more than this in human experience; it has played a central role in human consciousness, particularly our strongly held beliefs, feelings and emotions, often described as "heartfelt". This use of the word heart continues today despite the fact that we now know emotion and reason originate in the brain. It seemed to me important, therefore, to begin with a discussion of the origins of the concept of heart. How did it come to play an important role in how we think about and communicate our emotional lives? How did this role for the heart arise? Is it a feature of some but not all cultures or is it something innately human? These questions are important to everyone. Some have argued that advances in biomedical sciences and a deeper understanding of cardiac biology have contributed to displacing metaphorical meanings previously attributed to heart and as a result have diminished the emotional aspects of human experience [1], leading to a battle for the human heart [2]. Consequently, understanding the origins of "heart" is even more important for those of us steeped in the biomedical world in which the physical heart has such a prominent place.

Heart in European Languages

Romance and Germanic

Language may be considered a system of shared meanings within a culture that supports communication. Within such systems, metaphors are devices which help us to understand one concept in terms of another. Abstract thoughts are unique to humans and as interior conceptions they are exceptionally difficult to communicate. Metaphors help us to do this by providing a system of accepted meanings in terms of more accessible concepts. One of the most widely used examples of this is the concept of love as a journey. Love is a complex human experience; a journey on the other hand is a simpler and a more accessible one. And so, expressions such as "go our separate ways", "marriage on the rocks", "a relationship foundering", and "being at a cross roads" can all convey ideas about love in terms of travel. These metaphors speak about the condition or state of love within a relationship. When it comes to expressing a sense of what love or other emotions feel like, we often use heart metaphors, for example, heartache, broken-hearted, soft-hearted, big-hearted, and strong of heart. These are all based on the concept that our emotions are contained within the heart. Even though we now understand that emotions originate in the brain and not in the heart, we continue to use these metaphors because we accept and value their utility as a system of shared symbolic meanings. Just as love and a journey are quite distinct concepts, we find the latter helpful in expressing a sense of the former. In the same way, it is not necessary that emotions should originate in the heart for it to be metaphorically useful in conveying a sense of them.

In an elegant study [3], Regina Gutiérrez Pérez examined cross-cultural uses of heart metaphors in five European languages; three Romance (French, Spanish and Italian), and two Germanic (English and German). The use of heart metaphors to express a wide range of concepts was compared, such as expressions of emotion like love, sympathy, sadness, kindness, anxiety, desire and courage. This study also examined how heart metaphors are constructed in each language. These included attributing various qualities to the heart; for example, material, such as "gold-hearted" *or* "stone-hearted"; size, "big-hearted" or "small-hearted"; and temperature, "warm-hearted" or "cold-hearted". The study also examined the use of the heart as a living organism as in, "my heart tells me"; as the seat of intellect, "learning by heart"; or as the centre of things, "the heart of the city". Perez found that conceptualisations of the heart were similar in all the languages she studied, the most frequent of which was the concept of the heart as the seat of emotions. Some differences were also noted, which mainly reflected idiomatic expressions in each language, of which a few English examples are "cross my heart", "change of heart" and "his heart is in his boots". However, for most heart metaphors there were corresponding examples in all the five languages studied. This might suggest that they originated from some shared human experience and evolved in each language separately. An alternative possibility is that they originated as linguistic devices in an earlier language from which all of them derived. Although the study was based on languages derived from two sources, Romantic and Germanic, all share a common earlier root. Furthermore, all are likely to share cross-cultural influences from Greek.

Uralic and Basque languages

Europe has rich mix of distinct cultures that embody a strong sense of national identities. Languages are one component of this and while most European languages share a common route, Finnish, Estonian, Hungarian and Basque are distinct from that. Finnish and Hungarian are widely accepted to have originated from a proto-Uralic language around the Ural Mountains and are members of the Uralic family of languages. They are distinct from and have no shared origins with Germanic or Romance languages and yet they too contain metaphors based on their word for heart which are widely used to convey meanings related to emotion and feeling.

The Finnish word for heart is *sÿdan* and it is used in many metaphors which parallel those in English and other languages. For example, *sydänkäpynen* is used for sweetheart or intimate friend, *sydänsuru* means heartache, sadness, and heartbreak and *sydänveri* means to put one's heart into something or with heart and soul. The word for heart also appears in metaphors used to convey time as in *kesäsydän* meaning mid-summer and *sydänyö* meaning midnight. The concept of place can also be expressed using heart as in *sydänalue* meaning the centre of an area and *kynttilänsydän* meaning the wick of a candle. A more extensive list of Finnish examples is presented in Appendix I.

The Hungarian word for heart is *szív* and here also, the word appears in many metaphors which we would recognise in English. For example, *összetört szív* means a broken heart, *elnyeri a szívét* means to win someone's heart, and *a város szíve* can be used for the centre of the city. Hungarian also uses the heart to convey time as in *nyári szív* meaning the middle of summer and *téli szív* means mid-winter. A sense of character can be conveyed using *oroszlán szívű* meaning courageous (literally, heart of a lion) and *szívélyes* meaning kind-hearted. A more extensive list of examples is presented in Appendix II.

Basque, Euskara is spoken in the western parts of the Pyrenees in northern Spain and Southwest France and is believed to predate the arrival of Indo-European languages in Europe. It is an example of a language isolate, meaning one which has no discernable relation to any other language. So, it is not related to the Romance languages used in adjacent areas and therefore is of particular interest to explore whether it also uses heart metaphors.

The Basque word for heart is *bihotz*. Examples of its use for the physical heart include *bihotz taupadak* meaning beating heart and *bihotz-biriketako gaixotasunak* meaning heart and lung diseases. The word heart is also widely used as a metaphor in Basque. Examples which convey emotion include *ene bihotza* meaning beloved or dearest and *bihotz-erdiragarri* meaning heart breaking. Courage can be conveyed using *bihotz eman* meaning to encourage and the feeling of regret and remorse by *bihotz zimiko*. The concept of time can be conveyed using *neguaren bihotzean* meaning mid-winter. For a more comprehensive list of Basque metaphors, see Appendix III.

Once again, in these ancient European languages which arose independently, it seems clear that heart metaphors evolved to convey emotional experiences as well as describing the physical heart.

Heart in Non-European Languages

Japanese has two words for heart. *Shinzou* means the heart as a physical organ, while *kokoro* is used to link the heart with mind and spirit. Heart is also expressed in Japanese calligraphy, known as kanji. Japanese kanji is inherited from Chinese hànzi and both systems use the same character, 心, for this sense of heart. This has the same meaning as *kokoro* meaning heart, mind, and spirit. The corresponding Japanese syllabic word and character for the anatomical heart are *shinzou* and 心臓. The latter comes from 心 meaning heart and 臓 meaning organ. Not surprisingly this system lends itself readily to the use of heart metaphors of which there are many. Some examples are included here. A fuller list is given in Appendix IV. In each case, the syllabic form is given followed by the kanji calligraphic equivalent in parenthesis. Metaphors which convey emotion include *kokoro no soko kara* (心の底から), meaning from the heart and soul or sincerest or deepest feelings and *magokoro* (真心), meaning kindly feeling or cordiality. A sense of character can be conveyed with *futagokoro* (二心) [literally two and heart], meaning duplicity or double dealing, and the concept of location with *shin* (芯), the centre of something as in candlewick or lead of a pencil.

It is clear from these that heart metaphors have an important role in Japanese as we have seen in other languages. It is also evident that there are important cross-cultural influences between Japanese and Chinese and therefore some may have been inherited.

The Uses of Heart in Ancient Languages

Ancient languages provide a means to clarify for how long heart metaphors may have been in use. Ancient languages are generally considered older than fifth century CE and several are still spoken now, among them, Tamil, Korean and Hebrew. Languages are recognised as "living" in that they continue to evolve while in use by inheriting and absorbing influences from speakers in neighbouring regions. Consequently, ancient languages still in use are likely to contain usages derived from a variety of sources over time. Early writings avoid this; the expressions used in them were "frozen in time" when written. In the following, I have chosen to look at heart metaphors in writings from three regions, China, Mesopotamia and Egypt, where some of the earliest known writing has been discovered.

The Heart in Chinese

The Chinese language as spoken and written today is based on a continuous evolution of its calligraphy over more than 2,000 years [4]. In this respect, it differs from other ancient languages such as Sumerian cuneiform script and Egyptian hieroglyphs which have long been extinct. It is widely accepted that Chinese writing evolved from pictograms to modern abstract characters over many iterations, based on a variety of objects unearthed during archaeological excavations. Representation of the word for heart also evolved. Figure 1.1 illustrates the character for heart at some recognised stages during this process.

Jiaguwen: The earliest script form, known as *jiaguwen,* is found on tortoise shells and animal bones from the Shang Dynasty (1711–1066 BCE). However, the characters used appear mature, suggesting that they may have been the result of a significant period of earlier development. These are known as oracle inscriptions and were used for divination. It is estimated that 150,000 such artefacts have been unearthed. Figure 1.1(a) illustrates the word for heart written in *jiaguwen* script.

Jinwen: From the Shang Dynasty to the Zhou Dynasty (1046–771 BCE), *jinwen* script was used to inscribe bronze bells and pots, which were highly

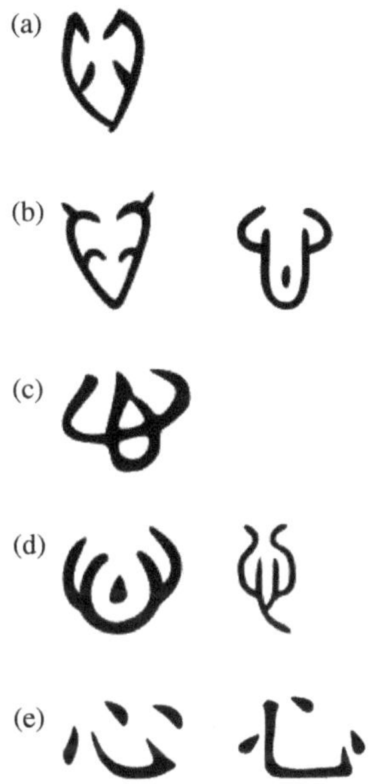

Figure 1.1: Heart pictographs (a–d) and the uniform abstract script (e) for heart used in ancient Chinese texts. (a) Heart pictograph used during the Shang Dynasty (1711–1066 BCE) for oracle bone and shell inscriptions. (b) Heart pictographs on bronzeware (*jinwen*) from the Shang (L) and Zhou (R) Dynasties (1711–771 BCE). (c) Heart pictographs as used on bamboo and silk script, which allowed the creation of longer texts. Bamboo scripts from the fifth century BCE have been unearthed but may have been in use much earlier. (d) Heart pictographs as used in large (left) and small (right) seal scripts, which were used from around 500 BCE–100 CE to confirm identity. (e) Examples of Chinese characters for "heart" in use for more than 2,000 years.

prized and played an important role in court life. Used as musical instruments and for divination, about 9,000 of these have been found which contain the most important preserved texts of the period. *Jinwen* pictograms are considered to have evolved from oracle bone script. Figure 1.1(b) illustrates the heart in *jinwen* script.

Bamboo and silk scripts: Recording more detailed events required larger writing surfaces and for this bamboo or wood strips and silk were used. Bamboo strips have been unearthed from the fifth century BCE but are believed to have been in use much earlier. These allowed the production of long texts on cultural, military and economic subjects. Figure 1.1(c) illustrates the character for heart as written in *bamboo* and *silk scripts*.

Seal script: During the Warring States Period, 457–221 BCE, a time of intermittent war and instability among several states, several scripts were in use. These are generally referred to as large and small *seal scripts*. Figure 1.1(d) illustrates the word for heart in large and small *seal scripts*.

Clerical script: Following the conquest of the other states and founding of the Qin Dynasty (221–206 BCE), all scripts in use were unified under Emperor Qin Shi Huang resulting in a simpler set of characters which were easier to write. Then during the Han Dynasty (206 BCE–220 CE) further refinements occurred with the introduction of "*clerical script*". These characters were abstract forms far removed from the earlier pictograms but were more convenient and faster to write. Figure 1.1(e) illustrates the character for heart in this abstract form and which is still in use today.

The invention of paper, first documented in China during the Eastern Han period (25–220 CE), together with the availability of a simple uniform script allowed for more and much larger documents to be written including the great cultural texts of ancient China and facilitated a wide dissemination of knowledge and bureaucratic regulation. It is interesting to note that early pictograms for heart bear a remarkable resemblance to the structure of the heart and may indicate some knowledge of anatomy as early as the 1700 BCE.

The Heart in Chinese Medicine

There is no evidence of formal studies of anatomy at that time and it seems more likely that knowledge of the heart could have been derived from hunting and warfare. The earliest evidence of human dissection in China is from the Han Annals in a biography of Wang Mang (45 BCE–23 CE). At that time dissection was still frowned on due to Confucian teaching which held that the human body is sacred; "our body with skin and hair comes from our parents. We must not mutilate it". There was nevertheless a real interest and desire to understand more about the internal structures of the body. This is evident from the book, *Huangdi Neijing*, also known as *The Inner Canon of the Yellow Emperor*, which is generally accepted to have been written between 475 and 221 BCE [5].

> It is beyond human capacity to measure the height of the sky or the extension of the earth. But it is easy to obtain surface measurements of a human being who is eight feet tall. After death his body may be dissected to get a general idea of the appearance, the size and the capacity of his viscera, to measure the length of his blood vessels, evaluate the quantity and condition of the blood and the Qi.

Various means were implemented to circumvent the Confucian restriction. Wang Mang did so in 16 CE by having the body of a rebel dissected by a skilled butcher while a court physician observed and noted details of the internal organs. The text tells us that

> Measurements were then made of the internal organs and bamboo rods were inserted into the blood vessels in order to discover where they begin and where they end for the purpose of a better understanding of how to cure diseases.

A similar approach was documented during the Northern Song dynasty (960–1127 CE). A robber known as Ou Xi-Fan and his followers were captured and executed. Within two days they were dissected, and drawings were made of the bodies. The drawings, known as *Ou Xi-Fan's Pictures of the Five Viscera* [6] have remained important for traditional Chinese medicine and beyond. They are likely to have been important for Japanese medicine also as indicated by the appearance of the drawings in 14th century Japanese texts [7]. Figure 1.2 shows a frontal view from these drawings and of interest here is the presence of the heart, clearly marked with the character 心 which has been in use for over two millennia. It seems clear therefore that early Chinese physicians were aware of the heart as a physical organ. Traditional Chinese medicine did not depend on knowledge of physiology and pathology as we understand them now.

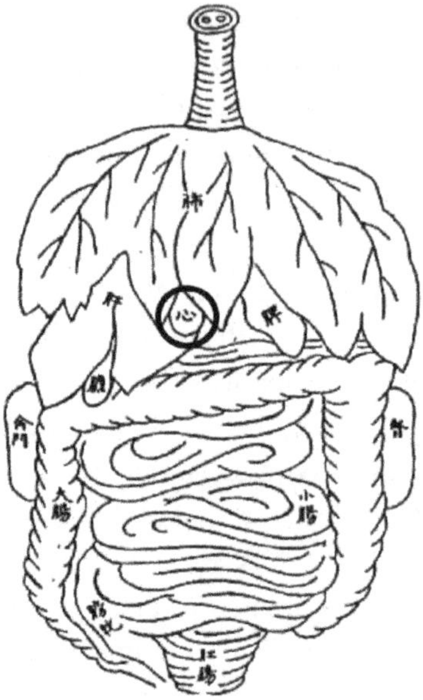

Figure 1.2: Drawing from Ou Xi-Fan's *Five Internal Organs* including the heart (circled) made following the dissection of a group of executed robbers during the Northern Song Dynasty (960–1127).

However, examination of the pulse was important and would have generated much interest in anatomy. We may be inclined to think of traditional Chinese medicine as markedly different from western medicine today however, the latter also began as a blend of science, religion and mysticism. What differs is that traditional Chinese medicine has been practised continuously alongside the growth of western medicine. Uniquely Chinese medicine relied on metaphors of the state to describe the functions of the heart and other organs. In the *Huangdi Neijing,* the work of the organs is considered analogous to that of the state, as in

> The heart is the king who directs the body, the lungs are the promulgators who carry out his orders. The liver is the general whose duty is to meditate carefully. The gall-bladder is the central legal officer, who makes judgments. The pericardium is the minister who brings happiness. The spleen is the officer of granaries who creates the five tastes. The large intestine is the officer of communications. The small intestine is the receiving office in

> which digestion is performed. The kidney is the officer of vigor or strength who serves through his intellect. The three burning spaces constitute the sewage system from which all the canals drain into the bladder where the fluid is stored; after having been acted upon by the air, it is finally passed out.

Although ancient Chinese physicians thought of functions of the body in these metaphorical terms, like their counterparts in Egypt and Mesopotamia, they were aware of the heart as a physical organ and knew of its relation to the pulse which was important in examining patients. Formal dissection for study purposes was restricted from the time of Confucius, however curiosity and a compelling sense of a need to improve treatments led to their circumvention and recording of anatomy.

The Heart in Chinese Culture

As Chinese characters evolved into simpler forms and larger media became available, longer texts began to emerge of which the *Analects of Confucius* and the *Book of Odes* are best known. The latter are believed to have been composed between the 11th and 7th centuries BCE and traditionally have been attributed to Confucius. They broadly consist of eulogies to leaders and ritual songs and have been an important part of classical Chinese education for over two millennia. The *Analects of Confucius* were written by followers of Confucius who wished to record his ideas and sayings. They are believed to have been originally written during the Waring States period (475–221 BCE) but copied subsequently with some amendments until the mid-Han (206–221 BCE). The analects have long been considered of great importance in education and discourse in China and beyond. They offer valuable insights into the use of heart metaphors in ancient China.

The concept of the heart and mind as one is deeply embedded in the *Odes* and *Analects*. In fact, the character 心 which is still used for the anatomical heart also means heart/mind in both texts. The Confucian master Dong Zhongshu (174–104 BCE) expresses this as, "the Odes tell of the aspirations of the heart and mind; therefore, it excels in substance" and Yang Xiong (56 BCE–18 CE) as, "what is divine is the heart/mind (xin, 心) [the single seat of the intellect and emotions]". Not surprisingly, heart metaphors are frequently found in both texts related to emotion and reason.

Heart Metaphors Expressing Emotion

In the *Odes* expressions of love and friendship often relied on heart metaphors and there are many examples of this. When Zhou Meng wished to express his thanks to officials whom he met while travelling on a diplomatic mission for a generous welcome he received, he did so by quoting a stanza from the "Lowland Mulberry" (Ode 228):

> My heart is full of love.
> Why should I not say so?
> The core of my heart will treasure him.
> What day could I forget him?

We are not told whether the author of these lines had intimate or social relations in mind, however the Odes have long been used for education in ethics and behaviour and their use in such a diplomatic manner would not be unusual. The following lines from the "Cypress Boat" (Ode 26) express the intimate feelings of a woman under pressure to marry one man while she loves another:

> My heart is not a mirror, for
> There's more than meets the eye.
> My heart is not a stone, for
> It cannot be turned.
> My heart is not a mat, for
> It cannot be rolled and put away.
> My sorrowing heart wastes away,
> Harassed by hordes of petty men.

And in Ode 147, longing for a lover is expressed as:

> That a mere glimpse of plain leggings
> Could tie my heart in knots!
> Enough! Let us two be one!

In simple terms the *Analects of Confucius* are concerned with imbuing its readers with a sense of ethics, integrity, good behaviour and wisdom.

Heart metaphors again are frequently used but mostly in conveying actions of the mind, reason and will. In the following example, *ren* refers to ethical virtues such as benevolence, integrity, and humaneness and Confucius advises they have great value and should be taken seriously or, as we might express it today, taken to heart.

> If one sets one's heart on ren, there will be none he hates.

Similarly, *Dao*, meaning effectiveness through acquired skills of the mind, is recommended in the following lines from Books 4 and 7:

> If a gentleman sets his heart on the dao but is ashamed to wear poor clothes and eat poor food, he is not worth engaging in serious discussion
>
> and
>
> Set your heart on the dao, base yourself in virtue, rely on ren, journey in the arts.

In the following lines from Book 9, the heart conveys what we think of as the will:

> One can seize the general in charge of the three army divisions, but one cannot seize the heart's intent of a peasant.

In each case the heart is the holder of knowledge and wisdom needed to lead a good life and understanding the power of the heart was considered important. The heart is in fact equivalent to the mind and this is consistent with the fact that the character 心 is still used for both. In ancient China then the heart was thought to be the centre of emotion and reason. Its use as a metaphor for functions of the mind is therefore to be expected and it is remarkable that their usage is quite easily understood today.

Mesopotamia

Sumerian was the language of Sumer, the earliest known civilisation in Southern Mesopotamia in modern South Iraq. Texts, first linear and later in the form of cuneiform writing, emerged there about 3300 BCE. It was also a language isolate, meaning it has no relationship to any other language. Akkadian was a Semitic language in use in Northern Mesopotamia and Sumerian–Akkadian bilingualism was common from around 3000 BCE. Sumerian was gradually replaced by Akkadian from about 2000 BCE. However, Sumerian persisted in sacred texts until the first millennium BCE and Akkadian texts adopted Sumerian cuneiform script. Both were written on wet tablets of clay using wedge-shaped incisions and left to dry in the sun. If intended for temporary use, they could be soaked in water and re-used. If important, they could be fired in kilns producing durable documents. Some have been preserved by unintended heating when buildings or cities were destroyed. At least 500,000 tablets, including many fragments, are estimated to have been excavated in Iraq with collections in many museums around the world. The British Museum holds the largest collection with about 130,000. Although cuneiform tablets were known to exist from the 15th century, deciphering them was not fully achieved until the mid-19th century. Akkadian kings built libraries to preserve these written texts as sources of knowledge, the best known of whom is Aššurbānipal (669–627 BCE). It was during the destruction of Aššurbānipal's palace by the Medes that many of these tablets were preserved by unintended firing. Around 30,000 tablets have now been translated and as a result understanding of Middle East History, which was formerly reliant on biblical texts, has been transformed.

The Heart in Mesopotamian Medicine

Part of that legacy is a rich source of medical texts, the earliest of which is written in Sumerian and dates from the period 2112–2004 BCE. This was not haphazard or fortuitous. These early healers were actively writing texts to preserve knowledge between generations and to make copies to teach students. Following the demise of Sumerian as a spoken language, medical texts continued to be produced by Akkadian speakers who also used cuneiform script derived from Sumerian. They give us a wonderful insight into the extent to which the early Mesopotamian healer (*āšipu*) had developed diagnostic and therapeutic methods, many of which have resonances with modern medicine. From the painstaking work of many scholars [8–10] we know that they had collections of medical instruments, including forceps tweezers, blades and needles. From the beginning of the second millennium BCE, texts began to include signs and symptoms and lists of therapeutic remedies. Contrary to some claims, these texts were not simply copied verbatim until the end of the Mesopotamian civilisation, rather they were amended, added to, and reorganised in specialist series as their knowledge developed.

Mesopotamian healers had no understanding of physiology, pathology or therapeutics as we know them now and omens and gods were likely to be invoked in healing. This has led to them being ignored or denigrated as "sorcerers manuals" and having nothing to teach us. However, the *āšipu* was not a shaman or a priest engaged in religious rituals. They were trained using diagnostic and prognostic manuals written by their predecessors over many generations. They knew about blood vessels but not that blood circulated or that the heart pumped blood. However, there is good evidence that these early healers were aware of the heart and pulse, which they observed carefully. They noted whether veins were engorged or collapsed. They observed irregular heart rhythms, changes in breathing patterns and the presence of oedema (swelling due to water accumulation beneath the skin). The absence of the pulse or a "sleeping" heart was associated with death. The examples given here are taken from the decipherments of Scurlock [10]. In the following, the heart that is awake is a sign of possible recovery:

> If the patient continually turns his head to the left, his hands and his feet are immobilised, he closes and rolls his eyes and spittle flows from his mouth and he makes mumbling noises. If when it afflicts him, his heart is awake, it can be removed. If when it afflicts him, he does not know who he is, it cannot be removed.

An irregular heartbeat is described as shuddering in the following example:

> If his chest is clear, (the blood vessels of) his temple have collapsed, blood incessantly flows from his nose/mouth, and his heart shudders incessantly, he will be worried and die.

There are also many descriptions of crushing chest pain, suggesting the occurrence of angina and coronary heart disease as in the following:

> If his heart continuously crushes him and shudders incessantly, the anger of god and goddesses is upon him.

Congestive heart failure would likely have been common in Mesopotamia. It would have been impossible to diagnose as a distinct illness without auscultation of the lungs and heart. However, Mesopotamians did recognise the prognostic significance of oedema and shortness of breath. Examples of ankle oedema are frequent:

> If a person's feet are continuously swollen
> If the lower parts are swollen.

In the following example the combination of shortness of breath and oedema is an ominous sign:

> If his breath becomes rapid, and his muscles are tense, he will die.

It is evident that ancient Mesopotamian physicians paid close attention to the pulse and examined it at several locations, including the temples, hands and feet. That they were all palpable was considered a good prognostic sign, as in the following example:

> If his head is hot and the blood vessels of his temples, hands and feet all are pulsating together — he will get well.

It seems likely also that the *āšipu* observed veins and particularly if they were collapsed which is referred to as having "gone". In the following example, this would have been a feature of dehydration, blood loss or shock from any cause:

> If he keeps asking for more and more water but his temperature is normal, the blood vessels in his hands "go" and he groans from the beginning of the night until the middle watch —

As these examples show, the primary goal was to recognise the condition in terms of past documented experience and make a prognosis. This has led some to conclude that medicine until modern times had little efficacy [11]. However, this misses the fact that all the achievements of modern medicine are based on past discoveries, including understanding diagnoses based on clinical observations. For example, we now know that a healthy diet, avoiding tobacco and excessive alcohol help to prevent coronary heart disease; but this knowledge is based on research which depended on understanding what coronary disease is and how to diagnose it based on clinical observations and investigations. The idea that modern therapies were developed independently of knowledge and experience gained in the past is unfounded. This includes ancient knowledge from Mesopotamia: modern medicine acknowledges its inheritance from early Greek physicians but Greek medicine may in turn have inherited much from nearby Mesopotamia, although how much is not yet clear.

Thus, Mesopotamians had a very similar concept of the physical heart as we have today. What then of their use of the heart in a metaphorical sense?

Heart in Mesopotamian Culture

The development of medicine was just one aspect of a much wider culture in Mesopotamia and the durable nature of cuneiform writing on clay tablets has left us an astonishing view of their civilisation. Deciphered tablets include accounts of Sumerian and Akkadian myths and epic tales, collections of laws, Akkadian treaties, legal documents, historical texts, rituals, hymns and prayers, proverbs and counsels, observations on life and the world order, theodicy, lamentations, sacred marriage texts and Sumerian love songs. These contain many uses of heart metaphors we can readily recognise today. Some examples are presented here, and a more comprehensive list with source references is given in Appendix V. All are based on translations from Pritchard's *Ancient Near Eastern Texts* [12].

Metaphors to convey emotion:

For him who is the honey of my eye, the lettuce of my heart[i]

---

[i]A love-song for Shu-Sin; Sumerian Sacred Marriage Texts

My precious sweet, having lain by my heart[ii]
Agga the king of Kish, restrained not his soldierly heart[iii]

Metaphors for the mind/soul:

On the road whose traveler returns not, how hath thy heart led thee[iv]
Their hearts having plotted evil[v]

Metaphors for location:

Let us not go down into the heart of the forest[vi]

It is clear from these examples that Mesopotamians had rich emotional lives and that they used metaphors freely. Heart metaphors were especially important in conveying thoughts and feelings [In the following, the sources for quotations discussed can be found in Appendix V]. It is surely remarkable to find such close correspondences with expressions still in use today, for example, "the honey of my eye" and "the apple of my eye". Although we might not say "the lettuce of my heart" we can recognise it as a message of friendship or love and "having lain by my heart" as indicating intimacy. A "joyful heart" or a heart that rejoices conveys happiness, while a heavy heart or an oppressed heart means sadness. A warm heart is used to indicate kindness or sympathy just as it is today. Agga's courage and determination in battle are indicated in his not restraining "his soldierly heart". When the young Gilgamesh claims great strength, courage and valour (then unproven) and that he will vanquish Huwawa, the elders reply with the line "Thou art yet young, Gilgamesh, thy heart has carried thee away". It is a remarkable insight to an intergenerational

---

[ii] The sated lover, *Set Me Free My Sister*; Sumerian Sacred Marriage Texts
[iii] Sumerian myths and epic tales; Gilgamesh and Agga (2000–1500 BCE)
[iv] Sumerian myths and epic tales; Inanna's Descent into the Nether World (2000–1500 BCE)
[v] Akkadian Creation Epic, Tablet I (2000–1500 BCE), Marduk the God of Babylon was created in the "heart" of the primal God Apsu
[vi] The Epic of Gilgamesh, Tablet VIII, Old Babylonian version (2000–1500 BCE)

dialogue from 4,000 years ago. The elders fear Gilgamesh's emotions have taken charge of his reason.

Heart is also frequently used to convey activities of the mind. When Enki is said to know the hearts of the plants planted by the Goddess Ninhursag, it conveys the idea that he has gained some forbidden knowledge. Creation epics frequently describe the act of creation as occurring in the mind (or heart) of the chosen God, as in "In the heart of holy Apsu was Marduk created". The act of thinking is conveyed by heart metaphors as in "hearts having plotted evil", "pondered in his heart" and "the plan conceived in his mind". The heart is also both the source of thoughts expressed and the recipient of advice from others as in "He told him what was in his heart" and "the woman's council fell upon his heart".

We can also recognise the use of heart metaphors to convey location, particularly as the centre of a place as in the "heart of the city" and "the heart of the forest". The "heart of Erech" refers to the centre of the old Sumerian city of Uruk.

There are remarkable similarities between these heart metaphors in use thousands of years ago and how we express ourselves today. We readily recognise the interior meanings they convey. One striking difference is the frequent attribution of emotions and reason to the heart itself. We frequently find the heart directly experiencing the emotion as in "joyful heart", "soldierly heart" or "heart rejoiced". In the case of reason also, the heart itself is often the actor as in "knew in his heart", "how hath thy heart led thee" or "hearts plotted evil". These quotations are from important poems and stories of the period and therefore may have employed a more poetic form of language compared to that in daily use. Nevertheless, in reading them, one is left with a sense that the writer may have believed that the heart was the actual source of these experiences.

It is clear from their writings that Mesopotamians understood the heart in real physical terms, while also able to conceive of it in a metaphorical sense just as we do now. Their epic tales, myths and poems suggest that they had rich emotional lives and enjoyed a strong aesthetic sensibility. But they also engaged in war and experienced battles which would have inflicted devastating personal injuries and there seems little doubt that they were aware of human anatomy by this means. This is evident in their written stories as in, "the arrow cut through her insides splitting her heart" and "when they had slain the bull they tore out its heart". Their medical texts also show that they had a keen awareness of the heart and pulses, which were closely observed in examining patients and that the heart rate and rhythm were important prognostic indicators. In other words, Mesopotamians were much like us today. We may have a much greater understanding of the biology of the heart and circulation, but our concept of the heart in physical and metaphorical terms are very similar.

Ancient Egypt

The discovery of wonderfully preserved mummies and decorated sarcophagi from ancient Egypt stunned the modern world. We are astounded by the achievements of their civilisation. We know much about how they lived and what they believed and understood. Most of this knowledge comes from hieroglyphic writing in tombs and burial goods. The earliest known writing is from the First Dynasty around 3100 BCE. Texts were also written on specially prepared rolls made from the papyrus plant. However, in contrast to the clay tablets used by Mesopotamians, papyri are much more fragile and relatively few remain. Nevertheless, a rich body of writing has been discovered and preserved. Deciphering Egyptian hieroglyphics was attempted as early as the fifth century, but was only achieved in the 1820s, building on previous work and the discovery of the Rosetta Stone. The latter proved crucial because it recorded the same decree from King Ptolemy V in 196 BCE written in three scripts, ancient hieroglyphs, a Demotic Script [developed around 650 BCE] and in ancient Greek. The stone was discovered during Napoleon's campaign in Egypt and Syria but was seized by the British army following the defeat of the French at Alexandria and transported to London in 1802, since when it has been on display at the British Museum. However, copies of it appeared in European museums soon after its initial discovery and led to its decipherment by Jean-François Champollion in Paris in 1822. The Akkadian language of Mesopotamia was known and studied in Egypt as indicated by the presence of cuneiform tablets dating from the 14th century BCE found at Tell El-Amarna. Some cross-cultural influences were therefore likely. Nevertheless, uses of the heart appear in much earlier Egyptian hieroglyphs and would therefore have evolved independently.

The Heart in Egyptian Medicine

Most of the known pharaonic doctors practised during the period of the Old Kingdom, from around 3100 to 2100 BCE. This was the time of the great pyramid building at Giza, which required great advances in intellectual, industrial and social organization. Although, the extant medical papyri were written during the period of the Middle Kingdom, 2040–1795 BCE, they are believed to have been copied from texts written during the Old Kingdom [13]. Medicine was already well developed in Egypt at the beginning of the third millennium BCE. The Ebers papyrus, named after George Ebers who acquired it in 1873, is the largest volume of medical writing from ancient Egypt, consisting of about 110 pages in a 200 m roll. It is currently displayed in the University of Leipzig. From this and other papyri a picture of how medicine was practised emerges.

Ancient Egyptians had no understanding of anatomy or physiology as we know them today. They describe "vessels" known as *metu* (plural of *met*) which they believed connected various parts of the body. *Met* could mean blood vessels and ducts as well as tendons and long slender muscles. They were thought to transport air, blood, mucus, semen and urine as well as harmful and beneficial spirits. Extensive lists of organs and body parts which *metu* connected are given in paragraphs of the papyri known as *The Vessel Book* [13]. These included the heart, throat, bladder, liver, lungs and spleen, testicles, nostrils, eye, anus and various parts of the head. Although these have no correlation with our understanding of anatomy, there are several references to all being connected to the heart. The number of *metu* varies from 12 to 52 in different papyri, possibly reflecting different traditions of practice.

The Egyptians used two words for heart, *ib* and *haty* which were often used interchangeably. The word *ib* was used to describe heart in an emotional context and featured in many compound words to describe feelings, as in *aw-ib*, meaning long heart or joyful heart; *haty* was used exclusively in describing the physical heart and never to convey emotion. The hieroglyph for *ib* was a simple picture of a heart (Figure 1.3). Compounds of *ib* were also used as anatomical terms reflecting the Egyptians' concepts of anatomy. For example, the stomach was written as *r-ib*, literally the mouth of the heart and the two were not always clearly distinguished. This may sound less strange to us now when we recall that the word *kardia* is still used as the anatomical term for the entrance to the stomach and the word stomach is derived from the Greek word for mouth, stoma. Thus, early Egyptians' understanding of anatomy differed greatly from ours today; nevertheless, their language describes both physical and emotional uses for heart.

There is no written evidence to suggest that Egyptians considered the brain as the source of thought, reason and emotion. Indeed, it is almost certain that they did not, bearing in mind their embalming procedure involved removal and discarding the brain. Alcmaeon of Croton, fifth century BCE, is thought to be the first to document the function of the brain [14]. This was well documented in the Hippocratic Corpus, named after Hippocrates but in reality a collection of medical texts from the end

Figure 1.3: The Egyptian ideogram for heart (*ib*) on the upper right beside that for the god Thoth.

Source: By Karen Green, licensed under the Creative Commons Attribution-Share Alike 2.0 Generic (CC BY-SA 2.0), https://commons.wikimedia.org/wiki/File:Flickr_-_schmuela_-_IMG_7153.jpg.

of the fifth and early fourth centuries BCE. In the section on the sacred disease, which we now know as epilepsy, we find:

> Men ought to know that from nothing else but the brain come joys, delights, laughter and sports, and sorrows, griefs, despondency, and lamentations. And by this, in an especial manner, we acquire wisdom and knowledge, and see and hear, and know what are foul and what are fair, what are bad and what are good, what are sweet, and what unsavory; some we discriminate by habit, and some we perceive by their utility [15].

The author goes on to contradict the earlier idea that thought resided in the heart as follows:

> Wherefore, I say, that it is the brain which interprets the understanding. Some say that we think with the heart, and that this is the part which is grieved, and experiences care. But it is not so; only it contracts like the diaphragm, and still more so for the same causes [15].

Earlier doctors' misconception that the heart was the seat of emotion becomes understandable when we realise that they did not practice human dissection but were familiar with the very palpable heart and breathing responses to emotion. It would be wrong to suppose from this that they were incapable of cool reasoning. That they were careful thinkers is clearly shown in *The Edwin Smith Surgical Papyrus* [16]. This was written about 1700 BCE but is also thought to contain copies of earlier texts dating as far back as 3000 BCE. It consists of a series of case descriptions of serious injuries, likely due to injuries sustained during pyramid building and perhaps on battlefields. For example, the corrugated surface of the brain in a case of skull fracture is noted and likened to those which occur on molten copper (case 6). In the same case, breaking of the "skin" covering of the brain results in leaking of fluid from the interior of the head very likely refers to rupture of the meninges and leaking cerebrospinal fluid. In case 8, a fracture of the skull is described in which the author notes that "his eye is askew because of it on the side of him having that injury which is in his skull; he walks shuffling with his sole having that injury which is in his skull".

The ancient Egyptians had no understanding of the circulation as we know it. They did however realise that the heart is connected to the pulses. The *Edwin Smith Papyrus* advises the healer to palpate the pulses by putting "his hands (or) his fingers [upon the head, upon the back of the] head, upon the two hands, upon the pulse, upon the two feet". The presence of a palpable pulse was clearly understood to be related to the heart and its absence noteworthy as in the following text from the Ebers Papyrus, "It means the non-pulsating of the heart, or that the vessels of the heart are silent (not pulsating), there being no indications of them (the "vessels") under thy two hands". In case 1 of the *Edwin Smith Papyrus*, which details the examination of a patient with a severe head wound, examining the pulse is likened to counting with the fingers to know the action of the heart. [16] Here the pulse is examined to learn about the condition of the heart, indicating that they understood that the heart and pulses were connected. Further evidence for this comes from the lines which followed:

> There are canals (or vessels) in it (the heart) to [every] member. Now if the priests of Sekhmet or any physician put his hands (or) his fingers [upon the head, upon the back of the] head, upon the two hands, upon the pulse, upon the two feet, [he] measures [to] the heart, because its vessels are in the back of the head and in the pulse; and because its [pulsation is in] every vessel of every member.

Transliterations of the *Edwin Smith Papyrus* also refer to "counting" in the context of palpating the pulse. Although there is no evidence that they would have had access to clocks to do this accurately, texts such as "counting with a bushel" or "counting something with fingers" may refer to assessing the strength of the pulse or perhaps estimating from experience whether it is fast or slow as a physician would do today. They knew that the action of the heart gives rise to the pulses as expressed in the Ebers Papyrus as the heart speaking: "As to: the heart (*ib*) weakens, it means the heart (*haty*) does not speak, or it means the vessels (*Metu*) of the heart (*haty*) are dumb". Here the word *ib* and *haty* are both used for heart and it appears that in later texts *haty* rather than *ib* was exclusively used for the physical heart. There are many references to the heart in sickness, which are often difficult to interpret based on present knowledge.

Comments such as "his heart is powerless and flooded" have been suggested to possibly refer to congestive heart failure but should be taken with caution.

Given the fragile nature of Papyrus scrolls, that so much has been preserved and translated is remarkable. They do not reveal fully the knowledge and experience of these early healers. However, they do show that they were capable of careful observation, deduction and reasoning. Their understanding of the function of the heart was limited but their concept of it as a physical organ was much the same as ours today. What then of their notions of the metaphorical heart?

Heart in Egyptian Culture

In contrast to the fragile medical texts from ancient Egypt, much of what we know about its culture was engraved in stone and is therefore far more durable. These contain the oldest known Hieroglyphic writings, the earliest of which is the Palermo Stone, written during the Fifth Dynasty (2500–2400 BCE) and contains a list of the kings of Egypt from the First to the Fifth Dynasty. The Pyramid texts consist of a collection of religious writings during the Fifth and Sixth Dynasties of the Old Kingdom from around 2400–2300 BCE and provide valuable information about Egyptian culture of the period. They were carved on the walls and sarcophagi of pyramids at Saqqara, the burial site of the ancient city of Memphis located about 19 miles from modern Cairo. In addition, there are also many extant texts recorded on papyrus, which together give an astonishing view of Egyptian life over many centuries. The style of writing in these is often poetic and rich in its use of metaphors. Of the latter, heart is one of the most frequently encountered if not the most.

Heart Metaphors in Ancient Egypt

Heart metaphors in ancient Egypt convey a wide range of meanings, which is not surprising when we recall that they believed that the heart was the seat of emotion and reason. Some of the earliest examples are found in the story of the creation by Atum of Heliopolis, which was carved around the 24th century BCE inside the pyramids of Mer-ne-Re and Pepi II (Nefer-ka-Re) during the Sixth Dynasty. The act of creation is ascribed to Atum by his "spreading wide his heart (in joy) at his begetting (you) in your name of the nine bows" [17]. It becomes clear just how important the heart was considered in ancient Egypt by the way in which it is embedded in the theology of the Old Kingdom at Memphis. This is documented in the Shabaka Stone:

> Ptah [...] who gave birth to the gods; There came into being as the heart and there came into being as the tongue (something) in the form of Atum. The mighty Great One is Ptah, who transmitted [life to all gods], as well as (to) their kas (=souls), through this heart, by which Horus became Ptah, and through this tongue, by which Thoth became Ptah. Thus it happened that the heart and tongue gained control over [every] (other) member of the body, by teaching that he is in every body and in every mouth of all gods, all men, [all] cattle, all creeping things, and (everything) that lives, by thinking and commanding everything that he wishes [17].

The concept here is that creation took place in the heart (=mind) of Ptah and was brought into being by his speech:

> Indeed, all the divine order really came into being through what the heart thought and the tongue commanded [17].

Some examples of the use of heart for metaphorical meanings in ancient Egypt are included here and a more comprehensive list is presented in Appendix VI.

Metaphors for emotion:

> My heart rejoices [18]
> My heart is distraught [18].

Metaphors for mind and soul:

> Her heart was craftier than a million men [19]
> One skilled in his heart, who can write with his fingers [20].

Although there is no doubt about Egyptians' understanding of the physical heart from the evidence of medical texts, such descriptions of the physical heart are rarely if ever found in non-medical Egyptian texts, perhaps suggesting a distinction between different elements of Egyptian culture and learning. It is as if, beyond the medical world, knowledge of the physical heart and perhaps of the interior of the body generally was limited. In this context, it has been suggested that extant medical papyri may have been copied from an earlier text, which was known to its original users as the "Secret Book of the Physician" [16, p. 10]. And yet preserving the heart at burial was considered important for a good burial and therefore its presence and location would have been well known to embalmers. The following reference makes clear that the heart was preserved, not because of its physical nature but because it was a record of whether the person had lived a virtuous life:

> His reward is a goodly burial for the heart
> which is satisfied with truth [21].

An important aspect of Egyptian culture was the belief that those who lived virtuous and wise lives would enjoy an afterlife. A large proportion of available texts relate to this and are specifically concerned with securing a good after-life for the dead. These are collectively known as the *Book of The Dead.* A particular feature of them is a list of denials of various wrongdoings, which gives a remarkable insight into Egyptian society. The heart played a special place in burial rituals in that unlike the other organs which were removed, it was left in place. This was because it was believed to hold a record of actions during life and would be weighed against the feather of Ma'at (the Goddess who personified Truth) to determine whether the deceased would be permitted to the afterlife (Figure 1.4).

For Egyptians therefore, the heart was of great importance to their concepts of life and death. As early as 3000 BCE they had developed remarkable insights into its role in the body and how to examine its function. Their understanding of its physiology and effective treatments was very limited. It is clear however that they valued the ability to diagnose and prognose patients and at the time these would have had great value.

Figure 1.4: Weighing of the Heart from the *Book of the Dead*. Left, the deceased and his wife enter as the god Anubis weighs the heart of the deceased against the feather of Maat, observed by the goddesses Renenutet and Meshkenet, the god Shay, and Ani's own *ba*. The monster Ammut (right) will devour the soul of the deceased if his heart is unworthy, while Thoth prepares to record the result. At the top are gods acting as judges.

Source: Public domain, https://commons.wikimedia.org/w/index.php?curid=10971230.

Alongside this knowledge, the heart was considered the centre of what was often most important in their lives, such as love, fear, virtue, courage and death. And of course, their ability to hold at one time these two distinct ideas of the heart — one physical and the other abstract — is just the same as we do today.

The Concept of Heart in 21st Century Culture

Advances in medical science have broadened and deepened our understanding of how the heart works. And while there is always much more to learn, cardiac medicine is built on a strong foundation of knowledge about the physical heart and circulation. Nevertheless, we do not need to look far to find evidence of continued interest in symbolic concepts of the heart. As I write, Oxford's famous Ashmolean Museum is holding an exhibition about the history of witches and magic. This has several exhibits related to the heart. For example, it shows actual animal hearts which had been pierced as a protection against evil. A human heart from the 12th century, which had been removed after death and buried separately in a lead box lined with silver reminded viewers of the belief that the soul resided there. There was also an installation by artist Katharine Dawson in which a heart sculpted in glass is used to reflect light in a darkened space [22]. The experience is intended to symbolise and recall the fact that actual hearts were pierced and left in places to protect against witches in the belief that special powers were endowed in it. These of course tell us about concepts and beliefs which were held in the recent and distant past.

Music is a form of language that is found in all cultures and has the capacity to affect us deeply at an emotional level. It enriches our lives and the pleasure it gives has led to interest in the possibility that it may also be beneficial for health and well-being and this has stimulated research to learn more about how its effects are mediated [23]. Music can have a calming or exciting effect on our emotions. Physiological studies have also confirmed that music can influence breathing, heart rate and blood pressure by mechanisms mediated through the autonomic nervous system [24]. The heart, which was once thought to be the source of our emotions, we now know is intimately connected to the brain via the autonomic nervous system and responds to changes in emotion and mood. Thus while the heart does not originate our emotions, it is closely involved in how they are expressed and therefore how we feel and perceive them.

I once attended a concert given by the Sacconi String Quartet in Oxford. A flyer handed to members of the audience mentioned a project called "Heartfelt" [25] which the quartet had performed on several occasions. The idea behind it was to give the audience a sense of what it feels like to be a musician as they performed Beethoven's string quartet in A minor. During the concert each player wears a device which records the heartbeat and each beat is transmitted by a wifi signal to devices that create a palpable pulse with each beat. Members of the audience are given one of these devices to hold during the concert and can "feel" how the heart rhythm of one of the players responds to playing Beethoven's quartet. Since music is primarily an auditory and to some extent a visual experience, awareness of the players' heart rhythm may not add greatly to it. However, music can have a powerful effect on how we feel and in this case the players are inviting the audience to share their experience of performing the quartet by feeling how their hearts responded to it. This is of special interest here because it relates directly to the question of why the heart became such a powerful symbol of emotion for all humans. The players evidently are moved by their playing of Beethoven and believe that changes in their heart rhythm are the easiest way to convey this to the audience. In other words, the heart responds to emotion in a way we can readily feel, which led our ancestors to believe it must be where it originates. Today, despite all we know of the brain as the centre of emotion, the heart's response to it is a fundamental part of what we feel and why we continue to think of the heart in a physical and metaphorical way.

Heart in Religion

Heart has long been used as a metaphor and ikon in religion. It is perhaps most widely known for its frequent use as a metaphor for human love in the Song of Solomon from the Hebrew Bible, as in the lines:

> Set me as a seal on your heart,
> as a seal on your arm;
> for love is strong as death
> [...]
> You have ravished my heart, my sister, my bride.
> You have ravished my heart with one of your eyes,
> with one chain of your neck
> […]
> My heart pounded for him.
> I rose up to open for my beloved.
> My hands dripped with myrrh,
> my fingers with liquid myrrh,
> on the handles of the lock.

Although attributed to Solomon it is widely accepted to have been written much later. Its language is Aramaic which gradually replaced Hebrew after the end of the Babylonian exile in the sixth century BCE and based on its style vocabulary and syntax is most probably from the third century

BCE. Although firmly embedded in Judeo-Christian theology, some have noted possible earlier influences from Mesopotamian love poetry [26], which may have arisen during the period of Babylonian captivity. For example, the use of the word "sister" for the lover in the Song of Solomon:

> You have ravished my heart, my sister, my bride.
> You have ravished my heart with one of your eyes,
> with one chain of your necklace.
> How beautiful is your love, my sister, my bride

is also found in the Sumerian Love Poem, *Set Me Free My Sister*; the sated lover:

> Set me free, my sister, set me free,
> Come, my beloved sister, I would go to the palace,
> You will be a little daughter before my father.

To this day, the most frequently used religious heart metaphor and ikon is that of the Sacred Heart of Jesus. Representing the love of Jesus for humanity, it is revered by some Christian denominations, but mostly notably by Roman Catholics. Its origin is attributed to St Gertrude, a 13th century mystic and prolific writer. It was later popularized in the 17th and 19th centuries in France and Portugal and today an image of the Sacred Heart can be found in most Roman Catholic Churches. That the heart rather than the brain of Jesus is the centre of this devotion of course raises the question as to whether it also arose out of the seemingly universal ancient concept of the heart as the seat of the mind and soul.

Another recent religious event also serves to remind us of this. St Laurence O'Toole was made Archbishop of Dublin in 1162–1180 and has long been the patron Saint of the city. He was an important reformer at the time and was canonised by Pope Honorius III in 1225. His heart was preserved as a relic in a lead box at Christ Church Cathedral and venerated for 800 years. However, it was stolen in October 2012 and only recovered in April 2018 after a lengthy investigation. Its return to the cathedral was marked by a joyous service which as the Cathedral's Dean put it, expressed "in some way the emotion and feelings felt by all of us on the return of such a precious object. I feel the heart has come back into this cathedral". Such events remind us that however much we learn from scientific discoveries metaphor and allegory remain highly significant in human experience.

Scepticism about what is unproven is a guiding principle of the methods of science whereas the nature of human experiences and why they occur are more prominent in religion. A recent case exemplifies the continuing importance of religion for many people and difficulties which may arise when religion and science interact. In October 2008, a consecrated communion wafer [and therefore believed by Roman Catholics to have become the body of Christ] fell during distribution. A week later it was found to have a discoloration on it and a little over two months later it was examined by two histopathologists who reported that the material had the features of human myocardium. Although not formally accepted by the Church as such, the event is now widely touted as a miracle [27] with the backing of science, to the delight of some and dismay of others. However, it fails to comply with even the most basic standards of science. The report was not peer-reviewed, the evidence has not been made public, there is no possibility to confirm it and the many questions which arise from it remain unanswered. If myocardium is present, DNA analysis would clarify whether it is human or animal in origin, what its gender was and its ethnic origin. Whatever the truth of this case, claims of scientific evidence to support it are unjustified. One might be tempted to speculate whether the stain appeared as heart muscle rather than brain or other tissue may somehow be linked to the ancient concept of the heart as the centre of our being; the physical realisation of the metaphor?

The Physical and Metaphorical Heart: Why We Need Both

There are important reasons why science, especially medical science, and religion might benefit from a closer engagement, not least because for many patients, faith is an important aspect of their lives. A recent joint commission of medical scientists and multi-faith religious leaders recognised this and the fact that each has much to learn from the other. Science has the possibility to remove superstition from religion while religion can help science to retain a sense of values in its search for truth [28].

My interest in exploring the origins of heart arose from claims [1, 2] that medical science is undermining the quality of human experience by its emphasis on the heart as a physical organ rather than a symbol of the mind, emotion and human feelings, qualities which are linked to faith. It is clear however that the concepts we attribute to the heart today both physical and metaphorical originated independently in ancient cultures and their meanings have not changed irrespective of scientific advances. What has changed of course is our understanding and perception of the world around us and how we live in it. But the kinds of inner experiences we have of the world, our reactions to it and to each other have not changed and for these we still find heart metaphors most appropriate. However, as scientific knowledge and understanding has expanded and re-shaped the environment it has done so in a way which requires much greater technical attention. We spend more time attending to the physical world and have less time and space to reflect on what we feel and think about it. As rational and emotional beings we have need for both and what has been described as the battle for the human heart [2] and a conflict between the physical and metaphorical meanings of heart [1] are, I suggest, themselves metaphors for that tension.

The advances in medicine which have transformed the treatment of many diseases in the past two centuries have to a large extent focused on physical pathology. As I write, mental illness is often seen as the Cinderella of health care with poor funding, long waiting lists and poor outcomes. The lived experiences of illness and approaching death are of immense importance for patients and their families but can all too easily get lost in the busy technical environment of modern medicine. Loneliness and isolation are increasingly recognised as important contributors to ill health and increased mortality. They are also associated with higher rates of heart disease [29]. These aspects of ill health frequently involve experiences which we might well use heart metaphors to describe such as broken-hearted, down-hearted and cold-hearted. It is important for all who work in healthcare to be aware of such expressions as part of the journey all patients must take and to cultivate the skill of reading how others feel [30, 31]. This includes those expressions most often described by heart metaphors. Indeed the physical and metaphorical heart is important for us all, we need both.

Conclusion

Heart metaphors seem to have been present in all human cultures since ancient times and used for largely similar purposes. This means that they are unlikely to have been inherited from a single source but rather to have emerged universally and independently at different times. It also suggests that the popularity and widespread use of heart metaphors were not due to their value in expressing feelings and emotions alone, but because they originated in a shared human response to these experiences. The heart's reaction to emotion is a sensation we all recognise and offers the best explanation for our ancestors' belief that their feelings arose from it. Thus although the heart is not the originator of our emotions but it gives life to them in how it responds to them. Heart metaphors therefore help to remind us that we are emotional creatures and that being conscious of this in ourselves and others is important and especially for those who work in healthcare.

References

[1] Alberti, FB. Heartfelt emotions. *Lancet*. 2009;374:520–521.

[2] Hoystad OM. The battle for the Western heart. In: *A History of the Heart*. London: Reaktion Books Ltd; 2007.

[3] Pérez RG. A cross-cultural analysis of heart metaphors. *Revista Alicantina de Estudios Ingleses*. 2008;(21):25–56.

[4] Wei Be. The origin and evolvement of Chinese characters. Available from: http://www.ejournals.eu/pliki/art/3137/. (Accessed on 24/1/2018).

[5] Schnorrenberger CC. Anatomical roots of acupuncture and Chinese medicine. *Schweiz Z Ganzheitsmed*. 2013;25:110–118.

[6] Shiying J. Five internal organs. [Wuzàng tù kao]. *Chinese Journal of Medical History*. 1994;24(2):68–77.

[7] Ogawa T. The beginnings of anatomy in Japan. *Okajimas Folia Anatomica Japonica*. 1975;52:59–72.

[8] Kinnier Wilson JV. Organic diseases of ancient Mesopotamia, and mental diseases of ancient Mesopotamia. In: Brothwell D, Sandison AT, editors. *Diseases in Antiquity: A Survey of the Diseases, Injuries and Surgery of Early Populations*, Chaps 15 and 56. Springfield, Illinois: Charles C Thomas; 1967.

[9] Finkel IL. Adad-apla-iddina, Esagil-kin-apli, and the series SA.GIG. In: Leichty E, Ellis M, Gerardi P, editors. *A Scientfic Humanist: Studies in Memory of Abraham Sachs*. Philadelphia: The University Museum; 1988, pp. 143–159.

[10] Scurlock JA, Anderson BR. *Diagnosis in Assyrian and Babylonian Medicine*. University of Illinois; University of Illinois Press, Champaign, IL, 2005, p. 2.

[11] Wooten D. *Doctors Doing Harm Since Hippocrates*. Oxford University Press; Oxford University Press, Oxford, 2007.

[12] Pritchard JB. *Ancient Near Eastern Texts*. Princeton University Press; Princeton, NJ, United States, 1969.

[13] Nunn J. *Ancient Egyptian Medicine*. The British Museum Press, British Museum, London, 1996.

[14] Gross CG. Neuroscience, early history of. In: Adelman G, editor. *Encyclopedia of Neuroscience*. Basel: Birkhauser; 1987, pp. 843–847.

[15] Hippocrates, on the sacred disease. Translated by Francis Adams. Available from: http://www.public-library.uk/ebooks/84/26.pdf. (Accessed on 15/10/2018).

[16] *The Edwin Smith Surgical Papyrus*. Translated by JH Breasted. Available from: https://oi.uchicago.edu/sites/oi.uchicago.edu/files/uploads/shared/docs/oip3.pdf. (Accessed on 15/10/2018).

[17] J A Wilson. The theology of Memphis, Egyptian myths, tales, and mortuary texts. In: Pritchard JB, editor. *Ancient Near Eastern Texts*. Princeton University Press; Princeton, NJ, United States, 1969.

[18] J A Wilson. The contest of Horus and Seth, heroic tales; exploits of gods and human beings. In: Pritchard JB, editor. *Ancient Near Eastern Texts*. Princeton University Press; Princeton, NJ, United States, 1969.

[19] J A Wilson. The repulsing of the dragon, heroic tales; exploits of Gods and human beings. In: Pritchard JB, editor. *Ancient Near Eastern Texts*. Princeton University Press; Princeton, NJ, United States, 1969.

[20] J A Wilson. The legend of the possessed princess, heroic tales; exploits of Gods and human beings. In Pritchard JB, editor. *Ancient Near Eastern Texts*. Princeton University Press; Princeton, NJ, United States, 1969.

[21] J A Wilson. Hymns to the Gods as a Single God, Egyptian Hymns and Prayers. In Pritchard JB, editor. *Ancient Near Eastern Texts*. Princeton University Press; Princeton, NJ, United States, 1969.

[22] Birkwood, K. https://www.museumsassociation.org/museums-journal/reviews/2018/12/01122018-spellbound/.

[23] Music and wellbeing. Available from: https://digitalmedia.sheffield.ac.uk/channel/Music+and+Wellbeing+Research+Centre/82867491. (Accessed on 5/11/2018).

[24] Sleight P. A historical perspective on peripheral reflex cardiovascular control from animals to man. *Experimental Physiology*. 2014;99(8):1017–1026.

[25] Available from: http://www.heartfelt.org.uk/. (Accessed on 2/11/2018).

[26] Exum JC. Song of songs. In: Newsom CA, Ringe S, Lapsley J, editors. *Women's Bible Commentary*. Westminster John Knox Press; 2012. ISBN 9780664237073.

[27] Available from: http://www.therealpresence.org/eucharst/mir/english_pdf/Sokolka1.pdf and http://www.therealpresence.org/eucharst/mir/english_pdf/Sokolka3.pdf. (Accessed on 8/11/2018).

[28] Horton R. The fact of faith. *Lancet*. 2017;389:724.

[29] Leigh-Hunt N, Bagguley D, Bash K, Turner V, Turnbull S, Valtorta N, Caan W. An overview of systematic reviews on the public health consequences of social isolation and loneliness. *Public Health*. 2017;152:157–171.

[30] Comer Kidd D, Castano E. Reading literary fiction improves theory of mind. *Science*. 2013;342:377–380.

[31] Marchalik D, Jurecic A. Novel remedies. *Lancet*. 2015;386:1223.

Appendix I

Finnish Metaphors

Metaphors which use the Finnish word for heart, "sÿdan".

To convey emotion:

sydämensivistys	wisdom, maturity, cultured, educated
sydänsuru	heartache, sadness, heartbreak
sydänkäpynen	sweetheart or intimate friend
sydänystävä	soul mate, bosom friend, life-long friend
särkynyt sydän	broken heart, sense of loss and grief
sydämenasia	strong feelings about something, close to the heart, best interests at heart
sydänveri	[literally "heartblood"], to do with supreme effort

To convey concepts of time:

kesäsydän, sydänkesä and sydänsuvi	mid-summer, the heart of summer
sydäntalvi and talvisydän	mid-winter or the heart of winter
sydänyö and yösydän	midnight or middle of the night

To convey concepts of place or location:

sydänpuu	heart wood of a tree
sydänhalkeama	a shake or crack in the heart wood
sydänalue	the centre of an area
pähkinänsydän	kernel or the edible part of a nut
kynttilänsydän	the wick of candles

Appendix II

Hungarian Metaphors

Metaphors which use the Hungarian word for heart, "sziv".

To convey emotion:

szívfájdalom	heartache
szívszorító	heartbreaking
szívből jövő	heartfelt
összetört szív	broken heart
elnyeri a szívétwin	someone's heart
torkában dobog a szíve	heart beating in one's throat, meaning fear
elszorul a szíve	one's heart tightens, meaning sadness
valami nyomja a szívét	something is pressing your heart, meaning guilt
elnehezedik a szíve	one's heart becomes heavy, meaning sad

To convey emotion as material:

aranyszív	heart of gold, meaning very kind, amiable
kőszív	heart of stone, meaning without feeling
uvegszív	heart of glass, meaning very unsympathetic

To convey place, location:

a város szíve	the city centre

To convey qualities of character:

oroszlán szívű	heart of a lion, meaning brave
szívélyes	kind-hearted
jégszívű	cold-hearted, meaning indifferent, unsympathetic
melegszívű	warm-hearted, meaning kind
szívélyes	warm-hearted, meaning kind, generous
szívtelen	heartless, meaning cruel, unkind
szívéből szól	speak from the heart, meaning sincerely, honestly

a szívem azt súgja	my heart whispers to me, meaning deeply felt
szívből jövő	comes from the heart, meaning sincerity

To convey concepts of communicated emotion:

szívébe vési	to carve it in his heart, meaning to be deeply affected by something
szívhezszóló	it speaks to the heart, meaning it makes a deep impression

Appendix III

Basque metaphors

Metaphors which use the Basque word for heart, "bihotz".

To convey emotion:

ene bihotza	beloved or dearest
bihotz-bero	lover
bihotzezko maitasun egiazkoa	true love or from the heart
bihotz-erdiragarri	heart breaking
bihotz-berri zaitezte	happy, rejoicing
bihotz-bigun	soft-hearted or kind
bihozbera	sentimental
poematxo bihotz-ukigarria	poem that touches the heart
bihotzezko	deeply felt, genuine

To convey courage:

bihotz-altxagarri	encouraging
bihotz eman	to encourage
bihotz-galdu	losing courage
bihotz hartu	take courage

To convey concepts of location or time:

lurraren bihotzeraino	literally the heart of the earth or heart of the matter
neguaren bihotzean	literally the heart of winter or mid-winter

To convey feelings of regret, guilt, remorse:

bihotz damu	to repent, to be remorseful
bihotz-garbi	pure heart, without guilt
zorionekoak bihotz-garbiak	blessed are the pure of heart
bihotz zimiko	remorse, regret

Appendix IV

Japanese metaphors

Metaphors using the Japanese word for heart, "kokoro" and the calligraphic equivalent, 心 in parenthesis.

To convey emotion:

kokoro no soko kara (心の底から)	from the heart or soul; with sincerest or deepest feelings
kokoro o komete (心を込めて)	with all one's heart
kokoro kara yorokonde (心から喜んで)	from the heart, with pleasure
magokoro (真心)	[literally "to think" and heart] meaning kindly feeling, cordiality, heartiness
mi no kokoro mo (身も心も)	with all one's affections, one's whole being

To convey qualities of character:

futagokoro (二心)	[literally "two" and "heart"] meaning duplicity, double dealing
ryooshin (良心)	[literally "good" and "heart"] meaning moral sense, conscience
kokoro no soko de (心の底で)	one's innermost thoughts and feelings; a person's character; secretly; in reality
nesshin da (熱心だ)	to be deeply interested in something

To convey concepts of location:

chuushin (中心)	the innermost part or central part of anything
shin (芯)	the central part of something e.g., wick, marrow; lead for pencil
jushin (樹心)	heartwood
kakushin (核心)	essence, essential, vital

Appendix V

Metaphors used in Mesopotamian culture based on translations from Pritchard's *Ancient Near Eastern Texts* [12].

Metaphors for emotion — joy, fear, courage, anxiety, happiness:

"For him who is the honey of my eye, the lettuce of my heart"[i]
"My precious sweet, having lain by my heart"[ii]
"with a joyful heart opened the door of the house"[ii]
"Agga the king of Kish, restrained not his soldierly heart"[iii]
"Who has heart, let him stand up"[iv]
"At the word of the men of the city his heart rejoiced, his spirit brightened"[iv]
"In my city man dies, oppressed is the heart,
Man perishes, heavy is the heart"[iv]
"Then did the heart of Gilgamesh take pity on"[v]
"Wide his heart, warm his sympathy"[v]
"Thou art yet young, Gilgamesh, thy heart has carried thee away"[vi]

Metaphors for mind/soul:

"Of the plants, [Enki] decreed their fate, knew their heart"[vii]
"On the road whose traveler returns not
How hath thy heart led thee"[viii]
"In the heart of holy Apsu was Marduk created"[ix]

---

[i] A love-song for Shu-Sin; Sumerian Sacred Marriage Texts

[ii] The sated lover, *Set Me Free My Sister*; Sumerian Sacred Marriage Texts

[iii] Sumerian Myths and Epic Tales; Gilgamesh and Agga (2000–1500 BCE)

[iv] Gilgamesh in the Land of the Living (2000–1500 BCE)

[v] Akkadian Creation Epic, Tablet VI (2000–1500 BCE)

[vi] The Epic of Gilgamesh, Tablet III, Old Babylonian version (2000–1500 BCE)

[vii] Sumerian Paradise Myth of Enki and Ninhursag (2000–1500 BCE). Enki tasted the plants, which the Goddess Ninhursag planted and "new" them. Here the implication is that by tasting them Enki hoped to gain a deeper knowledge of Ninhursag

[viii] Sumerian Myths and Epic Tales; Inanna's Descent into the Nether World (2000–1500 BCE)

[ix] Akkadian Creation Epic, Tablet I (2000–1500 BCE). Murdak the God of Babylon was created in the "heart" of the primal God Apsu

"Their hearts having plotted evil" [ix]
"Having pondered in his heart[x]
He told him what was in his heart"[x]
"Thou hast charged thine own heart to stir up conflict"[xi]
"To impart the plan he had conceived in his heart" [xii]
"He hearkened to her words, approved her speech;
The woman's council fell upon his heart"[xiii]
"He released the arrow, it tore her belly, It cut through her insides, splitting the heart" [xi]
"When they had slain the Bull, they tore out his heart,
Placing it before Shamash"[xiv]
"He touched his heart, but it does not beat"[xiv]
"and their hearts began to pound, terror fell upon them"[xv]

Metaphors for location:

"The purified palace … the heart of Erech" [xvi]
"Let us not go down into the heart of the forest"[xvii]
"If a slave-girl or slave of a man has fled into the heart of the city"[xviii]

---

[x] Akkadian Creation Epic, Tablet II (2000–1500 BCE)
[xi] Akkadian Creation Epic, Tablet IV (2000–1500 BCE)
[xii] Akkadian Creation Epic, Tablet VI (2000–1500 BCE)
[xiii] The Epic of Gilgamesh, Tablet II, Old Babylonian version (2000–1500 BCE)
[xiv] The Epic of Gilgamesh, Tablet VI, Old Babylonian version (2000–1500 BCE)
[xv] Assyrian Historical Texts of the Reign of Sargon II (721–705); The Fall of Samaria
[xvi] Sumerian Myths and Epic Tales; The Death of Gilgamesh (2000–1500 BCE)
[xvii] The Epic of Gilgamesh, Tablet VIII, Old Babylonian version (2000–1500 BCE)
[xviii] Collections of Laws from Mesopotamia and Asia Minor Lipit–Ishtar Law code (circa 1900 BCE)

Appendix VI

Metaphors used in Ancient Egyptian culture based on translations from Pritchard's *Ancient Near Eastern Texts* [12].

Metaphors for emotion — joy, fear, courage, anxiety, happiness:

"his heart was very wretched"[xix]
"My heart rejoices"[xix]
"Their Hearts were very pleased"[xx]
"My heart is distraught"[xix]
"Hearts were in mourning"[xxi]
"Their Hearts were very sad"[xx]
"Sing to him, do not let his heart take on cares"[xxii]
"when the princes of Kadesh see him,
the terror of him is in their hearts"[xxiii]
"then we shall fight against these foreigners,
then we shall not trouble our hearts [about] the rear of our army"[xxiv]
"Their battle cry in their hearts was as one"[xxv]
"Valiant warrior in the heart of the battle"[xxv]
"She was beautiful in the heart of his majesty,
and he loved her more than anything"[xxiii]

Metaphors for mind and soul:

"Her heart was craftier than a million men"[xxvi]
"I lifted up my heart and collected myself"[xxi]
"I found favor in his heart"[xxi]
"it was not planned, it was not in my heart"[xxi]

[xix]The Contest of Horus and Seth, Heroic Tales; Exploits of Gods and Human Beings
[xx]The Story of the Two Brothers, Heroic Tales; Exploits of Gods and Human Beings
[xxi]The Story of Si-Nuhe, Heroic Tales; Exploits of Gods and Human Beings
[xxii]The Journey of Wen-Amon to Phoenicia, Heroic Tales; Exploits of Gods and Human Beings
[xxiii]Beth-Shan Stelae of Seti Iand Ramses II, Egyptian Historical Texts
[xxiv]The Annals in Karnak, The Asiatic Campaigns of Thut-mose III, Egyptian Historical Texts
[xxv]Texts from the Tomb of General Hor-em-heb, Egyptian Historical Texts
[xxvi]The Repulsing of the Dragon, Heroic Tales; Exploits of Gods and Human Beings

"One skilled in his heart, who can write with his fingers"[xxvii]
"Perception is his heart, Command is his lips"[xxviii]
"Thou shouldst not express thy (whole) heart to the stranger,
to let him discover thy speech"[xxix]
"heart's affliction from a very great evil"
"He is the herdsman of all men.
Evil is not in his heart"[xxx]
"His reward is a goodly burial for the heart
which is satisfied with truth"[xxxi]

[xxvii] The Legend of the Possessed Princess, Heroic Tales; Exploits of Gods and Human Beings

[xxviii] Hymn to Amon-Re, Egyptian Hymns and Prayers

[xxix] The Instruction of Ani, Egyptian Instructions, Didactic and Wisdom Literature

[xxx] The Admonitions of Ipu-Wer, Egyptian Oracles and Prophecies

[xxxi] Hymns to the Gods as a Single God, Egyptian Hymns and Prayers

2

Origins of the Heart

The heart is the first organ to develop during human embryogenesis and it occurs at a very early stage. At around 18 days following fertilisation of the ovum, cells which will grow to form the heart have appeared and by 21–22 days a primitive heart has begun to pump. In another sense, the heart could be said to originate millions of years ago during the evolution of our ancestors. It has long been observed that during the development of an embryo, known as ontogeny, it has similarities with intermediate forms of animals at various times during the evolution of our ancestors, known as phylogeny. For example, during its life in the womb, a human embryo adopts transient features reminiscent of gills and a tail. This led to the idea that embryogenesis is a speeded-up version of evolution and during the 19th century this theory was widely held and known by the phrase "ontogeny recapitulates phylogeny", coined by Ernst Haeckel. However, it became clear during the early 20th century that evolution did not progress in a linear manner but involved many branching divisions and so no single path could reflect embryogenesis. Nevertheless, the biology of modern humans incorporates the evolutionary innovations that occurred in ancient creatures millions of years ago and which we have inherited through our genes. Our chemistry and physiology are encoded in the DNA received from our parents as they and their parents did in turn over countless generations. Elements of the DNA which control human development including our hearts are also found in other animals, even in the simplest and oldest life forms such as bacteria. In that sense, our hearts can also be said to originate in ancient times during the evolution of life. Both origins, that of our own lives in the womb and our evolutionary inheritance, are encoded in the DNA which we received at our conception.

Starting Life with a Clean Slate

At conception we received a copy of each of our parent's DNA, the genetic code which controlled their development in the womb and their lives up to the point at which they conceived us. In this way, we too received all the information needed to make a new human being. However, the DNA we received also contained information related to experiences our parents had during their lives. For example, exposure to extreme environments, infections or cancers can have secondary effects on DNA. These are known as epigenetic effects and some of them can be harmful. Therefore, it is important that such information is not passed on to future generations. This requires that each new individual starts life with DNA from which these epigenetic effects have been erased.

DNA is composed of nucleotides which bind together to form long strand-like molecules. Four nucleotides are found in DNA: cytosine, guanine, adenine and thymine, each linked to a sugar called deoxyribose and a phosphate group. These long molecules adopt a helical structure and bind with a second strand to form a double helix. This is configured with the nucleotides facing inward and with specific pairings so that adenine only links with thymine and cytosine with guanine. These are known as base pairs and human DNA consists of around 3 billion base pairs forming a molecule around 1.8 m long. Fitting this into a nucleus would be equivalent to packing eight miles of thin thread into a three-inch ball. To achieve this, DNA is wrapped around a core which consists of nucleosomes, as in Figure 2.1. Nucleosomes are composed of proteins known as histones. Each nucleosome is composed of nine histone molecules arranged in four pairs and a fifth histone which binds them together. DNA is wound around them in an arrangement of 1.8 windings per nucleosome in a manner which facilitates packing our DNA in very compact sequential helices.

DNA contains the coding information to make the proteins we need for life. These sections are our genes and we have around 19,000 of them, each containing the code to make a specific protein. This is facilitated by transcription factors which bind to DNA at specific locations and enable

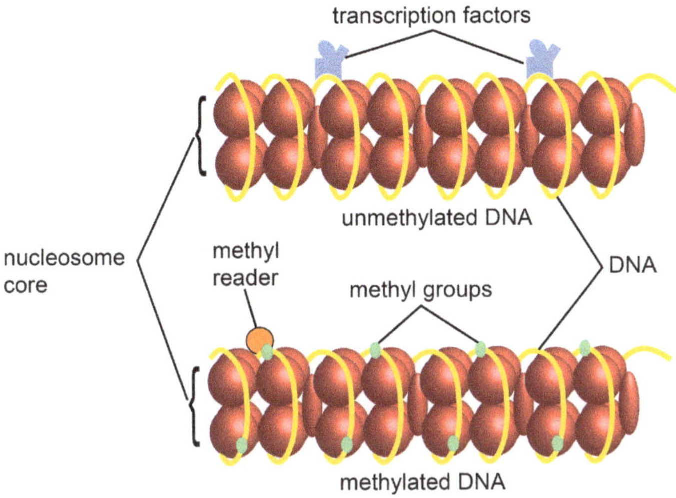

Figure 2.1: Diagram illustrating DNA (yellow) wrapped around a core which consists of nucleosomes (red), each composed of nine histone molecules arranged in four pairs and one which binds them together. DNA is wound around them in an arrangement of 1.8 windings per nucleosome. DNA contains the coding information to make the proteins we need for life and this is facilitated by transcription factors (blue) which bind to the DNA at locations which contain the code to produce specific proteins. Some genes are not expressed for much of our lives while others are expressed in some but not all tissues. One way to achieve this silencing is by methylation of our DNA and is known as epigenetic regulation. This consists of binding methyl groups to cytosine nucleotides on DNA (green). This causes DNA to be wrapped more tightly around the nucleosome core in a manner that makes it less accessible to transcription factors. In addition, methyl binding proteins (orange) which "read" the presence of methyl groups in the DNA bind to them and induce further changes in the nucleosome core which also inhibit transfer factor binding to DNA (after Dow and Cedar [1]).

RNA and protein synthesis related to the DNA code at that point. How our genes are used varies greatly depending on the function they regulate. When DNA code is read and used to synthesise a protein, it is said to be expressed. Some parts of our DNA are never expressed while others are expressed only at one time in our lives, for example, in regulating growth during a particular stage in our development in the womb. Some sections are expressed in almost all cells in our bodies, while others are expressed in some organs and not in others, for example, only in liver or heart cells. In addition, how we respond to the environment or diseases may involve temporary or long-term activation and expression of certain genes.

In all our cells, genes are annotated in a manner which determines whether they are active or quiescent [1]. This involves the addition of a methyl group (one carbon with three attached hydrogen atoms, –CH3) to cytosine-containing nucleotides along the DNA strand (see Figure 2.1, bottom panel). Known as methylation, the presence of methyl groups causes DNA to be wrapped more tightly around nucleosome cores in a manner which inhibits transcription factor binding to the DNA. This in turn blocks transcription of that section of DNA and consequently production of the protein it encodes. DNA in all cells contains both methylated and unmethylated segments. Some genes are unmethylated (and therefore active) in all cells while others are only unmethylated in certain tissues and not in others. In this way, functions such as energy generation needed by all tissues are available in all cells while gene transcription needed for specific brain functions can be restricted to these cells only, and similarly for other organs. The pattern of methylation in our DNA is remarkably stable throughout our lives, however disturbances in it are also believed to be responsible for some cancers. This can occur due to abnormal demethylation leading to inappropriate gene activation causing abnormal cell growth or alternatively through failure to suppress abnormal tissue growth due to methylation of growth suppressing genes.

In order to avoid transmitting such epigenetic effects to future generations, almost all methyl groups are removed from both parents' DNA within days of fertilisation of an ovum and before it is implanted in the womb. Although this prevents epigenetic transmission, it does not alter transmission of inherited conditions as these are due to abnormalities in the DNA code itself. Following implantation at around nine days after ovulation a second methylation stage occurs, whereby almost the entire genome undergoes *de novo* methylation except for specific islands of DNA which are protected and remain unmethylated. These unmethylated segments among other things provide the code needed to initiate and regulate the development of the embryo (embryogenesis) and the growth of each organ including the heart.

In this way, the genetic code needed to create each unique person is transmitted. The epigenetic information from each parent is erased and a new map is established for the embryo. Embryogenesis is an immensely complex process with many stages, some of which involve the expression of genes which will never be used again during adult life. In these cases, a secondary stage of methylation occurs which results in their permanent suppression. The pattern of methylation or the epigenetic map established at this very early stage in development remains remarkably stable throughout life and is thought to be a key factor in long lived animals.

Development of the Heart

It is no accident that the heart is the first organ in the embryo to develop. In humans, the primitive heart tube, which begins to pump at Day 22, is already contributing to further development and growth of the embryo.

The development of the heart begins with the designation of some future heart cells called progenitor cells in the very early embryo. This takes place during a stage known as gastrulation, which describes an initial folding of the early embryo. At the beginning, the embryo consists of a single layer of cells surrounding a yolk sac which provides nourishment for the early embryo. Folding of the embryo then occurs so that what had been a ball-shaped structure known as the blastula becomes cup-shaped and known as a gastrula. The gastrula is also lined with a single layer of cells, the outer of which is known as the epiblast and the inner the hypoblast. These two layers will form the three germ layers, ectoderm, mesoderm and endoderm (see Figure 2.2). Initially cells in the epiblast layer migrate through its centre and spread out between it and the hypoblast layer to form the mesoderm layer. When that has occurred, the remaining cells in the epiblast layer become the ectoderm and the hypoblast becomes the endoderm (Figure 2.2, bottom panel). These three germ layers will give rise to all tissues and organs of the foetus as it develops (Figure 2.3). The mesoderm gives rise to the heart and blood vessels. Cells from the mesoderm will become specialised to form the outer membrane of the heart (pericardium), heart muscle (myocardium), inner lining (endocardium), and the blood and lymphatic vessels.

This process begins with the migration of cells designated as future heart tissue from the epiblast through its centre to the mesoderm below (Figure 2.4, left panel). From there they migrate towards the cranial end (where the head will develop) of the embryo and spread laterally to form a horseshoe shaped pattern known as the heart fields (Figure 2.4, middle panel). Around Days 18–19 following fertilisation, the heart fields grow

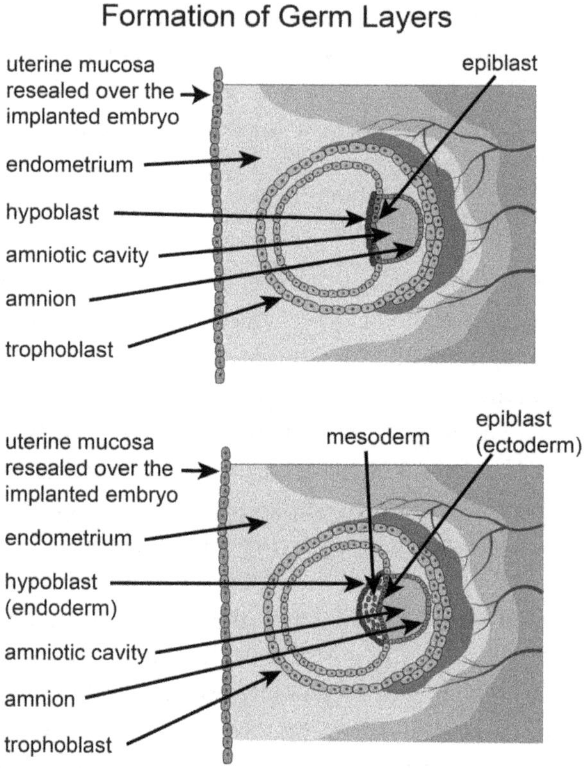

Figure 2.2: Formation of germ layers during the gastrula stage. Once the fertilised ovum becomes implanted in the uterus it evolves into a disc composed of two cell layers known as the epiblast and hypoblast, upper panel. These layers will give rise to the three germ layers which form all tissues and organs of an adult. Initially cells from the epiblast layer migrate through its centre and spread out between it and the hypoblast layer to form the mesoderm, bottom panel. Once that has occurred, the remaining epiblast cells transform into ectoderm and the hypoblast become endoderm. These three layers will give rise to all tissues of the growing foetus.

Notes: Endometrium: Uterine tissue in which the embryo is embedded; Amnion and amniotic cavity: The sac in which the embryo and foetus will develop surrounded by the amnion layer; Trophoblast: Outer lining of the embryo.

to form primitive blood vessels on each side of the embryo (Figure 2.4, right panel). These will later coalesce to form tubes known as endocardial tubes. At this point, the embryo undergoes a side to side folding which brings the endocardial tubes in close contact and they fuse to form a single primitive heart tube around Day 20 after fertilisation. At this time, regions

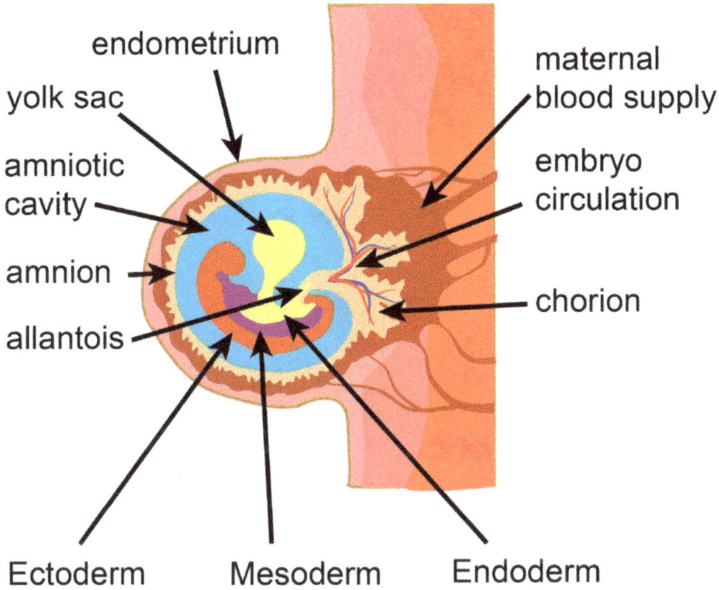

Figure 2.3: As the embryo develops, the three germ layers, ectoderm, mesoderm and endoderm, will give rise to all its tissues and organs. The mesoderm forms the heart and blood vessels. All these processes are regulated by the DNA in each cell. All cells originate from the same cell type but during the very early phase of development they are differentiated to take on the specific roles of the three germ layers. This involves specifying the function of each cell by permitting certain genes to be expressed during its development and silencing others not required for the role it has been designated to. This is done by the process of methylation which creates a profile of active and silent genes. In this way, the developmental path of cells in the three germ layers is designated very early in embryogenesis.

Notes: Endometrium: Uterine tissue in which the embryo is embedded; Amnion and amniotic cavity: The sac in which the embryo and foetus will develop surrounded by the amnion layer; Yolk sac: Provides nourishment for the early embryo; Allantois: Assists nutrition and excretion in the early embryo; Chorion: The outermost part of the embryo formed from the trophoblast and amnion. Ectoderm gives rise to the skin, hair, nails and nervous system; Mesoderm gives rise to the heart, blood vessels, muscles, skeleton, joints and lung structure; Endoderm gives rise to the digestive tract, liver, pancreas and lungs (lining of airways).

of the primitive heart are already designated (Figure 2.5). The sinus venosus at the lower end of the heart tube conveys blood from the placenta and yolk sac into the primitive atrium which is contiguous with the primitive ventricle and in turn with the conus cordis (which will form the outflow tracts of the right and left ventricles) and the truncus cordis (which will form the pulmonary and aortic roots). At this time, the primitive heart

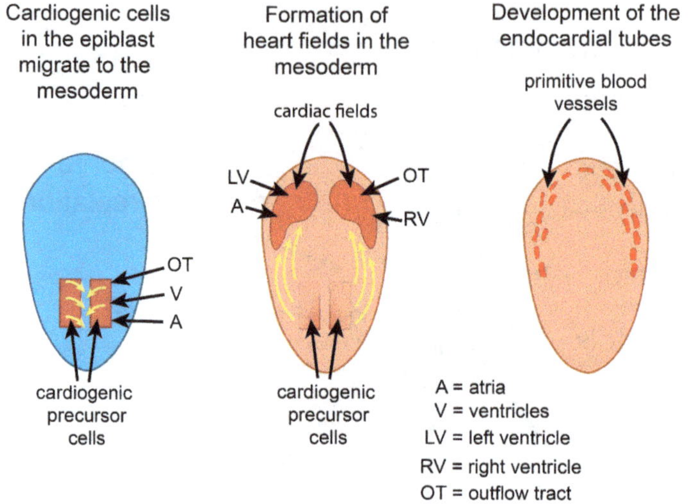

Figure 2.4: The epiblast stage occurs at around nine days after ovulation and by then has formed two layers of cells which will give rise to the embryo proper. These are named the epiblast (blue, left panel) and hypoblast and they will form the ectoderm and endoderm, respectively. Cells that will give rise to the heart (cardiogenic cells) are present in the epiblast and already designated to form specific parts of the heart (left panel). These cells migrate through its midline to the mesoderm layer below (middle and right panels) where they then migrate to the cranial area (where the head will develop) and form the cardiac fields (middle panel). Over the next several days, cells in the cardiac fields coalesce to form primitive blood vessels on each side of the embryo (right panel). These in turn will come together as a result of transverse folding of the embryo, initially forming a cord and later hollowing to form an endocardial tube in each side by Day 18.

begins pumping blood which enters through the sinus venosus and is ejected through the early aortic roots.

During Days 20–23, the heart begins a rotation process which will bring it close to the adult chamber configuration. Initially the sinus venosus and primitive atrium rotate to the left and upward, bringing them alongside the ventricle (Figure 2.5, left and middle panels). Then a second rotation takes place but this time taking the sinus venosus and primitive atrium to the rear and then upward behind the ventricles (Figure 2.5, middle and right panels). By Day 28 the atria and ventricles have also begun to divide into primitive left and right chambers. Later, division of the atria and ventricles will be completed to form the left and right sided chambers and the heart valves will be formed.

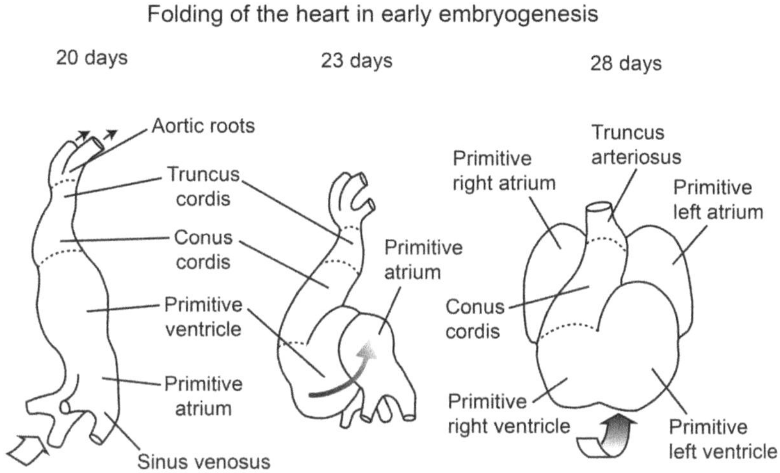

Figure 2.5: Diagram illustrating folding of the primitive heart tube (left panel). At this stage, the regions which will form the chambers of the adult heart are already designated; the sinus venosus receives blood from the placenta and yolk sac, which then passes consecutively through the primitive atrium and ventricle and then to the conus cordis and truncus arteriosus (left panel). This begins to pump spontaneously around Day 20 following fertilisation, blood entering through the sinus venosus below and being ejected through the aortic roots at the top. Over the next 2–3 days, the heart tube begins a rotation process which will result in the adult chamber configuration. Initially the primitive atria rotate in a manner which moves them about 90° to the left (middle panel). Then a further rotation takes place but this time the primitive atria rotate to the rear and then up behind and the heart (right panel). The result of this is that by Day 28 following fertilisation the atria and ventricles are close to their adult positions.

Genetic Control of Heart Development

Like all our functions, the instructions which regulate development of the heart are encoded in our DNA. The four-chambered heart is a remarkably efficient pump and perfectly adapted for our needs. However, many of the genes which regulate the early growth and development of the heart are ancient. For example, some are also present in primitive life forms such as *Drosophila* (fruit flies), even though their evolution diverged from that of vertebrates around 300 million years ago. Fruit flies also have a circulation driven by a pump in the form of a contracting tube, reminiscent of the primitive heart tube in humans at Day 22. The *Drosophila* heart tube circulates a fluid called haemolymph in an open circulation, meaning it lacks direct connections between arteries and veins. Haemolymph like blood carries oxygen, but it does so using haemocyanin, a protein containing copper rather than the iron which is found in haemoglobin. Haemocyanin is colourless when deoxygenated but turns blue when it binds oxygen. Haemoglobin is a much more efficient oxygen carrier but is toxic if freely circulating and for this reason, it is confined to red blood cells. In contrast, haemocyanin is non-toxic and circulates free in haemolymph. *Drosophila* also periodically reverses the direction of flow to pump haemolymph via a caudal opening into the abdominal cavity (see Figure 2.6(b)). This is achieved by closing the four caudal-most ostia and relaxing the heart valves [2]. Fruit flies do not have lungs. Air enters through tiny pores in its abdomen and passes through multiple channels by which it exchanges oxygen and carbon dioxide with haemolymph. By periodically reversing heart flow, haemolymph is pumped into the abdominal space to where gas exchange takes place. Although *Drosophila* does not have the chambers found in vertebrates, the contractile mechanisms by which its heart pumps have many similarities to ours. In addition, the genes which initiate its development have similarities to ours.

Drosophila Hearts

Drosophila melanogaster is a small fruit fly around 2–3 mm in length which has played an important role in our understanding of how our hearts develop during the early stages of life in the womb. *Drosophila* and all vertebrates including ourselves shared a common evolutionary path until around 300 million years ago. As a result, we also share a significant genetic heritage. Around 75% of genes which cause disease in humans are also found in *Drosophila* and many newly discovered genes in humans are also found in *Drosophila*. For this reason, studies of how the genes of *Drosophila* function have also shed light on how genes in vertebrates work. For example, studies of *Drosophila* genes which regulate the early development of the heart led to the discovery of genes and factors which do so in vertebrates [3]. In *Drosophila*, a protein known as "tinman" is one of these. Tinman is a transcription factor, meaning it is a protein folded in such a manner that it binds to DNA at a specific location and facilitates

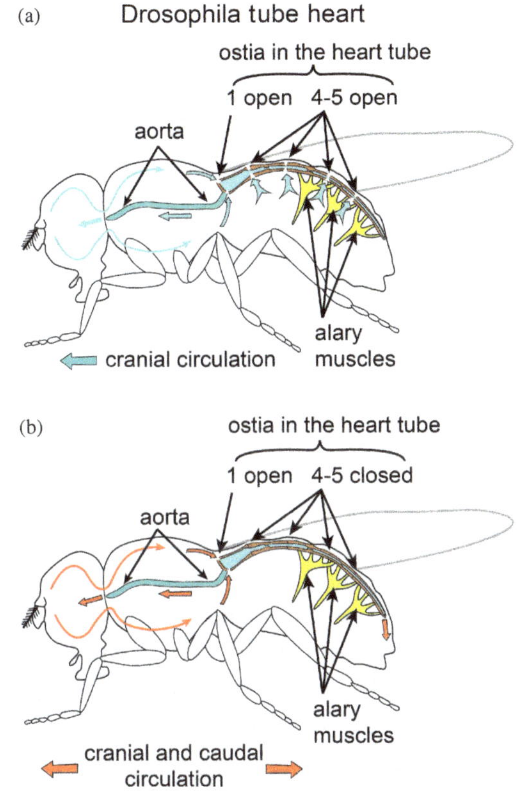

Figure 2.6: The fruit fly (*Drosophila melanogaster*) is about 2.5 mm in length and has a tube-shaped heart (shown in orange) along the back of its abdomen. This pumps haemolymph (shown in blue) to the head and thorax (chest) via a thin aorta into an open circulation, meaning its blood passes through spaces between its tissues to nourish them; it does not have a capillary and venous circulation as vertebrates have. Haemolymph re-enters the heart directly through five ostia. This is assisted by the alary muscles which are attached to the stiff carapace at one end and "pull" the heart tube to increase the size of its lumen, thereby sucking haemolymph into it to aid filling. The heart tube is also fitted with valves which compel forward flow to the head and thorax as shown in (a). Fruit flies periodically reverse the direction of flow to pump haemolymph via a caudal opening into the abdominal cavity (b). This is achieved by closing the four caudal-most ostia in the heart tube and relaxing the heart valves. Fruit flies do not have lungs. Instead air enters through tiny pores in its abdomen and passes through multiple channels by which it exchanges oxygen and carbon dioxide with haemolymph. Its periodic reversal of heart flow facilitates this. Vertebrates also have a tube-shaped heart at an early stage during intrauterine development and the genes which initiate heart development in *Drosophila* are homologous to those in vertebrates. (Blue arrows indicate normal flow direction and red arrows indicate reverse flow direction.)

transcription of the code from that point to make RNA and a protein based on that sequence of DNA. Tinman itself is of course the product of another gene also called the "tinman gene" and expression of the tinman gene is crucial for heart development in *Drosophila*: without it the heart fails to develop. The name tinman comes from the Tin Man in *The Wonderful Wizard of Oz*. In the story, Tin Man loses his limbs and heart due to the spells of the Wicked Witch and when the tin smith, Ku-Klip, replaced his limbs with tin versions, he neglected to replace his heart. Although crucial for heart development, tinman is just one of a collection of proteins and their genes needed for normal development of the heart.

Expression of the tinman gene in *Drosophila* is maintained by a protein known as decapentaplegic (DPP) (see Figure 2.7(a)) which is produced by cells in the ectoderm layer and diffuses through the adjacent mesoderm [4]. DPP is a member of a group of proteins known as bone morphogenetic proteins (BMP). These play an important role in the development of bone and cartilage as well as heart development. In the context of the latter, it acts as a morphogen, meaning it plays a key role in designating sections of the early embryo which will later form the limbs, organs and other structures of the adult. There are 15 of these segments, known as "imaginal discs" and if DPP is absent, they fail to develop normally. Hence the name decapenta (15) plegic (paralysis) or DPP for short. DPP works by acting on a cell receptor on the surface of mesodermal cells known as "thickvein" (named by scientists who found that it was associated with thickened veins in the wings of *Drosophila*). Activation of the thickvein receptor in turn activates a signalling pathway by which events occurring outside a cell can be transmitted to the cell nucleus and influence the expression of DNA. When DPP binds to the receptor thickvein, it leads to the activation of a protein complex inside the cells known as MEDEA. MEDEA is a member of a family of transcription factors known as SMAD4 (named by scientists as an acronym for "mothers against decapentaplegic" because if absent normal segment development fails to occur). MEDEA then travels through the cell cytoplasm and into the nucleus where it binds to a specific segment of DNA and facilitates its expression — in this case, tinman. Thus, MEDEA is another transcription factor and is the end of a signalling pathway that begins in the ectoderm with the secretion of DPP which in turn diffuses into the mesoderm, acting

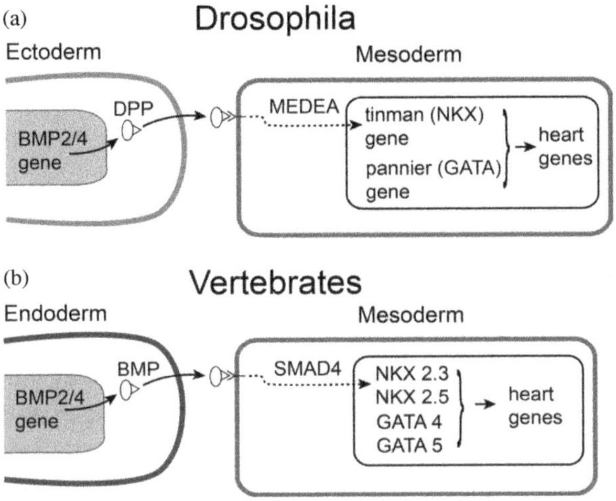

Figure 2.7: Diagram illustrating the genetic mechanisms which initiate heart development in fruit flies (*Drosophila melanogaster*) and vertebrates. (a) Heart cell designation and development in *Drosophila* occurs in mesoderm cells and is stimulated by a protein known as tinman, which is produced by the tinman gene. Maintenance of tinman activation is dependent on the presence of a signalling protein, DPP (decapentaplegic), which is produced in *Drosophila* by cells in the adjacent ectoderm layer. DPP acts on a receptor called "thickvein" on mesoderm cells, causing activation of a transcription factor known as MEDEA which travels into the nucleus and leads to the expression of the tinman gene and production of tinman protein. A gene known as pannier also plays an important role in heart development in *Drosophila*. (b) In vertebrates, heart development depends on activation of the NKX 2.3 and 2.5 genes which are homologous with tinman in *Drosophila* and their activation is stimulated by BMP, which is homologous with DPP in *Drosophila*. BMP is produced in endoderm cells in vertebrates and acts on mesoderm cells causing activation of SMAD4 and expression of NKX 2.3 and 2.5. GATA genes 4 and 5 also play an important role and are homologous with pannier in *Drosophila*. All these genes are members of gene families found in both *Drosophila* and vertebrates. DPP is a member of the BMP (bone morphogenic protein) family, tinman of the NKX gene family and pannier is a GATA gene. In addition, MEDEA is a member of a family of transcription factors known as SMAD4 (after Chen *et al.*) [3].

on cells there to cause expression of the tinman gene. In biology, there are many such pathways by which external events including the environment can impact cell functions including gene expression. In *Drosophila*, tinman works in conjunction with a second gene known as "pannier" (see Figure 2.7(a)). Once again, the name pannier was coined by scientists in the 1980s who observed that if the gene was absent or altered, it led to early *Drosophila* embryos having a shape resembling a pannier. Pannier is a member of the "GATA" family of genes. These are so named because the proteins they code for are transcription factors which bind to DNA at sites which have the nucleotide sequence Guanine–Adenine–Thymine–Adenine or GATA.

Development of Vertebrate Heart

While studies of fruit flies might initially seem a long way from humans, the benefits of such research are illustrated by the way they led to the understanding of vertebrate heart development. As we have seen, the tinman gene is crucial for heart development in *Drosophila*. Tinman is an example of a "homeobox" gene and the DNA sequence which codes the tinman protein is known as a homeobox domain. The term homeobox is derived from the Greek word "homeo" (meaning the same) and comes from the word "homeosis", coined by biologist William Bateson in 1894 who used it to describe changes in body structure where one part was replaced by another or is duplicated, as, for example, polydactyly (extra fingers) in humans or when a limb is replaced by an antenna in bees. In genetics, the term homeobox is used to describe genes involved in such malformations.

Recognition of the role of tinman in *Drosophila* heart development led scientists to look for similar genes in vertebrates. This was done using the DNA sequence of the tinman homeobox domain obtained from *Drosophila* as a probe to detect similar DNA sequences in vertebrates such as mice and frogs. Essentially short strands of nucleotides corresponding to the beginning (upstream) and end (downstream) of the tinman homeobox domain are synthesised. Known as primers, they bind to DNA at specific complementary sequences if present. These are then added to DNA obtained from vertebrate hearts such as mouse or frog together with the enzyme DNA polymerase. This enzyme then initiates a chain reaction which can produce thousands of copies (clones) of the DNA corresponding to the segment bound by the upstream and downstream primers. Any such clones are then extracted and analysed. In this way, potential homeobox genes and domains can be detected in other species. Several genes have been found in vertebrates with very similar DNA sequences, known as homologues, to tinman. These are known as NKX2.5, NKX2.3,

NKX2.6, NKX2.7 and NKX2.8 and they all play important roles in vertebrate heart development. However, unlike *Drosophila* none individually plays such a crucial role as tinman. Defects in any one of them can produce abnormal development of the heart but not complete failure, thought to reflect overlapping roles among them. Like tinman expression in *Drosophila*, NKX expression in vertebrates depends on secretion of a DPP-like protein in adjacent endoderm (Figure 2.7(b)), which acts on mesoderm cells leading to the activation of the transcription factor SMAD4 and which in turn causes expression of the NKX gene [3]. In addition, homologues of the pannier gene in *Drosophila* are also involved in regulating early heart development in vertebrates. In mice, for example, three GATA genes are found in precardiac mesoderm of which GATA4 and GATA6 are involved in early heart development. Thus, despite 300 million years since our evolutionary paths diverged, genetic regulation of our heart development at a very early stage in our embryonic lives has remarkable similarities to that of *Drosophila*. This is an example of conservation of evolutionary innovations and is well known to be widespread in biology. All evolutionary changes are implemented through alterations in DNA coding and these are retained in the genome of species long after their evolutionary paths have diverged.

Evolution of the Four-Chambered Heart

A simple tubular heart first appeared in a bilaterian (meaning animals which have symmetrical left and right sides) ancestor 600–700 million years ago. This also features as a step in early embryonic development of invertebrate and vertebrate hearts and, as discussed previously, forms the main circulatory pump in insects like *Drosophila*. Vertebrates had to evolve further in order to sustain life as air breathing, upright, warm-blooded animals. Buoyancy in water greatly reduces the effect of gravity and standing upright on land removes this effect. As a result, blood must be pumped against gravity to an extent that depends on the height of the animal and especially the brain. Blood pressure is recorded as systolic (the peak level during ejection) and diastolic (the level when the ventricles are relaxed and filling). Humans normally have a blood pressure of around 130/75 mmHg which is significantly higher than in aquatic and amphibian animals, typically around 40/20 mmHg or less. But for tall animals like the giraffe, a blood pressure of up to 260 mmHg is needed to maintain blood perfusion of the brain. The ability to maintain a stable body temperature over a range of environmental temperatures has advantages in surviving and procreating under varied conditions but also comes at a price of requiring higher rates of energy production. This is achieved in part from more efficient metabolism and enzyme kinetics, but also requires more efficient oxygen delivery. To achieve this, blood must be pumped around the body at higher pressure, while maintaining a separate circulation to the lungs at a far lower pressure to allow gas exchange. Mammals can do this having evolved a four-chambered heart with valves to control flow direction and an electrical system to regulate the timing of the pumping action.

Insects such as *Drosophila* have a simple tube heart which pushes haemolymph around its open circulation, contracting 300–600 times per minute. This is sufficient to maintain a pressure of 20/10 mmHg. In addition, some insects have additional accessory hearts at the wing bases, antennae and limbs to aid flow. For insects, flow is relatively slow and efficient diffusion of gases in these open circulations is limited to short distances. This limits the size to which animals can grow and organisms larger than about 10 cm have evolved more efficient pumps.

Molluscs

Molluscs are soft bodied animals which comprise the second largest group of invertebrates. About 85,000 species are still extant and as many as 100,000 are found in the fossil record. They include gastropods (slugs and snails), which make up 80% of the phylum, and cephalopods (such as octopus and squid). Molluscs like vertebrates evolved from *Bilateria* but diverged from the path of vertebrate evolution more than 550 million years ago.

Gastropods have a two-chambered heart (Figure 2.8(a)), an inflow chamber (atrium), outflow chamber (ventricle) and an atrio-ventricular valve between them. This has distinct mechanical advantages compared with the tube hearts of arthropods. The atrium acts as a reservoir of venous haemolymph and its ability to contract actively primes the ventricle during its filling phase (diastole). This priming stretches the ventricular muscle which induces it to contract with greater force. This provides an elegant system which matches demand to performance: with increasing activity, venous return is greater and this induces the ventricle to pump more forcefully and eject a larger volume. The atrio-ventricular valve closes during pumping phase (systole) which prevents back-flow. Gastropod ventricles have an intrinsic rhythm with an average contraction rate of around 20 per minute maintaining a pressure of 10–20 mmHg. The heart pumps directly to the gills and the oxygenated haemolymph continues from there to the organs and tissues.

Cephalopods (octopi and squids) are more evolved molluscs. They are capable of much greater activity and consequently have higher energy needs. Their circulatory systems have evolved to meet this in a number of ways. Unlike gastropods, they have a closed circulation (meaning its outflow and return vessels are connected by a network of small capillaries) which achieves a much more efficient delivery of oxygen and nutrients. Cephalopods such as octopi have a systemic heart and two additional branchial hearts which push haemolymph through the gills (Figure 2.8(b)). They are more active than gastropods and consequently have greater energy needs. To facilitate this, the systemic heart develops a pressure of 50–70 mmHg, which provides more effective perfusion of organs and tissues. However, that pressure would be unsuitable for gas exchange. All animals need oxygen to facilitate metabolism and to expel the CO_2 produced. In molluscs, gas exchange takes place through very thin-walled vessels in gills which carry haemolymph on one side and surrounding water on the other. Such thin-walled vessels could not withstand the higher pressures generated by cephalopod hearts. The evolutionary solution for this has been to place the gills on the inflow side of the circulation. In contrast to gastropods, the systemic heart in cephalopods pumps directly to the organs and tissues of the body and from there it continues to the gills. However, pressure dissipates with distance from the heart and cephalopods have evolved two additional accessory hearts to perfuse the left and right gills at <20 mmHg to allow efficient gas exchange. These mechanisms allow cephalopods like octopus to adapt their cardiac output (volume ejected/minute) to meet rapid movements by increasing their heart rate (from 40 to 100 per minute) and stroke output (volume ejected/beat) by two to threefold.

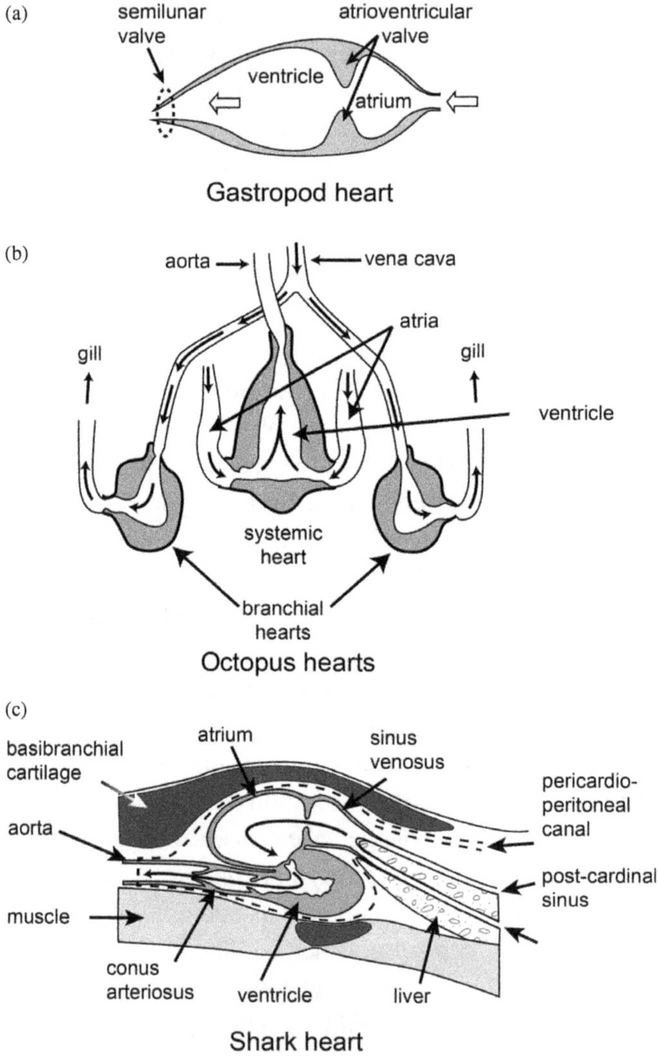

Figure 2.8: Diagram illustrating different heart structures. (a) Gastropods (slugs and snails) have a two-chambered heart, comprising atrium and ventricle, separated by an atrio-ventricular valve, which is suitable for an animal of low metabolic needs. (b) Cephalopods such as octopi have three hearts. A systemic heart receives haemolymph from the gills in left and right atria. These drain into a single ventricle which pumps haemolymph through the body tissues. From there haemolymph returns towards the gills, but at this point there is insufficient momentum to carry flow through the gills. For this, octopi have two additional branchial hearts which perfuse the gills and return haemolymph back to the heart. (c) Fish also have a two-chambered heart, but reconfigured

Fish

Fish were among the first vertebrates which appeared about 540 million years ago during the Cambrian explosion, a time of massive diversification of life forms on earth which lasted about 25 million years. Fish evolved substantial modifications of the heart and circulation compared with molluscs. Although they also have a single circulation with two main chambers, fish have a closed circulation with capillaries allowing continuous flow between arteries and veins. The capillaries are permeable which provides the means for delivery of nutrients and oxygen to the tissues and removal of carbon dioxide and by-products of metabolism. An important innovation in the vertebrate circulation was the introduction of the lining of all vessels with endothelial cells. These are thin flat cells which form a continuous layer on the inner surface of blood vessels. They have important functions in regulating nutrient transport, blood flow and coagulation. Vertebrates have haemoglobin as the oxygen carrier, the genes for which can be traced back to their divergence from invertebrates over 650 million years ago [5].

The linear two-chambered heart found in molluscs evolved a rotation in fishes such that the atrium came to lie alongside the ventricle rather than in line with it and the atrio-ventricular valve was placed at 90° to the direction of inflow (see Figure 2.8(c)). As a result, flow into the atrium is much less impacted by ventricular contraction than in molluscs (Figure 2.8(a)). In addition, a more continuous flow pattern is achieved which maintains flow momentum and preserves energy. The sinus venosus appeared as a ballooning of the venous return at its junction with the heart (Figure 2.8(c)) providing a reservoir and improving blood flow to the heart. A further modification of the outflow from the heart evolved in teleost fish, the largest fish group which includes salmon, haddock and

Figure 2.8: (*see figure on facing page*) in an "S", which confers important mechanical advantages. They also have an additional reservoir chamber (the sinus venosus) which assists venous return and an outflow chamber (bulbus arteriosus) which acts to dampen systolic ventricular pressure. In elasmobranches like sharks, the pericardial and peritoneal cavities are connected by a canal (dashed line) which allows transfer of fluid between them. This acts to modulate chamber filling during diastole and provides a mechanism to greatly increase cardiac output in times of greater energy demands.

around 96% of all fish species. This consists of a dilated elastic structure between the ventricle and the aorta called the conus arteriosus (Figure 2.8(c)). It acts to absorb and store some of the kinetic energy of the contraction of the ventricle. This has two benefits; first, the gills are the first organs to be perfused by the heart and are the organs where gas exchange takes place across thin-walled vessels. The damping effect of the conus arteriosus on the heart's pumping pressure helps to protect them. Second, by storing some of the kinetic energy developed during ventricular contraction, it assists flow during diastole.

Elasmobranch fishes, a group that includes sharks, rays and sturgeons, have evolved unique modifications of the pericardium and ventricular outflow which improve their ability to adapt to sudden increases in energy demands. They have a stiff cartilaginous pericardium (the outer lining of the heart), Figure 2.8(c), and the pericardial cavity (between the pericardium and the heart) is normally filled with fluid. When the ventricle empties during systole, it develops a negative pressure creating a suction which aids filling the atrium. A further modification of the pericardium in sharks is a canal between the pericardial cavity and the peritoneum (the space containing the abdominal contents) called the pericardio-peritoneal canal. By shifting fluid from the pericardial cavity to the peritoneum, elasmobranches can increase filling of the atrium and ventricle and thereby increase stroke volume and cardiac output by around 40%. Mammals increase cardiac performance by activation of sympathetic nerves which stimulate the heart. This mechanism is not developed in elasmobranchs and so this pericardium to peritoneal fluid shift provides an important mechanism to enable rapid bursts of activity. A consequence of the negative pericardial pressure which occurs during systole is that the ventricular outflow tends to be decompressed; this is prevented by the structure of the conus arteriosus (Figure 2.8(c)), which is thicker and stiffer than the bulbus arteriosus found in most fish. It also contains a one-way valve to prevent back-flow.

Amphibians

Amphibians have evolved lungs to enable them to live both on land and in water. Unlike lungfish which evolved primitive lungs while retaining gills, the latter have been lost by amphibians. In amphibians such as frogs, most of their respiratory function takes place in the lungs but they can also achieve significant gas exchange cutaneously (through their moist skin). Indeed some toads and salamanders have no lungs and achieve all their gas exchange through the skin. When diving underwater, frogs derive a significant amount of oxygen cutaneously and this is sufficient for long periods of hibernation in cold water. During their larval stage as tadpoles, frogs have hearts similar to fish. A two-chambered heart pumps blood through the gills and then around the body in a single circulation. As adults they lose their gills and develop lungs.

They also develop a three-chambered heart with two atria and a single ventricle (Figure 2.9).

The right atrium receives deoxygenated and carbon dioxide rich blood via the sinus venosus from around the body. Oxygenated blood from the lungs flows to the left atrium via the pulmonary veins. Both atria empty into the single ventricle but the blood from each maintains fairly separate flow streams. The single ventricle then ejects both into the single outflow tract. This has a longitudinal spiral valve, which acts as a baffle directing the oxygenated blood into the systemic arteries and the venous blood to the pulmonary arteries. The pulmonary artery gives off branches to the skin which divert around 25% of pulmonary flow. The blood oxygenated in the lungs returns to the heart via the pulmonary veins, while oxygenated blood from the skin drains into the systemic veins and returns to the heart via the sinus venosus. When swimming under water, the circulation undergoes marked changes. The pulmonary arteries vasoconstrict which reduces blood flow to the lungs. However, systemic vessels also perfuse the skin and this source increases about threefold [6] so that oxygen diffusion through the skin becomes the major source of oxygen uptake. A single ventricle also means that blood pressure is the same in the pulmonary and systemic circulations. This is possible in amphibians because the ventricle generates fairly low systolic and diastolic pressures of around 40/20 mmHg. This means that their thin-walled pulmonary capillaries have an appropriate pressure for gas exchange.

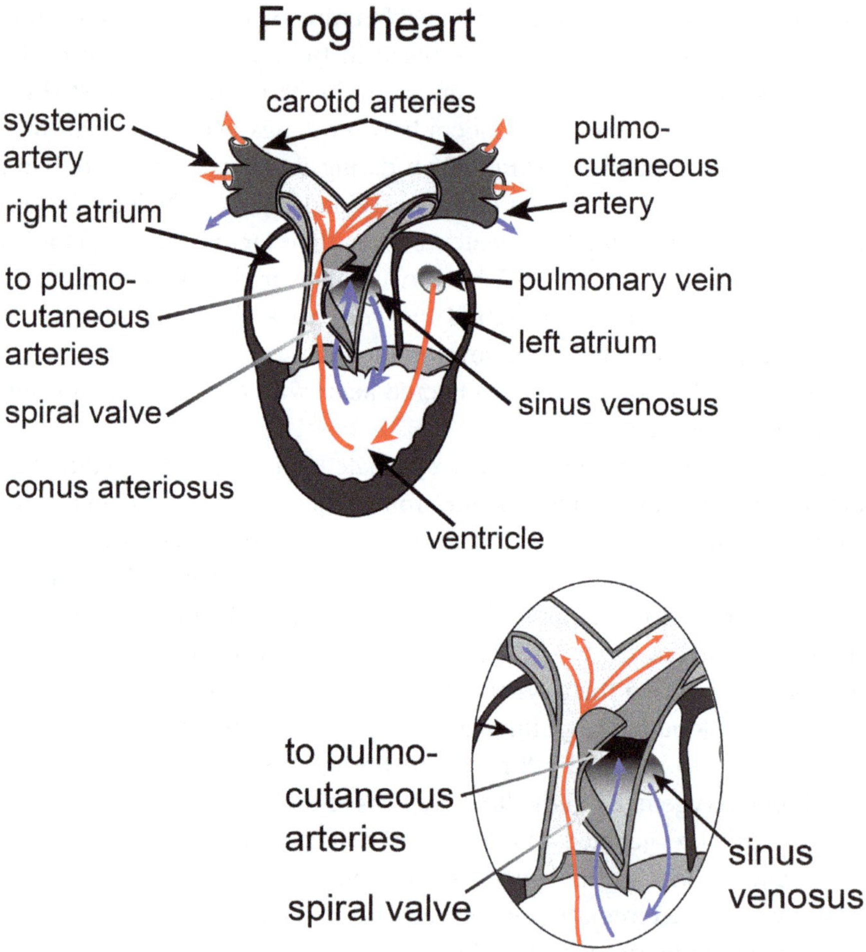

Figure 2.9: Frog hearts are three-chambered, the single ventricle ejecting into a common outflow tract. However, left and right atrial blood flows maintain separate parallel streams through the ventricle and this together with a longitudinal spiral valve in the outflow tract (see inset) effectively separate them so that oxygen depleted venous blood from the body goes to the lungs and skin via the left and right pulmo-cutaneous arteries (blue) and oxygen rich (red) blood from the lungs to the body.

Reptiles

Reptiles represent an intermediate evolutionary stage between the two-chambered heart and single circulation of fish and the four-chambered heart and double circulation of mammals. They are tetrapod (four limbed) animals which evolved over 300 million years ago and include snakes (snakes evolved from four-limbed ancestors), lizards, turtles and crocodiles. The non-crocodiles have a partially divided single ventricle: an interventricular septum but one that is incomplete and consequently allows some shunting of blood between their pulmonary and systemic circulations. Their ventricle generates fairly low pressures of around 40/20 mmHg which is adequate for animals with low energy demands. This also means that most of these animals can function with similar pulmonary and systemic blood pressures. Exceptions include pythons which have evolved the ability to rapidly increase their metabolic rate and oxygen consumption over sixfold [7] after eating a large meal or when incubating their eggs in response to a drop in ambient temperature. These responses place an extra load on the cardiovascular system which they cope with by two unique mechanisms. First, they can rapidly increase the size of the ventricle by about 40% (within 48 h of eating) for the period of digesting a meal, thereafter, regressing to normal [7]. This is an extraordinarily large and rapid change and much faster than anything observed in mammals. Second, although their interventricular septum is incomplete like other non--crocodile reptiles, the residual defect is effectively sealed during systole by the contraction of the ventricle. The result is that they have functionally separate pulmonary and systemic circulations like mammals and generate high systemic and low pulmonary pressures independently. This greatly increases their ability to increase tissue perfusion and metabolism [8]. This also explains why they can maintain high levels of oxygen in the systemic circulation and increase oxygen consumption in times of increased energy demand.

Crocodiles are tetrapod semiaquatic reptiles which emerged about 95 million years ago. They along with birds are the only living species from ancestors that included the dinosaurs. Although ectothermic, they are highly effective predators capable of short periods of rapid activity. These characteristics are facilitated by their cardiovascular system which has a four-chambered heart and two separate circulations. This arrangement is similar to the four-chambered heart in mammals and quite different to the three-chambered heart of amphibians. The evolutionary explanation for this is that crocodiles evolved from endothermic dinosaurs which would have had a higher metabolic rate and energy needs, for which a four-chambered heart was required [9]. Blood pressure in extant crocodiles is similar to amphibians and lower than in mammals and like amphibians they evolved a means for shunting blood between the pulmonary and systemic circulation when diving. However, the mechanism for this differs from amphibians and evolved subsequent to their four-chambered heart.

Crocodiles have retained a right and left aorta, the right receiving blood from the left ventricle, while the left aorta is connected to the right ventricle (Figure 2.10). The main pulmonary artery is also connected to the right ventricle and during low levels of activity receives its output. Under these conditions, the left aorta also receives blood from the right aorta via a connecting hole between them, called the foramen of Panizza located close to their origins (Figure 2.10, left panels). Within the pulmonary conus (just below the pulmonary valve) there is an additional valve composed of tough "cog teeth" connective tissue which under special conditions can close off the outflow to the pulmonary artery (Figure 2.10, right panels). When this occurs, flow is directed instead to the left aorta. During intense activity and diving, the cog teeth valve closes providing a mechanism which shunts the right ventricular output from the main pulmonary artery to the left aorta. This arrangement increases cardiac output to the systemic circulation providing additional tissue perfusion.

Shunting of pulmonary blood flow to the systemic circulation affords crocodiles an additional advantage. Like other carnivorous reptiles they consume large meals consisting of whole or parts of animals. In the case of crocodiles, these often include entire limbs and large bones. For this, they have evolved a unique adaptation of the cardiovascular system. The evolutionary advantage of this relates to the importance of stomach acid for digestion. The left aorta supplies blood to the stomach [10], which as in vertebrates produces hydrochloric acid and this requires hydrogen (H^+) ions. Blood returned to the right atrium and ventricle normally has a high carbon dioxide content, which when dissolved in water produces carbonic acid. Although in itself weakly acidic, the H^+ ions, it makes available, can be used by the stomach to produce a more powerful concentration of hydrochloric acid. Thus by closing the cog teeth valve, carbon dioxide rich blood is shunted from the pulmonary artery to the left aorta (Figure 2.10, right panels) providing a source of H^+ ions to assist acid production and digestion in the stomach. The fact that this shunting is

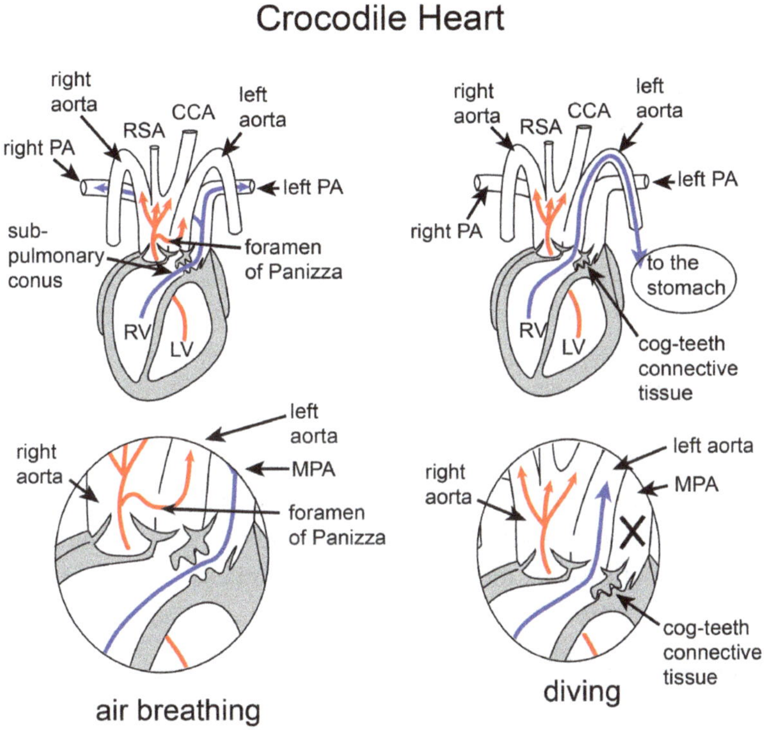

Figure 2.10: Crocodiles have four-chambered hearts. Under resting conditions, the right ventricle gives off the pulmonary artery and a left aorta. The left ventricle pumps blood into the right aorta some of which is passed to the left aorta via the foramen of Panizza. Crocodiles have an actively controlled mechanism to switch right ventricular output between the pulmonary artery and left aorta. Under normal conditions, right ventricular blood (blue) is pumped into the pulmonary artery (left panel). After feeding and diving, this is diverted to the left aorta which supplies the stomach by closure of the cog teeth valve below the pulmonary valve (right panel). This carbon dioxide rich blood (blue) provides a source of H^+ ions which allows the stomach to secrete a stronger acid solution and facilitates digestion of large bones.

responsive to adrenergic control indicates that is an actively controlled mechanism to aid digestion [11].

Development of the Four-Chambered Heart in Mammals

The heart structures of the animals described above suggest a progressive evolution from the simple tube hearts of insects to the four-chambered hearts of crocodiles. Their evolutionary pathways are more complex than this, however. Although these species share a common ancestor, their pathways diverged in ancient times. The evolution of insects, molluscs and cephalopods diverged from that of vertebrates over 550 million years ago. Reptiles and mammals shared a common ancestor until around 320 million years ago and primates emerged around 65 million years ago. Biped early humans are thought to have appeared around 5.5 million years ago although this is still debated and is subject to new discoveries. Despite these divergent paths, genetic and molecular biological research has clarified that evolutionary adaptations are frequently conserved through different evolutionary routes, as exemplified in the similarities between the early development of hearts in *Drosophila* and vertebrates described above. Indeed this conservation is the basis of all progressive evolution. It is not surprising therefore that embryological development of the four-chambered heart in humans recapitulates some of the intermediate evolutionary hearts of insects, molluscs, fish and reptiles.

Division of the atrio-ventricular canal

As we have seen, a primitive heart tube forms in humans around 18 days after fertilisation and is reminiscent of the heart tube found in insects. It then begins to fold during the middle of the fourth week, taking on an "S" shape not unlike the two-chambered heart in fish (see Figure 2.5). At the same time, the heart undergoes a complex set of divisions which will lead to formation of the four-chambered heart. The first step is the division of the atrio-ventricular junction. This process begins at the middle of the fourth week as the primitive heart tube is folding. At this stage, the atria and ventricles are continuous single chambers. The first change is the partial division of these two chambers by the growth of tissues which extend inwards at the front and back of the atrio-ventricular (A-V) junction (Figure 2.11, middle panel). These are known as cushions which consist initially of a gelatinous material called cardiac jelly which separates the inner (endocardium) and outer (myocardium) layers of the heart at this stage. At the same time, some cells in the endocardium transform into mesenchymal cells which facilitates their migration into the extending jelly to continue their growth. This transformation involves reversion

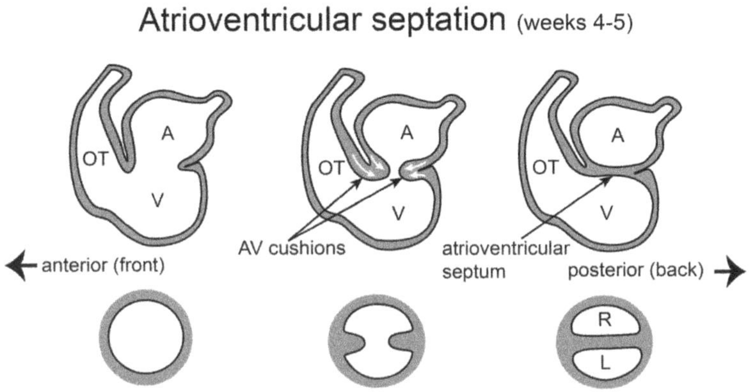

Figure 2.11: Lateral (viewed from the left side) diagrams of the heart (upper panels) during folding show the primitive atrium (A) and ventricle (V) and outflow tract (OT). The first step in forming the four-chambered heart is division of the atrio-ventricular (A-V) junction into left and right canals. This begins with the appearance of ridges at the anterior (front) and posterior (back) parts of the A-V junction (middle panel). Known as cushions, these are caused by expansion of cardiac jelly and migration and proliferation of mesenchymal cells from the local endocardium. These grow, extending into the junction and eventually fuse to form a septum which divides the A-V junction into left and right canals (right panel). These will become the left and right A-V canals in which the mitral and tricuspid valves, respectively, will be formed to ensure forward blood flow. The lower panels show diagrams of the A-V junction as viewed from above as the A-V septum is formed.

to a more primitive type of cell, mesenchymal cells; they become more rounded, they lose their ability to attach to adjacent cells and become pluripotential, meaning they can develop into many different cell types. As the cushions are invaded in this way they grow, eventually meeting and then fusing to form a bridge across the A-V junction (Figure 2.11, right panel), thereby partially separating the primitive atrium and ventricle and creating left and right A-V canals during the fifth week. The mitral and tricuspid valves will later form in these canals to regulate flow between the atria and ventricles.

Division of the primitive atrium

In adults, the left and right atria are separate and no blood flows between them. The transition from the primitive single atrium (Figure 2.12, left

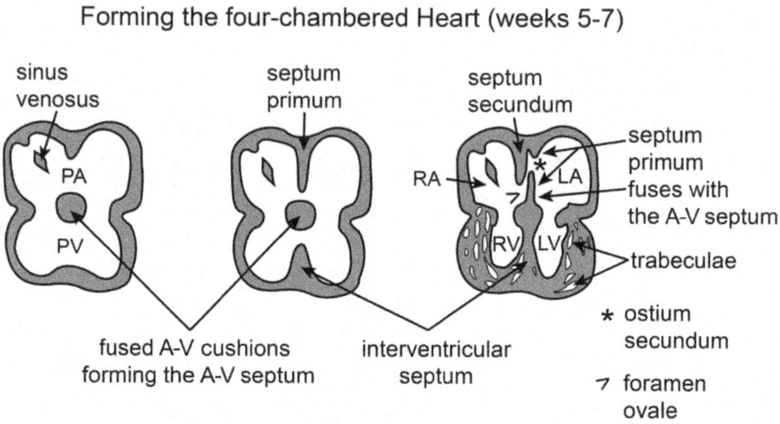

Figure 2.12: Ventral diagrams (viewed from the front after cutting away the front wall of the heart) of the heart during division of the primitive atrium (PA) and ventricle (PV). At this stage, the A-V septum is shown in cross section. The sinus venosus, which conveys blood from the placenta, is shown connected to the right atrium (RA). Division of the atria involves the formation of two septa. Initially the septum primum (middle panel) grows from the roof of the atrium and extends to fuse with the A-V septum. As it nears the latter, its upper part breaks down forming a hole named the ostium secundum (right panel). A second atrial septum, named the septum secundum, grows to the right of the septum primum and overlaps it but does not extend to the A-V septum (right panel). This arrangement is crucial for maintaining a viable circulation during intrauterine life and rapidly transitioning to postnatal life after birth. The lower part of the septum primum is relatively light and flexible and can be deflected by the momentum of blood from the right atrium (RA). This opening is known as the foramen ovale (right panel). Reverse flow is prevented because in the event of higher pressure in the left atrium (LA), the septum primum will be pressed against the stiffer septum secundum, effectively sealing the foramen ovale. Once the newly-born infant begins to breathe, the pulmonary circulation expands allowing the right ventricle to perfuse the lungs. Blood then returns to the left atrium (LA) via the pulmonary veins and increases left atrial pressure to a higher level than the right. This results in the septum primum being constantly pressed against the septum secundum to which it fuses shortly thereafter, sealing the interatrial septum. Division of the ventricles into right (RV) and left (LV) chambers results from the growth of a septum from the floor of the primitive ventricle which extends upwards towards the A-V septum with which it eventually fuses by the end of the seventh week. At this stage too, the ventricular walls have thickened with a spongy character (right panel).

panel) to this stage is complicated by the fact that throughout intrauterine life it is essential that blood continues to flow directly from the right to the left atrium. This is to allow oxygenated blood from the placenta to reach all parts of the growing foetus. Blood from the placenta returns to the right atrium and since the lungs are not in use it must bypass the pulmonary circulation and flow directly to the left atrium, ventricle and systemic circulation. This is achieved by maintaining an opening in the atrial septum until birth, which then closes soon afterwards. This occurs because of the way the atria are divided by the growth of two septa. The septum primum (Figure 2.12, middle panel) develops first from the top of the primitive atrium and eventually extends down to meet and fuse with the A-V cushion. As it does so, a process known as cell apoptosis occurs in the upper part of the septum primum. Apoptosis is a normal and highly regulated system of cell death initiated by machinery within the cells themselves, in this case triggered by chemical signals from the adjoining area. Here it results in holes appearing in the septum primum and formation of the ostium secundum which allows free blood flow to continue from the right to the left atrium (Figure 2.12, middle and right panels). At the same time, a second strong septum known as the septum secundum grows to the right of the septum primum and overlaps it by the end of the sixth week (Figure 2.12, right panel). The septum secundum is stiffer than the septum primum and this arrangement known as the foramen ovale enables the momentum of blood flowing from the right atrium to open the relatively light septum primum (to the left) and allow flow into the left atrium. In contrast, flow in the reverse direction is prevented because the septum primum is prevented from opening to the right by the stiff ostium secundum. This arrangement is crucial for the growing foetus and during the early transition to life after birth.

In the foetus, the pulmonary circulation is effectively closed because the lungs are compressed and will not expand until breathing can occur after birth. Therefore, the foramen ovale is essential to allow oxygenated blood from the placenta to reach the systemic circulation. Once the newly-born infant begins to breathe, the pulmonary circulation expands allowing the right ventricle to perfuse the lungs. Blood then returns to the left atrium (LA) via the pulmonary veins and increases left atrial pressure to a higher level than the right. This results in the septum primum being constantly pressed against the septum secundum to which it fuses shortly thereafter, sealing the interatrial septum.

Division of the primitive ventricle and outflow tract

As the primitive ventricle grows, it thickens and develops muscular ridges on its interior during the fifth week, known as trabeculations (Figure 2.12, right panel). These will later play an important role in forming the atrio-ventricular valves. At the base of the ventricle, a more prominent ridge appears and grows towards the atrio-ventricular septum, with which it will eventually fuse to complete the interventricular septum by the end of the seventh week.

In the adult heart, the right and left ventricles pump blood into the pulmonary artery and aorta, respectively. To achieve this, the outflow tract of the primitive ventricle must also undergo division, and this occurs during Weeks 5–7. This is the most complex three-dimensional change needed to complete the four-chambered heart. The outflow tract of the common ventricle consists of the conus cordis which is contiguous with the ventricle and the truncus arteriosus which joins the conus cordis to the aortic arches above (see Figure 2.5). Ridges appear on the left and right sides of both the conus cordis and in the truncus arteriosus during the fifth week and extend upwards and downwards, respectively, until eventually they meet, forming a continuous longitudinal ridge along the entire outflow tract [12]. However, the direction of this extension is not straight, but rather it extends in a spiral which completes a 180-degree rotation resulting in a helical ridge along the length of the outflow tract. The ridges on each side increase in size, extending into the lumen and eventually fuse during the sixth week. They also fuse below with the interventricular septum and atrio-ventricular septum. The effect of this change is that the outflow from the right and left ventricles is separated, becoming the pulmonary and aortic trunks, respectively. This process is reminiscent of the structure and function of the spiral valve found in frog hearts (see Figure 2.9), and it will not be surprising to find that genetic mechanisms which mediate it are also similar. The formation of the outflow tract and septation of the atria and ventricles require correct migration, proliferation and alignment of cells at several sites and for this reason is one of the most common areas in which developmental abnormalities can occur. These can take the form of failure to divide the outflow tract (common arterial trunk), double outlet right ventricle, interruption of the aortic arch, transposition of the pulmonary artery and aorta and ventricular septal defects.

Formation of the heart valves

The heart valves regulate the movement of blood through the heart ensuring one-way flow. Blood is not a "normal" fluid in the general sense.

It contains millions of cells which transport oxygen (red blood cells), counter infections (white blood cells), plug leaks in the circulation (platelets), as well as many proteins. It performs best in conditions of low turbulence. When a doctor or nurse listens to the heart with a stethoscope, normally nothing is heard except a noise of the valves closing. This reflects the absence of turbulence and requires that when "open" the valves must move out of the path of blood flow completely so as not to impede flow and yet while in the "closed" position they must be capable of withstanding a range of pressures reflecting varying physiological conditions. Man-made heart valves have saved many thousands of lives and are great achievements of medical technology. However, the challenge to match the function of our natural valves is formidable. A range of designs have been developed. A "ball and cage" valve ensures one-way flow, but the ball sits in the path of blood flow and causes turbulence which may damage some red blood cells or activate clotting mechanisms which can be harmful. Valves consisting of a restrained "wing-like" disk which "flies" in the stream of flowing blood offer much less impedance to flow but have posed design and construction challenges. Natural valves obtained from animals are also used. These are sterilised and "cured" to avoid allergic reactions and can function very well. However, they are not "living" tissues and are not nourished as our natural tissues are and consequently do not have the same longevity. That said, I should point out that artificial valves have been very successful. Patients in their sixties who have a new valve to replace a diseased aortic valve, once past the initial recovery stage, have an excellent long-term survival [13].

The heart contains four valves. The aortic and pulmonary valves regulate one-way blood flow at the outlets of the left and right ventricles, respectively. The mitral valve ensures one-way inflow from the left atrium into the left ventricle and the tricuspid valve has the same function between the right atrium and ventricle. As outflow valves for the ventricles, the aortic and pulmonary valves have similar structures and function in the same way. The mitral and tricuspid valves control ventricular filling and also have similar structures with the exception that the mitral valve consists of two rather than three leaflets. Blood flow and pressures differ significantly during filling and emptying of the ventricles and consequently the structure of the mitral and tricuspid valves is quite different from that of the aortic and pulmonary valves.

Mitral and tricuspid valves

The heart fills and empties in cycles. Emptying is achieved by contraction of the ventricular muscle, that is, the ventricular muscle shortens, squeezing blood out of the chamber. During contractions, pressure in the ventricles rises sharply and then falls away equally sharply as it relaxes. Filling takes place between contractions when the muscle is relaxed, and pressures are low. The structure and formation of the valves must reflect these conditions. Typically, left and right ventricular filling takes place at low pressures of about 3–5 mmHg. At such levels, the valves need to be very light and be capable of moving easily. During contractions, on the other hand, pressure rises to 130 mmHg and 30 mmHg in the left and right ventricles, respectively. At this stage, the mitral and tricuspid valves are closed and must be capable of withstanding such pressures. This dictates structures which are very light but robust in the face of high pressures.

Formation of the mitral valve begins during the fifth week and takes place concurrently with division of the ventricles and outflow tract (Figure 2.13, left panels). The two primordial mitral valve leaflets are formed by growth of tissue (a) at the point of fusion of the atrial septum primum and the atrio-ventricular septum (illustrated in Figure 2.12), supported by thickening of the ventricular muscle below and (b) on the opposing left ventricular wall (Figure 2.13, left upper panel). The latter expands to compose two-thirds of the developing valve and is also supported by a thickened ventricular muscle below. These thickened areas of spongy muscle below each cusp transform into muscular columns which support the ends of the forming valve cusps. They subsequently become increasingly compact, forming papillary muscles attached to the interior of the ventricle below and narrow tendons called chordae tendineae attached to the valve leaflets above (Figure 2.13, left lower panel). The leaflets also become thinner and finer, akin to curtains attached at one end to the circumference of the atrio-ventricular junction. This allows them to fall freely and quickly into the cavity of the left ventricle during ventricular filling and thereby allowing unrestrained blood flow. During contraction when pressure rises sharply, the leaflets are forced back towards the atrium, however tensioning by the papillary muscles and chordae ensure that the free ends are held in place with the opposing margins perfectly aligned to prevent back-flow.

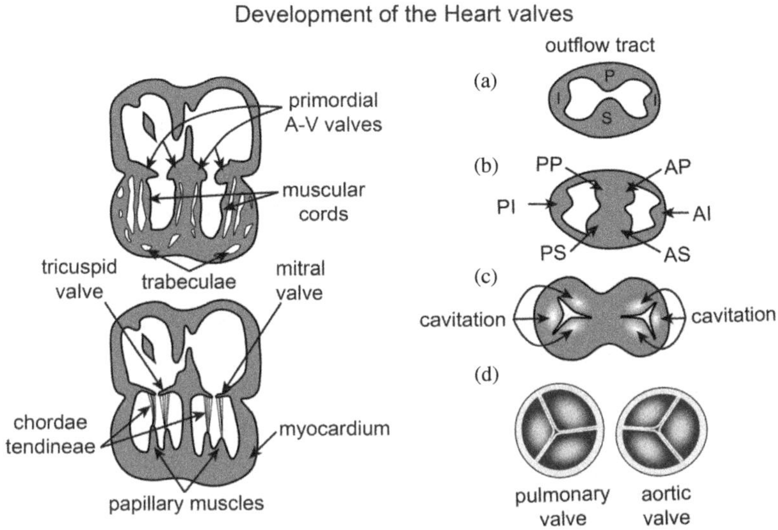

Figure 2.13: Diagram showing the formation of the heart valves. The mitral and tricuspid valves (left panels) develop from cushions which appear on the free walls of the left and right ventricles and from the fused atrio-ventricular septum and the atrial septum primum. The primordial leaflets are supported initially by thickened ventricles. The latter grow with a spongy character which then compacts to form muscular columns (left upper panel) supporting the valves and compact further to form the papillary muscles and chordae tendineae (left lower panel). The latter provide a tensioning apparatus which plays a crucial role in maintaining alignment of the valve margins under varying physiological conditions. The aortic and pulmonary valves (right panels) are formed at the junction of the primitive ventricle and the outflow tract. Initially two cushions appear (a); a septal cushion (S) forms on the side of the outflow tract where septation is occurring and the parietal cushion (P) on the opposite free wall. Somewhat later, two further swellings, named intercalated cushions, develop on the opposite quadrants (I). The septal and parietal cushions expand and fuse forming two canals in which the pulmonary and aortic valves will be formed (b). Further expansion of the fused junction occurs creating separate septal and parietal cushions in each canal, so that each now has three-cushions (b) and as they grow the cushions produce a three-cornered star shaped opening in each canal (c). At the same time, the outflow tract narrows at the points of the septal and parietal cushions and eventually divides into separate ventricular outlets (c and d). Each of the cushions then begins to cavitate creating triradial cup-shaped depressions when viewed from above (d). The fully formed valves each contain three cusps that are attached to the circumference of the ventricular outlet with internal margins of two sides at an angle of about 120°. The margins of each cusp abut against the other two to prevent back-flow when closed (d).

Notes: P = Parietal; S = Septal; I = Intercalated; PP = Pulmonary parietal cushion; PS = Pulmonary septal cushion; PI = Pulmonary intercalated cushion; AP = Aortic parietal cushion; AS = Aortic septal parietal cushion; AI = Aortic intercalated cushion.

Formation of the tricuspid valve has similarities with that of the mitral valve. As in the case of the mitral valve, a cushion develops on the free wall of the right ventricle (Figure 2.13, left panels). Concurrently, the right atrio-ventricular canal expands as the right ventricle develops and so does the newly formed cushion. This provides the tissue from which the two larger leaflets are formed. As with the mitral valve, they are initially supported by a thickened and trabeculated (spongy) right ventricle, which then becomes compacted forming the papillary muscle and chordae. The third leaflet and its tensioning apparatus are formed as a new layer which splits off the septum (Figure 2.13, left lower panel). The tricuspid valve is formed later than the mitral, the two larger leaflets appearing during the seventh week and the septal leaflet during the tenth. Completion of the valve occurs during the 12th week.

A remarkable feature of the atrio-ventricular valves is their ability to adapt to changing physiological conditions. During exercise or any condition requiring an increase in performance, heart rate and the force of contraction also change substantially. This involves marked changes in speed of contraction, reduction in the emptying and filling periods as well as a marked increase in the force of contraction and pressure generated in the ventricles. It is essential that the valves perform correctly under these conditions and this can be achieved because the papillary muscles and chordae which control the alignment of valve margins during closure are part of the ventricular muscle. This means that they undergo the same changes in the rate and force of contraction: if the ventricles contract more rapidly and forcefully, then the papillary muscles exert correspondingly faster and greater tension on the valve margins ensuring correct apposition and closure. This arrangement of tethered margins is also well suited to the relatively larger surface areas of the mitral and tricuspid valves (around 1.5 times larger than the aortic and pulmonary valves), a feature which is necessary for achieving rapid ventricular filling in the context of much lower pressures than that occur during emptying.

Aortic and pulmonary valves

The aortic and pulmonary valves, also referred to as semilunar valves because their cusps resemble half-moons, are formed at the junction of the primitive ventricle and the outflow tract (see Figure 2.11, left panel). During the sixth week, division of the outflow tract to form the aortic and pulmonary trunks is well advanced and below them four new cushions develop where it connects to the primitive ventricle which will form the aortic and pulmonary valves. The septal cushion forms on the side where atrioventricular septation is occurring and on the opposite free wall, the parietal cushion develops. Two further swellings, named intercalated cushions, develop on the opposite quadrants, shown in cross section in Figure 2.13(a). The septal and parietal cushions expand and fuse forming two canals in which the pulmonary and aortic valves will be formed (Figure 2.13(b)). This fusion is accompanied by further expansion creating separate septal and parietal cushions in each canal, so that each has three cushions. Each of these expands further producing a three-cornered, star shaped opening in each canal (Figure 2.13(c)). Simultaneously, the outflow tract narrows at the points of the septal and parietal cushions and eventually divides into separate ventricular outlets in conjunction with reconfiguration of the ventricles which is also taking place. As the aortic and pulmonary cushions expand further, their centres cavitate and evolve into triradial cup-shaped depressions (Figure 2.13(c) and (d)). The adult valves each contain three cusps or leaflets that are attached to the circumference of the ventricular outlet with internal margins of two sides at an angle of about 120°. The margins of each cusp are composed of stiff fibrous tissue and abut against the other two to prevent back-flow when closed (Figure 2.13(d)).

Formation of the aortic and pulmonary valves is broadly similar. The pulmonary valve has a slightly larger area than the aortic, reflecting that it must permit the same flow rate as the aortic valve at lower pressure (25 mmHg compared with 130 mmHg). Both valves also function in the same way. Viewed from above as in Figure 2.13(d), the cup-shaped cusps of each valve act as a trap preventing back-flow. Blood ejected from the ventricles would regurgitate once pressure falls at the onset of relaxation. This rapidly fills the cup-shaped cusps of each valve and exerts pressure on their margins forcing them tightly together and thereby sealing the valve. The valve cusps viewed from below have a convex rounded surface and are easily pushed upwards by the pressure of blood flow during ejection. Importantly too, the valve cusps lie close to the outflow walls during systole offering minimal impedance to blood flow.

Malformations of the heart

Abnormal formation of the heart (congenital heart disease) occurs in almost one in every hundred babies born. Most occur during formation of the valves or when the primitive heart tube is being divided to form the four chambers. About half of such babies have severe abnormalities for which there are no effective treatments available at present. The outlook for the remaining has improved greatly during the past century due to the development of surgical techniques and better medical management during and after surgery. Such abnormalities may be simply categorised in terms of those which occur outside the heart itself, mainly effecting the large vessels to which it is connected and abnormalities within the heart. They are also grouped depending on whether they cause cyanosis (blue babies). Recognition of abnormalities occurring outside the heart was important because they are more accessible and were the focus of initial surgical treatment. Cyanosis carries a more severe prognosis. This occurs when blood is unable to reach the lungs and therefore does not carry enough oxygen. Abnormalities which cause cyanosis are also more severe and more challenging to correct. Fallot's tetralogy is the most common form of cyanotic congenital heart disease in babies more than six months old. Babies with Fallot's have a narrow pulmonary valve which restricts flow to the lungs. Their right ventricles also grow larger and thicker than normal in response to greater effort required to eject their contents. Their interventricular septum is also incomplete with the result that their right ventricles partly empty into the left ventricle. The effect of this combination is that the lungs are largely bypassed, and the circulating blood is poorly oxygenated. Management of such cases remains challenging, however as a result of the extraordinary advances in surgical and medical techniques since the 1930s, they are now almost routine cases.

Surgical treatment of cardiac malformations

Initially surgery focused on abnormalities occurring outside the heart. Patent ductus arteriosus (PDA) was the first condition corrected surgically. In the womb, because the lungs and the pulmonary circulation are closed, blood pumped from the right ventricle must bypass them. This is made possible through a connection between the pulmonary artery and the aorta, called the ductus arteriosus, which diverts blood ejected by the right ventricle into the aorta. Normally this closes very soon after birth, but occasionally fails to do so, more often in preterm (about 8 in 1,000) than full term babies (2 in 1,000). In 1938, the first successful closure of PDA was undertaken by surgeons at the Children's Hospital in Boston. Having developed and demonstrated the technique, other surgeons soon followed. In 1944, surgeons in at the Karolinska Hospital in Sweden carried out the first successful repair of an abnormal narrowing of the aorta called aortic coarctation.

Cyanotic congenital heart disease proved a greater challenge due to its underlying complexity. Initial surgical efforts focused on getting more blood into the lungs, and in 1944 surgeons at John Hopkins Hospital in Baltimore introduced an operation to achieve this. Named the Blalock-Taussig shunt, its success was the result of several years of research. The initiative came from Helen Taussig, a cardiologist who cared for many patients with cyanotic congenital heart disease. She approached Alfred Blalock, a surgeon at John Hopkins Hospital, to enquire if something could be done to restore blood flow to the lungs in such patients. Vivien Thomas was an African American research assistant to Blalock and had been trained by him in experimental surgical techniques. Thomas developed a technique to anastomose the subclavian artery to the pulmonary artery and demonstrated that it could correct an experimentally induced cyanotic condition in dogs. After completing 200 such experimental procedures, demonstrating that it could be done, Blalock then repeated the procedure with him in dogs and was convinced it would work in humans. Together they undertook the first operation on a child in 1943. In time the operation became standard surgery for reversing cyanosis in the early management of Fallot's tetralogy. Although life-saving, it was still only palliative in that the underlying abnormality remained. Correction of the internal cardiac anatomy would need open-heart surgery.

The development of open-heart surgery proved a formidable challenge. It was evident in the 1930s that this would have to be done during cardiac arrest and it was known that stopping the circulation for more than three minutes produced permanent organ damage. Initial attempts focused on lowering body temperature and in 1952 surgeons successfully repaired an atrial septal defect under direct vision using moderate hypothermia. However, the safe open-heart time using this technique was too short to repair more complex lesions. To do this, surgeons began to explore means of extracorporeal support to substitute for the heart. John Gibbon at Massachusetts General Hospital developed the first artificial heart-lung machine consisting of a blood oxygenator and pump. His results demonstrated the feasibility of the technique, but initial outcomes were poor and discouraged its further use. However, they did encourage surgeons at the Mayo Clinic in Minnesota to investigate and modify his machine and they went on to use a heart-lung machine to develop the first successful program of open-heart surgical repair for congenital heart disease in 1955. Further refinement and simplification of blood oxygenation contributed to its rapid spread in the late 1950s and 1960s [14]. At that time, it was believed that surgery carried a higher risk in younger children and was therefore delayed where possible until age 5–7 years. In some cases, this required an additional palliative operation carried out earlier. With further knowledge and experience, it became clear that younger children could tolerate surgery. Furthermore, early correction allowed the heart to develop normally. It also avoided the need for palliative surgery and has become common practice since the early 1970s. Nevertheless, surgery at such a young age remains a significant challenge for patients and surgeons and this has stimulated further efforts to find less intrusive methods.

The concept of cardiac repair using catheters inserted through arteries or veins has been widely used for treating coronary artery disease since the late 1970s. Balloon angioplasty was originally introduced by Charles Dotter in 1964 to dilate large peripheral arteries. Andreas Gruentzig modified the technique to dilate coronary arteries in 1977. In 1967, Porstmann used a catheter technique to deliver a plug to close a patent ductus arteriosus [15] and Terry King successfully closed an atrial septal defect in 1975 using a catheter to deliver an umbrella shaped device [16]. Since then this has become the method of choice for treating many patients with coronary artery disease. Efforts to extend percutaneous approaches to repair other cardiac malformations are ongoing and no doubt will lead to improvements. More complex malformations are often unique and require detailed understanding of individual anatomy including 3-D reconstructions as part of the diagnostic workup. It is inevitable that surgery will continue to be needed for many cases of congenital heart disease.

At present, interest is focused on diagnosis before term and exploring possibilities of intrauterine intervention in cases where the malformation is severe and life threatening. Experience in this area is limited to some hundreds of cases globally. However, some remarkable successes have been achieved, for example, in dilating severely narrowed valves and in creating an opening in the atrial septum to improve blood flow to the foetus in cases of transposition of the aorta and pulmonary arteries a condition where their ventricular connections are reversed [17].

Genetics of the Four-Chambered Heart

Given the challenging nature of these conditions, efforts to understand the molecular and genetic origins of cardiac malformations have become the focus of intense research in recent years. Within a matter of days, the developing embryo designates certain cells as the cardiac progenitors which will form the heart, a process triggered by key genes and signalling molecules (Figure 2.7). Formation of the four-chambered heart from the primitive heart tube is immensely complex and depends on an ordered sequence of changes occurring at critical times. These too are triggered by timely expression of key genes and secretion of signalling proteins, which act singly and in conjunction with others in different regions of the developing heart at specific times [18]. The table is a simplified summary of their functions and interactions. TBX5 is a member of a large family of T box genes, of which currently 14 are known. These were first discovered in chromosome 17 of mice, mutations of which were associated with mice having short tails. TBX genes produce transcription factors which play a major role in limb and organ development, TBX5 being involved in heart development. In vertebrates, it has an important role on forming the atrial and ventricular septa and in a specialised area of the conduction system. GATA4 is a member of the GATA family which as discussed earlier plays important roles in early heart development (Figure 2.7). GATA4 also contributes to cushion formation and in conjunction with NKX2.5 to synthesis of contractile proteins in maturing heart muscle cells. NKX 2.5 is also important for division of the atria, ventricles and outflow tract and with

TBX5 for the conduction system of the heart. The various functions of these genes acting alone or in combination depend on the location and timing of their expression. Not surprisingly therefore a simple picture of single genes causing specific congenital heart defects does not emerge. Most cases occur sporadically with no known genetic link, but genetic associations are sometimes found, for example, alterations in the TBX1 gene are associated with tetralogy of Fallot and persistent truncus arteriosus (failure to divide the primitive outflow tract). In addition, some conditions such as atrial and ventricular septal defect are associated with abnormalities on more than one gene, TBX5, NKX2.5 and GATA4.

Transcription factor producing genes which regulate formation of the four-chambered heart	
TBX5	Formation of the atrial and ventricular septa
GATA4 + TBX5	Specialised areas of electrical conduction
GATA4	Formation and growth of cushions
GATA4 + NKX2.5	Synthesis of contractile proteins in heart cell
NKX2.5	Formation of the outflow tract, atrial and ventricular septa
TBX5 + NKX2.5	Formation of the heart conduction system, control of the heart chamber development

It is increasingly understood that blood flow patterns within the developing heart have an important impact on how it forms. Abnormal growth of the right ventricle in response to outflow obstruction as in Fallot's is one example of this and indicates that the earlier abnormalities begin in the developing heart, the more harmful they can be. In this context, the discovery that nitric oxide synthase (NOS) plays an important role in heart formation is important. NOS is an enzyme which converts L-arginine to L-citrulline, producing nitric oxide (NO) in the process. The latter is a small molecule which can rapidly diffuse through tissues and is an important signalling molecule in many physiological processes including the behaviour of blood vessels. All blood vessels are lined by a layer of endothelial cells which can sense the rate of blood flow. Increases in flow rate stimulate NOS in these cells to produce NO which rapidly permeates the vessel wall causing it to relax. Recent studies have shown that NOS is also involved in formation of the outflow tract and the atrio-ventricular canal in the developing heart. Its absence is associated with impaired cushion formation, leading to abnormalities in the aortic and mitral valves and incomplete septation of the atria and ventricles [19, 20]. These discoveries are of special interest because they raise the possibility that blood flow patterns in the developing heart may influence its formation through sensing mechanisms similar to those which regulate the adult circulation. Improvements in imaging techniques hold the promise of earlier detection of abnormal blood flow in the foetal heart and the future possibility of intervention to preserve its normal growth. An additional long-term challenge is the fact that artificial devices such as valves or systems to seal abnormal openings implanted in childhood cannot grow with the heart and often requires follow-up surgery with its attendant disadvantages. Availability of devices of a size and structure which matches the abnormality of individual cases would be a significant advance. The ability to fabricate such devices from engineered tissues derived from stem cells harvested at the time of birth holds the promise that they would grow physiologically with the child's heart.

Conclusion

The first tubular heart probably evolved in a bilaterian ancestor over 600 million years ago and its development would have incorporated genetic processes that evolved much earlier as essential molecular building blocks. Many of these have been conserved through subsequent evolution; for example, some of our enzymes involved in energy production are also found in bacteria and components of the contractile molecules in our hearts are also found in bacteria [21] and in yeasts [22]. Over millions of years, hearts of many different structures have evolved from simple pulsatile tubes to our own four-chambered heart. It would be misleading to suggest that our more complex hearts are superior or have reached a higher state of evolution. Rather, each arrangement seems uniquely and perfectly matched to the energy needs of the lifestyle they support. Diversification from an aquatic to a land-based life and bipedalism imposed marked increases in energy requirements and seems to have been the stimulus to further evolution including the four-chambered heart in mammals.

References

[1] Dor Y, Cedar H. Principles of DNA methylation and their implications for biology and medicine. *Lancet*. 2018;392:777–786.

[2] Wasserthal LT. Drosophila flies combine periodic heart beat reversal with a circulation in the anterior body mediated by a newly discovered anterior of ostial valves and venous channels. *Journal of Experimental Biology*. 2007;210:3707–3719.

[3] Chen JN, Fishman MC. Genetics of heart development. *Trends in Genetics*. 2000;16:383–388.

[4] Staehling-Hampton K, Hoffmann FM, Bayliss MK, Rushton E. *dpp* induces mesodermal gene expression in *Drosophila*. *Nature*. 1994;372: 783–786.

[5] Hardison R. Hemoglobins from bacteria to man: Evolution of different patterns of gene expression. *Journal of Experimental Biology*. 1998;201: 1099–1227.

[6] Moalli R, Meyelgs RS, Jackson DC, Millard RW. Skin circulation of the frog, Rana catesbeiana: Distribution and dynamics. *Respiration Physiology*. 1980;40:137–148.

[7] Andersen JB, Rourke BC, Caiozzo VJ, Bennett AF, Hicks JW. Postprandial cardiac hypertrophy in pythons. *Nature*. 2005;434:37–38.

[8] Jensen B, Nielsen JM, Axelsson M, Pedersen M, Löfman C, Wang T. How the python heart separates pulmonary and systemic blood pressures and blood flows. *Journal of Experimental Biology*. 2010;213:1611–1617.

[9] Seymour RS, Bennett-Stamper CL, Johnston SD, Carrier DR, Grigg GC. Evidence for endothermic ancestors of crocodiles at the stem of archosaur evolution. *Physiological and Biochemical Zoology*. 2004;77:1051–1067.

[10] Farmer CG, Uriona TJ, Olsen DB, Steenblik M, Sanders K. The right-to-left shunt of crocodilians serves digestion. *Physiological Biochemical Zoology*. 2008;81:125–137.

[11] Franklin CE, Axelsson M. An actively controlled heart valve. *Nature*. 2000;406:847–848.

[12] Anderson RA, Webb S, Brown NA, Lamers W, Moorman A. Development of the heart: (3) Formation of the ventricular outflow tracts, arterial valves and intrapericardial arterial trunks. *Heart*. 2003;89:949–958.

[13] Sharabiani MTA, Fiorentino F, Angelini GD, Patel NN. Long-term survival after surgical aortic valve replacement among patients over 65 years of age. *Open Heart*. 2016;3(1):e000388.

[14] Castañeda A. Congenital heart disease: A surgical-historical perspective. *Annals of Thoracic Surgery*. 2005;79:S2217–S2220.

[15] Porstmann W, Wierny L, Warnke H. Der Vershlouβ des Ductus arteriosus persistens ohne Thorakotamie. *Thoraxchirurgie und Vaskuläre Chirurgie*. 1967;15(2):199–203.

[16] King TD, Thompson SL, Steiner C, Mills NL. Secundum atrial septal defect: Nonoperative closure during cardiac catheterization. *JAMA*. 1976;235:2506–2509.

[17] Jaeggi E, Renaud C, Ryan G, Chaturvedi R. Intrauterine therapy for structural congenital heart disease: Contemporary results and Canadian experience. *Trends in Cardiovascular Medicine*. 2016;26:639–646.

[18] Kloesel B, DiNardo JA, Body SC. Cardiac embryology and molecular mechanisms of congenital heart disease: A primer for anesthesiologists. *Anesthesia and Analgesia*. 2016;123:551–569.

[19] Liu Y, Lu X, Xiang F-L, Lu M, Feng Q. Nitric oxide synthase-3 promotes embryonic development of atrio-ventricular valves. *PLoS One*. 2013; 8:1–13.

[20] Lee TC, Zhao YD, Courtman DW, Stewart DJ. Abnormal aortic valve development in mice lacking endothelial nitric oxide synthase. *Circulation*. 2000;101:2345–2348.

[21] Garner EC, Campbell CS, Mullins RD. Dynamic instability in a DNA segregating prokaryotic actin homolog. *Science*. 2004;306:987–989.

[22] Watts FZ, Miller DM, Orr E. Identification of myosin heavy chain in Saccharomyces cerevisiae. *Nature*. 1985;316:83–85.

3

Molecular Motors

As conscious beings, our thoughts are mostly concerned with the external world and our interactions with it. We observe the movement of other people, animals and objects to plan our safe interactions with them. In health, we are only occasionally aware of movements within our bodies, and yet internally we are constantly on the move. Blood flows continuously through our arteries and veins, food passes through our intestines and air flows through our lungs. At a microscopic level, particles and molecules are constantly moving to fulfil functions such as conveying signals between different parts of the cell or for structural maintenance. Most of this is beyond the reach of our consciousness. Were it otherwise, we would be overwhelmed by the number and complexity of the processes involved, which far exceed the capacity of our conscious minds to regulate. Nevertheless, we are aware of certain movements within us from time to time. Exertion and excitement can make us aware of the heart beating and respiration. Occasionally we become aware of intestinal movements or the "kicking" of a growing foetus. And although we have little or no conscious control over them, they are important elements of our experience. As discussed in Chapter 1, we frequently use a shared understanding of how the heart responds to emotion to convey how we or others feel, and the first "kick" of a growing foetus is a moment to remember. This chapter is concerned with how these movements are powered, and, although it is focused on the heart, the basic motor involved is very similar in all of them.

At its simplest, we think of the heart as a pump that receives blood at the end of its transit through the circulation and ejects it again, maintaining the momentum of flow. It is an intermittent pump, filling and emptying consecutively. The emptying phase generates a pulse of pressure which is transmitted as a wave through the arteries around the body and is the basis of the pulses. When we exercise or become aroused for any reason, we may become conscious of the heart beating. We may feel and sometimes see the pulse and we may be aware of the movement of the heart within the chest. However, for most of our lives, we are unaware of the heart, and yet it is astonishingly resilient and durable, for it never rests. Over the course of a day, an adult heart will beat around 100,000 times and pump about 8,500 L of blood, amounting to around 2 billion heartbeats and over 200 million litres of blood in an average lifetime.

This mostly silent working of the heart also hides a flexibility and adaptability which allows it to respond to the varying conditions of our lives. This requires a system of regulatory mechanisms which recognise the needs of our bodies minute by minute and communicate them to control the performance of the heart. At the centre of this is a molecular motor which powers the heart and its ability to modulate the force it generates and the speed with which it does so. This motor will be the main topic of this chapter, but initially a brief description of the regulatory mechanisms which control the heart and interact with its motor will be given.

Regulation of the Heart

The heart responds to the body's needs under varying conditions such as during exercise, increased mental activity or alarm. These responses are immediate and graded depending on the level of demand. Modest changes in our heart's performance track our needs minute by minute and this has two main advantages. It accurately provides for our needs and optimises energy utilisation by the heart. It also means that the heart is normally working well within its physiological range, which is important for durability. In other words, our body's energy needs are constantly being monitored and used to regulate the heart's activity. This avoids over or underuse of its pump function and keeps it within a safe operating range. The heart has a wide physiological range in which it can function safely and increased activity within that range as during exercise is beneficial to the heart and health generally. Indeed, there is abundant evidence that physical exercise is beneficial for the heart and our health generally. However, sustained exposure of the heart to abnormal demands can be harmful. High blood pressure is currently the most common risk for early cardiovascular disease in the world. It is harmful for several reasons, one of which is that high blood pressure requires the heart to pump against an abnormally high resistance over prolonged periods. The heart responds by growing so that its walls become thicker. This is helpful in meeting the demand in the short term but if prolonged is harmful. In contrast, intermittent exercise as in training also causes elevations in blood pressure, but these are temporary and although they may induce some growth of the heart, this is normal and reversible. Indeed, exercise benefits cardiovascular health. It follows that such close matching of cardiac function with demand means that its activity is carefully regulated. The mechanisms which provide this regulation can be divided into those which operate outside the heart and those which are intrinsic to the heart itself.

External regulation

In order to match the heart's performance to the needs of the body, we have sensors outside the heart which gather the information needed to regulate its function. Humans and all vertebrates have blood pressure sensors located on the main outlet from the heart (aorta), the carotid artery, large veins and pulmonary arteries. These are sensitive to stretch and, depending on their location, can detect both high and low blood pressure through distension of the vessel wall. They transmit signals through the glossopharyngeal and vagus cranial nerves to the brain which form the "input" side of a reflex that regulates blood pressure by controlling the performance of the heart and blood vessels. The "output" side of this reflex is composed of sympathetic and parasympathetic nerves, which have stimulatory and inhibitory effects, respectively, on the heart. They both enervate the sinoatrial node (the heart pacemaker), parasympathetic stimuli slowing the heart rate and sympathetic stimuli increasing it. At rest, the parasympathetic system is dominant, modulating heart rate at low levels of demand by increasing and decreasing stimulation and can induce changes in heart rate within a few beats. The sympathetic system is activated with higher levels of demand and provides a more powerful stimulus capable of driving the heart rate to much higher levels. Sympathetic nerves also enervate heart muscle (myocardium) and stimulation increases the force of contraction. We also have chemical sensors (chemoreceptors) located in the aorta and carotid artery, which sense levels of oxygen (O_2), carbon dioxide (CO_2) and acidity (pH). These also convey signals to the brain via the glossopharyngeal and vagus cranial nerves, providing additional input to the reflexes which regulate the heart through the autonomic nervous system. All of this neurological regulation of the heart occurs below our conscious level and is part of the autonomic nervous system.

Internal regulation

In addition to these external control systems there are several mechanisms within the heart which regulate its function.

Frank–Starling response

The best known of these is the Frank–Starling effect, named after Otto Frank and Ernst Henry Starling, and was formulated in 1914. It states that the heart responds to increases in filling by increasing its output. The degree of ventricular filling is an indication of the work the ventricle has to do and is known as *preload*, meaning the load *before* the aortic valve opens. The basis for the Frank–Starling response is that the ventricles can sense an increase in preload through mechanoreceptors which are sensitive to stretch and in response, induce an increase in the force of contraction and stroke volume. This is intrinsic to the heart muscle, requiring no external input and allows the heart to adjust its performance to changes in venous return to the heart. It is also an immediate effect such that it can be observed following even a single heartbeat in which the filling volume is increased. It is important for maintaining normal cardiovascular function. For example, when an athlete commences vigorous exercise such as running, the activity of his leg muscles squeezes out blood into the veins, and this leads to a sharp increase in blood flow returning to the heart. Without a prompt compensatory response by the heart, it would cause an increase in back pressure within the lungs and seriously hamper respiration. Sometimes we may become aware of this for a short period after lying down. Changing from standing to lying down alters the effect of gravity on the circulation and, for a short while, allows more blood to flow back to the heart from the legs and abdomen. This, in turn, increases ventricular filling and ejection volume and often results our being aware of the heart beating for a short while until the circulation readjusts.

The Anrep effect

Contraction of the ventricles commences at a time when the outlet valves are closed. Pressure within the ventricles rises quickly to a level that forces the outlet valves to open and ejection of blood begins. From the start of contraction until the aortic (outlet) valve opens, the contraction is referred to as isovolumic, meaning the volume of blood in the chamber remains unchanged. The main effect on the myocardium is an increase in wall stress due to the rising pressure. Once the outlet valve opens, ejection of blood commences and ventricular volume decreases. During ejection, the ventricle works against the resistance of the circulation, which can vary considerably as, for example, during exercise or with elevated blood pressure for any reason. This resistance is known as the *afterload* (meaning after the outlet valve opens). It has been known for over 100 years that the heart increases the force of contraction in response to increases in afterload. Known as the Anrep effect after Gleb von Anrep [1], it is intrinsic to the heart and is mediated by myocardial stretch in response to a higher resistance during ejection. Mechanoreceptors in the myocardium detect this and respond by signally an increase in the pumping force of the heart. This also has an important role in ensuring that the heart is capable of matching demand; without it the heart would fail to empty correctly under varying conditions with the risk of fluid back up and congestion.

Force frequency response

Increasing the cardiac output is readily achieved by speeding the heart rate and this is a daily occurrence in our lives. However, changes in heart rate have effects on the filling and emptying of the heart. For example, increasing the heart rate from 70 to 100 beats per minute shortens the period in which filling and emptying must occur with each beat. This means that if the contraction time remained the same, it follows that the ventricles would have less time for filling. The heart can accommodate for this by (a) increasing the force of contraction when the rate increases, (b) speeding up contraction so that it is completed in a shorter time and (c) speeding up the process of relaxation. In other words, the entire filling and pumping process is actively modulated in response to changes in heart rate, and this too is an intrinsic property of heart muscle and independent of all external factors. It is a vital property to allow the heart to respond normally to changes in heart rate; without it, increasing cardiac output at higher heart rates would be impossible due to inadequate filling.

In order to implement these control mechanisms, the pumping machinery in heart muscle needs to be sensitive and adaptable to changing conditions. The heart can do this because it has evolved a motor that operates at a molecular level and contains several features which enable it to respond to signals emanating from all these regulatory systems. It is an astonishing evolutionary achievement whose origins can be traced back to the beginnings of life on earth.

The heart is composed almost entirely of muscle and like all muscles, its power comes from its ability to shorten. In skeletal muscle, this shortening is most often used to move hinged bones. The heart chambers have the shape of hollow spheres so that shortening results in a reduction in their volume, which forces blood out. This property of shortening or contraction is due to the way in which the tiny motors which power the heart operate. At a molecular level, they consist of parallel filaments which slide along one another by an active process that generates force through the expenditure of energy.

Structure of the Heart

The heart is mostly composed of cells called myocytes that have the ability to contract along their length. They are roughly cylindrical in shape and around 200–240 μm in length. Adjacent cells are firmly attached at their ends. Heart muscle is regulated to ensure that all these interconnected cells contract in unison which results in substantial contraction of the entire organ. This is the essence of how the heart is able to contract and pump blood out of the ventricles. It also implies of course that the contraction of heart cells is highly organised to achieve uniform contraction and this will be the subject of Chapter 4. There are about 2–6 billion of these myocytes in an adult human heart.

Figure 3.1 is an electron micrograph showing parts of three myocytes in a longitudinal section (the long axis of the cells is from top right to

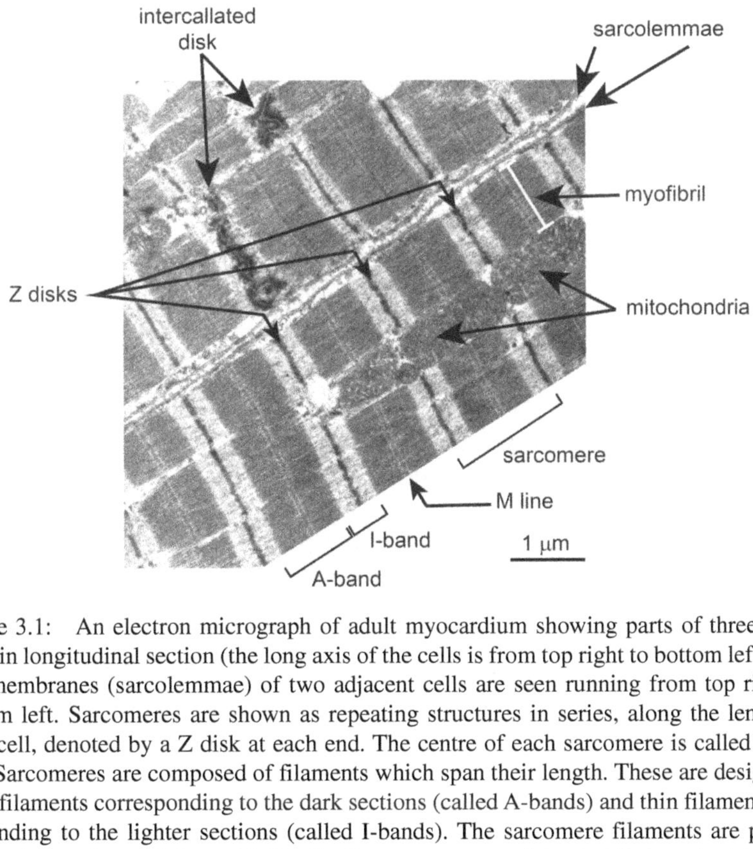

Figure 3.1: An electron micrograph of adult myocardium showing parts of three myocytes in longitudinal section (the long axis of the cells is from top right to bottom left). The cell membranes (sarcolemmae) of two adjacent cells are seen running from top right to bottom left. Sarcomeres are shown as repeating structures in series, along the length of each cell, denoted by a Z disk at each end. The centre of each sarcomere is called the M line. Sarcomeres are composed of filaments which span their length. These are designated thick filaments corresponding to the dark sections (called A-bands) and thin filaments corresponding to the lighter sections (called I-bands). The sarcomere filaments are packed closely in bundles called myofibrils. Cardiac myocytes are joined at their ends at intercalated disks, seen in the upper left section. These are formed by fusion of the terminal Z disk of each cell and afford strong junctions between adjacent myocyte so that the myocytes function as a continuous series of contracting units. The energy required for contraction is generated in the form of high energy phosphate compounds in mitochondria, from oxygen and nutrients.

bottom left). Immediately apparent is the regular appearance of the cellular contents. The repeating structures are called sarcomeres (from the Greek word *sarx* for flesh or muscle and *meros* for part), designated as the section between two Z disks, from the German *zwischenscheibe* (meaning disc between). The Z lines also provide a strong junction between adjacent sarcomeres, forming a contiguous series along the length of the myocyte. The light and dark sections of each sarcomere are called I-band and A-band, respectively. The "A" comes from anisotropic, meaning that its properties are dependent on its orientation. In early research based on light microscopy, it was found that A-bands refracted light differently depending on the angle of incidence, hence the termed anisotropic. The lighter sections of sarcomeres, which lack this property, were termed I-band, for isotropic. Each sarcomere is composed of filamentous proteins, which span its length attached at each end to the Z disk. The centre of each sarcomere is called the M line and is simply descriptive from *Mittelscheibe,* meaning middle. Sarcomere filaments are packed closely in bundles called myofibrils which give the repeating structure seen in Figure 3.1 and led to cardiac and skeletal muscle being described as "striated" on light microscopy. The nature and form of the filaments will be described in more detail below. Myocytes also contain mitochondria which produce the energy required for contraction, nuclei containing the cell's DNA, sarcoplasmic reticulum (which acts as a store for calcium ions) and other structures required for the activation of contraction. The cell membrane (called sarcolemma) of two adjacent cells is seen running diagonally (top right to bottom left) in Figure 3.1. In addition, intercalated disks, by which adjacent cells are joined at their ends, are shown in the upper left section. These are formed by fusion of the terminal Z disk of each cell. The intercalated disks afford strong connections between cells to support the forces generated by the contraction of each cell.

The filamentous proteins which span sarcomeres consist of "thick" and "thin" filaments and a very large protein called "titin". Together they make up around 47% of the volume of myocytes. The arrangement of the filaments within sarcomeres is illustrated in Figure 3.2, each sarcomere is composed of two equal halves. Taking the left half of Figure 3.2, five pairs of thick filaments with projecting structures (called myosin heads, discussed in more detail below) are shown abutting on

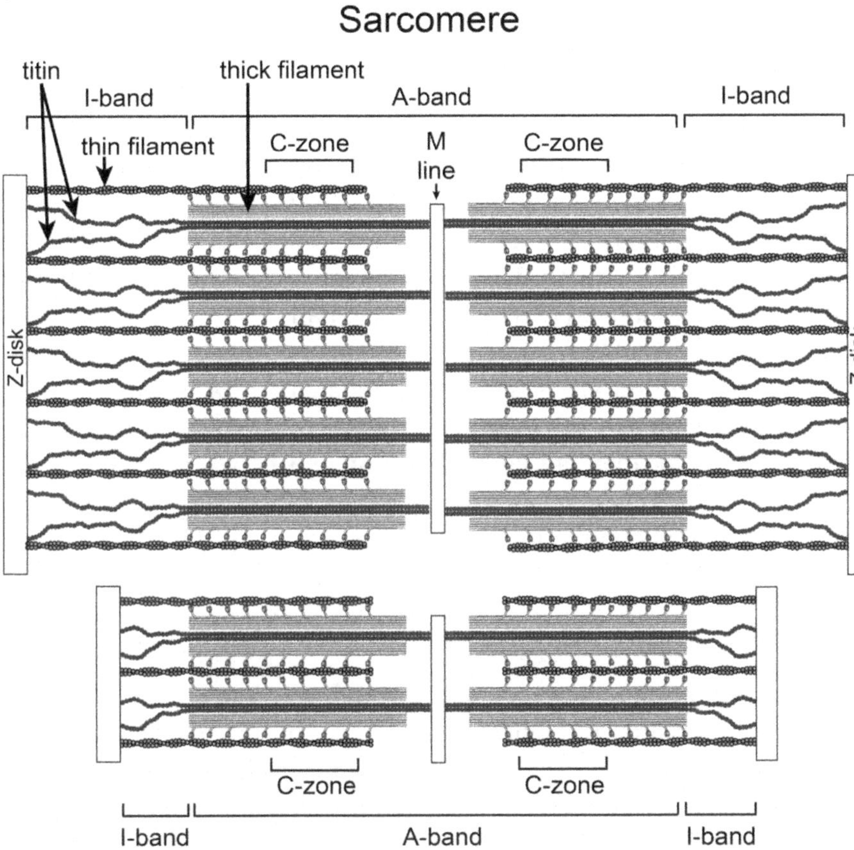

Figure 3.2: Diagram to illustrate sarcomere changes during contraction. The A-band relates to the thick filament region, I-band to thin filament section and the C-zone to the location of myosin binding protein-C. The midline of the sarcomere is called the M line and the sarcomere ends are attached to Z disks. During contraction (lower panel), cyclic actin myosin interaction propels the thin filaments towards the centre of the sarcomere sliding between thick filaments, resulting in shortening of the sarcomere.

adjacent thin filaments. This is repeated in mirror image form on the right side. At each end of the sarcomere, thin filaments are attached to a Z disk. Contraction involves thin filaments sliding along the thick filaments towards the centre of the sarcomere and therefore reducing its length, illustrated in the lower panel of Figure 3.2. The sliding motion is generated by interaction between the protruding myosin heads of the thick filaments and the thin filaments. In the relaxed state, sarcomeres are about 2 μm long and contract to around 1.6 μm. Since adjacent sarcomeres are joined at the Z disks along the length of myocytes and the intercalated disks afford strong junction between the ends of myocytes, it will be apparent that the contraction process from the level of sarcomeres upwards is functionally continuous. Cardiac myocytes are around 200–240 μm in length, so that we may consider them to have 100–120 sarcomeres in series along their length. From an estimated 2–6 billion cells in an adult human heart, a rough calculation based on the number of cells and the volume fraction of myofilaments suggests that there are about 12 quadrillion (10^{15}) sarcomere filaments in a human heart. In other words, although the contraction of each sarcomere filament is a mere 0.4–0.5 μm, they are additive within and between myocytes and this together with their connection in series and in such vast numbers results substantial contraction of heart muscle.

Molecular Motor Structure

The filaments in sarcomeres are called "thin" and "thick" reflecting their microscopic appearance (Figures 3.2 and 3.3). The thin filaments are composed of two strand-like polymers of actin molecules held together in a double helix that repeats about every 38 nm. The individual actin molecules are proteins consisting of 374 amino acids. The thick filaments are composed of myosin. Myosin consists of two proteins known as "heavy chains" and "light chains".

Myosin heavy chains

Myosin heavy chains have a head, neck and tail region and are arranged in pairs with the tail of each pair held together in a double helix about 150 nm long (Figure 3.4). They combine in pairs in a repeating pattern so that there are three pairs of heads, necks and tails at regular intervals of 43 nm along the thick filament. The thick filament itself is formed by the combined tails of all myosin heavy chains. The myosin head and neck can project about 20 nm from the thick filament towards the adjacent thin filament. Each head is capable of interacting with actin molecules on the adjacent thin filament in a manner which moves the thick filament along

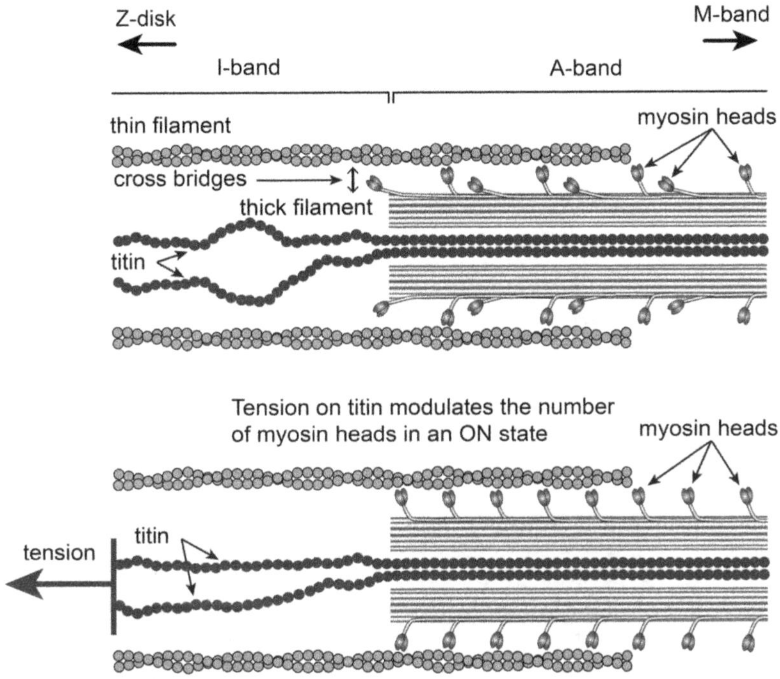

Figure 3.3: Diagram illustrating the arrangement of thick and thin filaments in sarcomeres. Thin filaments are composed of strings of actin molecules arranged in pairs as a double helix. The thick filaments are composed of myosin. These contain head, neck and tail sections. Myosin heads attach to and detach from actin molecules of the thin filament in a cyclic process which moves the thin filament towards the centre of the sarcomere resulting in shortening. The neck region acts as a lever which amplifies the motion generated by the head. The angle of the necks determines the availability of the heads to interact with actin; when myosin heads are lying flat the thick filaments are in an off state and when more perpendicular are in an on state. The myosin tail sections, together with those of other myosins, form the backbone of the thick filaments. Titin is a giant protein which spans the length of each half sarcomere and is closely aligned with a thick filament. They afford an important spring function in the sarcomere and when stretched (lower panel), they interact with myosin resulting in more myosin heads arranged perpendicular to the thick filament and thus promoting the on state of myosin. This is an important element of a mechano-sensing mechanism by which passive stretch induces stronger contractions. Known to physiologists for more than a century as the Frank–Starling response, it enables the heart to eject with greater force in response to increased filling.

it by a process known as cross bridge interaction. Figure 3.4 illustrates this arrangement in more detail with myosin heads in relation to active sites on actin with which they interact. It also illustrates the coiled pairs of heavy chain tails which form the thick filament backbone.

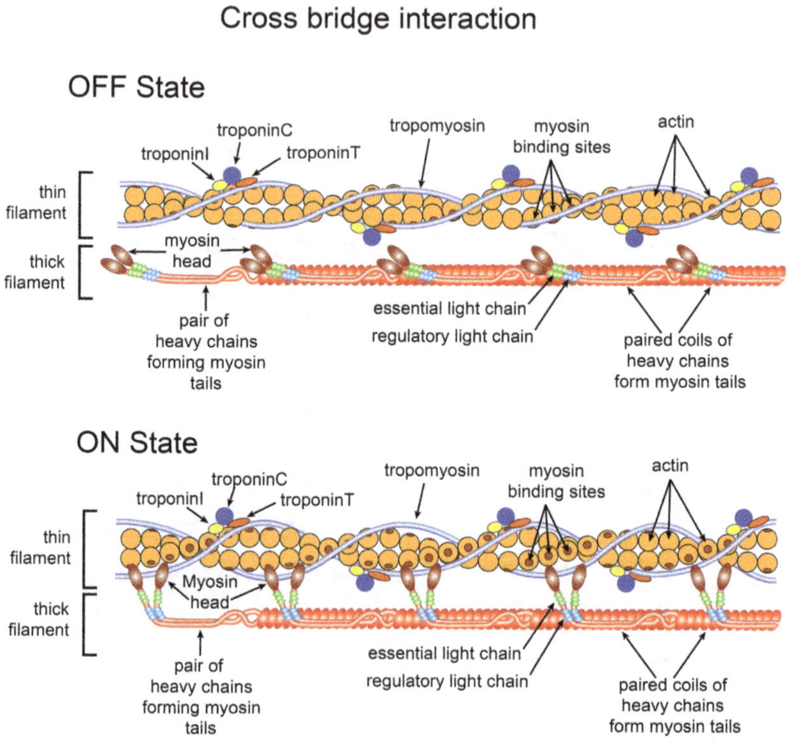

Figure 3.4: Diagram of thick and thin filaments in greater detail. Thick filaments are composed of pairs of myosin molecules with heads that project towards the actin molecules of the thin filaments. The tail sections of each pair of myosins are coiled together and with others form the back bone of the thick filaments. Myosin light chains are attached to the neck region of myosin. Thin filaments are composed of two strings of actin molecules arranged in a double helix. Troponin and tropomyosin are attached to the thin filaments and regulate contraction and relaxation in heart muscle. The upper and lower panels illustrate filaments in on and off states, respectively. Thin filaments are in the off state (upper panel) when tropomyosin tightly surrounds actin in a manner which blocks the myosin binding sites on actin molecules, preventing cross bridge interaction. Myosin heads lying along the thick filament are also less able to interact with actin, conferring an off state to thick filaments. The on state is illustrated in the lower panel. The rise in Ca^{2+} induced by an action potential increases Ca^{2+} binding to troponin C on thin filaments which causes a conformational change in troponin I. This acting through troponin T moves tropomyosin away from the myosin binding sites on actin molecules, making them available for cross bridge interaction with myosin heads. The angularity of myosin necks on thin filaments determines the ability of the heads to interact with actin. This is influenced by several factors; phosphorylation of myosin regulatory protein and myosin binding protein-C both increase perpendicularity. In addition, stretch of titin (not shown here) also does so. These provide important mechanisms for regulating contraction in response to changes in heart rate, filling and afterload (for more details, see text).

Myosin light chains

Each myosin neck is closely associated with two smaller proteins called myosin light chains. These are called essential light chains (ELC) and regulatory light chains (MyRLC). ELCs have a role in stabilising the neck region; MyRLCs have a regulatory role in controlling the angle of the neck region with respect to the thick filament backbone and thereby the ability of the head to interact with actin.

Origins of Actin and Myosin

Actin

Thin filaments are composed of pairs of actin polymers arranged in a double helix. Actin is a protein with an ancient heritage. It is found in all eukaryotic cells (cells that hold their DNA within a nucleus) and it is therefore widely present in nature, including plants, fungi, moulds, as well as mammals. The ancestor of all eukaryotic cells is thought to be around 1.8 billion years old [2]. This means that actin itself has also been a functional protein for this length of time. Most of the roles it performs, including those of the heart, depend on actin's ability to form polymers creating strands, known as fibrillar actin (or F actin). The ancient heritage of actin is consistent with the wide range of functions it is known to carry out in cells. Example includes building the cytoskeleton of cells which maintains their structure, as in axons (nerve fibres) where regular rings of actin maintain their shape, the formation of reinforced F actin rods to extend cell membranes facilitates the movement of cells such as amoebae. In addition, F actin strands provide paths along which particles attached to myosin molecules are transported as cargo within cells. For a detailed review of actin functions, see Ref. [3]. The ability of actin to fulfil specific roles is determined by the way actin molecules are polymerised to form F actin, so that a unique strand is formed for each role. Thus, the strands of actin in cardiac thin filaments are unique to that role and a key element of the contracting machinery, accounting for around 20% of the protein in cardiac myocytes. The polymerisation of actin molecules to form fibrillar actin is an active process; it requires energy input in the form of ATP hydrolysis (adenosine triphosphate is the most widely used source of chemical energy in biology) and other facilitating molecules.

Myosin

Myosins are also ancient motor proteins. Like actin they evolved during the origin of eukaryotes and therefore are of a similar age. They interact with actin to generate force in exchange for the utilisation of energy. They are involved in many cellular processes, such as cell division and transporting other molecules and particles within cells. Mutations in the genes that encode myosins are known to cause a wide range of abnormalities and not just in the heart. For example, humans require an isoform of myosin to transport melanin in retinal cells and for maintaining appropriate tension in stereocilia (sound detecting hair cells) in the inner ear. Mutations in genes which encode these functions can cause vision loss and deafness. As shown in Figure 3.4, myosins have head, neck and tail domains. The heads have sites specific for binding to actin and ATP, the latter providing the energy to support contraction. The tail domains vary depending on the function they provide, for example, to bind and transport particular cargos. At least 37 classes of myosins have been identified, reflecting a great diversification of their roles [4]. Mammals are known to have 40 genes [5] producing myosins of varying structure for different functions. They may exist as single molecules which transport cargo, such as large proteins or vesicles (intracellular material surrounded by a membrane), within cells. These have a region in the tail domain which binds to cargo, while the head and neck region interacts with an actin strand in a process akin to "walking" from one actin unit to the next. These transporting motors are also specific in terms of the direction they travel along the actin strand. For example, they may play a role in transporting vesicles to the cell surface from the interior. To achieve this, actin provides the path along which the myosin travels. The two ends of actin are distinct; one end attaches to the cell membrane and the other within the cytoplasm. In this way, the structure of actin determines the direction of travel of the myosin motor and its cargo. During its long history of evolution, a large number of myosins have evolved from the original, reflecting the wide variety of functions it fulfils. Myosin II is responsible for contraction in muscle cells and forms the thick filaments in skeletal and cardiac muscle. Thus, the motor which drives heart muscle is comprised of actin and myosin molecules. They react together in a process known as cross bridge interaction, which generates movement and causes the thin filaments to move along the thick filaments towards the centre of the sarcomere and thus reducing its length.

Anatomy of Cross Bridge Interaction

In 1954, two research papers changed the way muscle contraction was thought to occur. Until then it was believed that muscles contracted because a change in the shape of proteins in myocytes reduced their length. Based on observations of microscopic images of resting and contracted muscle, it was realised that the region corresponding to the I-band (Figures 3.1 and 3.2) is shortened in contracted muscle [6, 7] and this led to the idea that the thick filaments must be moving between thin filaments. Since then, it has been accepted as the basis of muscle contraction and is known as the sliding filament mechanism. During subsequent decades, researchers have revealed more and more about the way this occurs and how it is controlled and regulated. To understand this, we need to describe the components involved in the interaction between actin and myosin in greater detail. Figure 3.4 illustrates a thin filament with a series of myosin heads projecting from the side of a thick filament (for simplicity, the bulk of the thick filament is omitted). It shows the two strands of actin molecules (light brown) linked together in a double helix with myosin heads projecting toward the binding sites (darker brown) on actin. There are, however, a number of important additional components.

Tropomyosin

Surrounding the actin there is an additional pair of protein strands, shown in blue in Figure 3.4. These are composed of tropomyosin. This is an elongated protein containing 284 amino acids and in cardiac muscle it is found in pairs which form a double helical strand of around 1 nm. In resting heart muscle, these strands fit in the grove between seven adjacent actin molecules on each side of the thin filament. They are also joined with others, end to end, forming continuous strands along each side of the thin filament. As with actin and myosin, there are several isoforms of tropomyosin responsible for a range of functions. Cardiac muscle contains two isoforms of tropomyosin, α-Tm and β-Tm, in a ratio of 5:1. In resting muscle, their position places them directly over the myosin-binding sites on actin and prevents interaction between the myosin heads and actin and ensures that the contractile machinery is in an "off state" (Figure 3.4, upper panel). This is important, first, because it prevents unnecessary energy usage at rest and second in the heart it ensures that the muscle is fully relaxed and filling is unimpeded during diastole. This also implies that there must be a mechanism for rapidly switching the contractile system to an "on state". Tropomyosin does this in consort with troponin.

Troponin

Troponin is a protein that plays a crucial role in regulating muscle contraction in both skeletal and cardiac muscle. Different isoforms are found in both types of muscle but with only minor differences in their amino acid composition. It is perhaps best known as a test used in patients suspected of having heart attacks. This is because it is normally confined within muscle cells but leaks from damaged cells so that if cardiac troponin is present in blood samples, it indicates cell disruption due to injury and is therefore useful in early diagnosis of heart attacks. In cardiac myocytes, troponin is located on thin filaments attached to both actin and tropomyosin (Figure 3.4). Troponin is made up of three subunits: Troponin C (TnC), Troponin I (TnI) and Troponin T (TnT).

- TnC (dark blue in Figure 3.4) is a protein of 160 amino acids which has two functional domains; one end is highly sensitive to ionised calcium (Ca^{2+}) and the other is firmly bound to the thin filament.
- TnI (yellow in Figure 3.4) is known for its inhibitory action on contraction. TnI interacts with both TnC and TnT and in resting muscle is bound to actin in a manner which holds the troponin-tropomyosin complex in close contact with actin, blocking actin–myosin attachment.

- TnT (brown in Figure 3.4) is bound to both TnC and TnI and is also tightly attached to and holds the whole complex anchored to tropomyosin.

The onset of contraction in heart muscle is signalled by a rapid and transient rise in Ca^{2+} within the myocytes, the mechanisms for which will be described later. TnC is highly sensitive to and binds Ca^{2+}. This induces a change in the confirmation (the shape) of TnC. As a result of this change, TnC then interacts with TnI causing it to detach from actin. This, in turn, has the effect of releasing the whole troponin-tropomyosin complex from its tight position in the actin grove and thereby exposing the myosin-binding sites on actin to myosin heads (Figure 3.4, lower panel). Thus, tropomyosin acts as a switch mechanism along the entire length of thin filaments which can block and unblock cross bridge interaction to control the contraction and is capable of responding rapidly to facilitate the onset of contraction in as little as 10 ms [8]. This movement of the position of tropomyosin on thin filaments in response to Ca^{2+} has been observed using 3-dimensional reconstruction of electron microscopic images [9].

Myosin light chains

As discussed earlier, myosin is composed of a pair of heavy chains which consist of a head, neck and tail, the latter forming a double helix which is embedded in the body of the thick filament. In addition, two smaller protein molecules known as light chains are linked to the neck region of the heavy chains, shown as green and blue in Figure 3.4. The term light chain reflects the fact that they are around ten times smaller than the heavy chains. These are called myosin regulatory light chains (MyRLCs) and essential light chains (ELCs) and they have important roles in regulating contraction process.

- ELCs stabilise the neck domain of myosin. They also form a link to actin during contraction in a way that strengthens the actin–myosin cross bridge interaction and force generation. Removal of ELCs in isolated experiments causes a reduction in actin filament motility and in contractile force. In addition, some mutations of cardiac ELC in man are associated with hypertrophic cardiomyopathy, a condition which causes a marked increase in the thickness of the left ventricle and disorganisation of its microscopic structure. One ELC mutation causing this condition has been shown to alter the way in which myosin heads bind to actin during contraction.

- MyRLCs contain a serine amino acid that can be phosphorylated by a specific enzyme, myosin light chain kinase (MyRLCK). This simply means that a phosphoryl group (phosphorous and three oxygen atoms) is added to the protein at this location. Although a small addition, this alters the conformation of MyRLCs in a way that effects the dynamics of contraction. MyRLCs are located at the neck of the myosin heavy chain, and phosphorylation induces a change in the angle of the myosin neck and head away from the backbone of the thick filament and towards the thin filament. This brings the myosin heads closer to actin binding sites and increases the probability of cross bridge interaction. At a physiological level, it contributes to increased contractile force.

Myosin binding protein-C (MyBP-C)

MyBP-C regulates contraction in both cardiac and skeletal muscle. It is a long protein molecule which if fully stretched would be around 50 nm in length; however, in living cells, it exists in the form of a string of folded globular units, of which there are 11 in cardiac and 10 in skeletal muscle. These are called Ig and Fn domains of which cardiac MyBP-C has eight of the former and three of the latter (these are common folding patterns of proteins; the names come from proteins in which they were first observed, Ig refers to immunoglobulin and Fn to fibronectin). These elements are important for the way it binds to titin, an important structural component of sarcomeres (Figures 3.2 and 3.3). MyBP-C consists of 1,273 amino acids and is located on the thick filament (Figures 3.5 and 3.6). Electromicroscopic studies using antibodies to MyBP-C have shown it to be confined to the "C zone" of sarcomeres which corresponds to around half the length of the thick filaments on each side of the M-band (Figure 3.2). These studies have also shown them to be distributed at intervals of 43 nm along the thick filament, which corresponds to the repeating

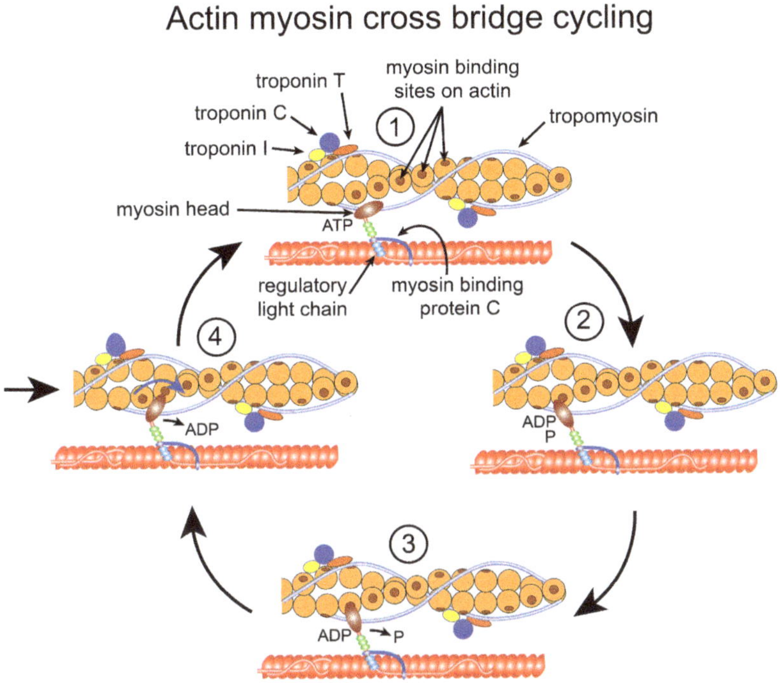

Figure 3.5: Diagram illustrating cyclic actin myosin interaction. 1. At the end of the power stroke, ATP binds to the myosin head. This changes the configuration of the head and forces the head to detach from actin. 2. Once detached, the head hydrolyses the ATP to ADP and a phosphate group (P) and undergoes another change in configuration which brings it close to actin again. 3. The myosin head binds weakly to actin and releases the P product of ATP hydrolysis. 4. This results in myosin binding strongly to actin and undergoing a configuration change referred to as the power stroke, which pulls the thin filament towards the centre of the sarcomere. The myosin head then releases the ADP product of ATP hydrolysis leaving the ATP site vacant, returning it to state (1) and the process can be repeated.

pattern of myosin heads. Thus, the structure of MyBP-C and its position on thick filaments are strongly suggestive of an essential role in contraction.

Titin

Titin is the largest known protein, with an atomic mass of 3 million Daltons (a Dalton is the mass of a single neutron or proton). It gets its

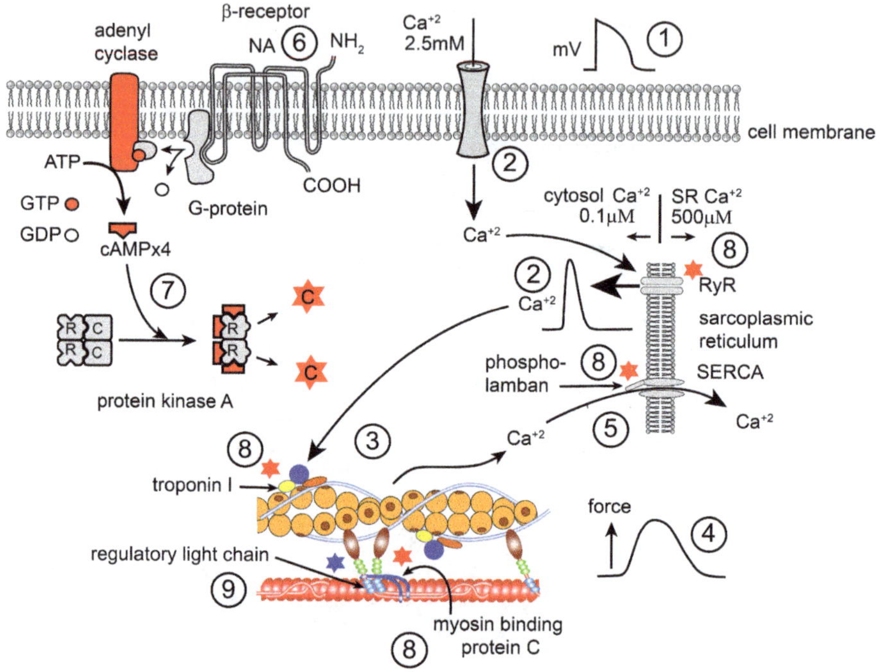

Figure 3.6: Schematic of cardiac muscle activation. (1) An action potential is induced from an impulse conducted from the pacemaker. The resulting depolarisation of the cell membrane opens voltage-gated calcium channels (2) in the sarcolemma and in turn an influx of Ca^{2+} ions into the cell. The rise in Ca^{2+} activates ryanodine (RyR) receptors on the sarcoplasmic reticulum (SR) which then releases more Ca^{2+} into the cytoplasm causing a rapid rise in its concentration (2). This increases binding of Ca^{2+} to troponin C on sarcomere thin filaments (3), which induces a conformational change in troponin I. This shape change, acting through troponin T, moves tropomyosin so that it no longer covers myosin binding sites on actin and making them available for interaction with myosin. Cyclic actin–myosin cross bridge interaction then follows which moves the thin filament towards the centre of the sarcomere resulting in shortening and force generation (4). The contraction is terminated by the rapid removal of Ca^{2+} from the cytoplasm by SERCA (5). Sympathetic nerve stimulation (the fight or flight response) is one mechanism for regulating cardiac muscle contraction. This causes release of noradrenaline (NA) at the nerve ending which binds to β-receptors (6), which then activate a closely coupled G-protein and in turn the enzyme adenyl cyclase which converts ATP to cyclic AMP. cAMP is a major intracellular signalling molecule which activates the enzyme protein kinase A (PKA) (7). PKA phosphorylates several key proteins involved in the contraction process (8 and red star). (a) Phosphorylation of RyR receptors increases Ca^{2+} release from the SR increasing contractile force. (b) Phosphorylation of troponin I increases exposure of myosin binding sites, increasing contractile force, but reducing sensitivity to Ca^{2+} which reduces the duration of contraction. (c) Phosphorylation of phospholamban disinhibits its effect on

name from Titan, which in Greek mythology refers to the ancestor of a family of giants. Titin is half a sarcomere in length (~1 μm) and is attached at one end to the Z disk and at the other to the M line (Figures 3.2 and 3.3). As we saw in the case of MyBP-C, titin is also folded in the form of Ig and Fn domains, containing 112 of the former and 132 of the latter. However, the distribution of these differs in the regions corresponding to the I- and A-bands. Titin within the I-band consists mainly of Ig domains, but with two sections which are rich in the amino acids, proline, glutamate, valine and lysine (named PEVK, the symbols for these amino acids). These sections are naturally folded but readily unfold when stretched. This unfolding gives titin within the I-band section a spring-like quality as sarcomeres are stretched and is a main contributor to the passive stiffness of heart muscle and its distensibility during diastole and chamber filling. Titin between thick filaments in the A-band is composed of alternating Ig and Fn domains and arranged in segments that repeat at 43 nm intervals. This, as we have seen, corresponds not only to the repeating pattern of three pairs of myosin heads but also to the spacing of MyBP-C in the C-region of thick filaments.

This relationship between titin, myosin and MyBP-C led many researchers to suspect that titin is more than a structural element of sarcomeres, but may also have a functional role. Recent research has begun to shed light on this, revealing a remarkable role for this giant protein. Mutations in the gene which encodes for cardiac titin are known to cause cardiomyopathy (a term used to describe impaired cardiac muscle function). Mutations resulting in the formation of incomplete titin molecules result in heart failure characterised by dilated ventricles and myocytes lacking in

Figure 3.6: (*see figure on facing page*) SERCA, resulting in more rapid Ca^{2+} reuptake into the SR, also curtailing the duration of contraction. (d) Myosin binding protein-C has a tethering effect on myosin heads which limits and slows the rate of cross bridge interaction. Phosphorylation by PKA reduces this effect resulting in greater force generation and more rapid relaxation. Myosin regulatory light chains are not targets for PKA but are phosphorylated by a specific enzyme, myosin regulatory light chain kinase (MRLCK) in response to raised Ca^{2+} (9 and blue star), causing myosin necks to be more perpendicular to thick filament backbone. This facilitates cross bridge interaction, resulting in greater force generation and is believed to be involved in the force frequency response.

Figure 3.7: Electron micrograph of growing myocardium. The formation of new sarcomeres (sarcomerogenesis) occurs at the cell margin as myocytes expand. Here partly formed pre-myofibrils (white double headed arrow lines) are shown at an angle to mature sarcomeres. They are closely associated with darker sections of the sarcolemma known as Z bodies which will later form the Z disks of mature sarcomeres. For a detailed review of sarcomerogenesis, see Ref. [34].

sarcomeres [10, 11]. It has been known for some time that sarcomeres are produced by a process in which the filaments are synthesised and then assembled into sarcomeres. Figure 3.7 shows an example of this; sarcomeres are seen in the process of being assembled from where they are synthesised close to the sarcolemma (cell membrane) in growing myocardium. Recently it has emerged that transporting and correctly positioning filaments within sarcomeres utilises the motor functions of myosin. Myosin transports components of the sarcomere into place during assembly. Full-length titin is also essential for moving components of the Z disks into place. Recent studies also suggest that as thick filaments are assembled, titin is wrapped around them and forms a terminal ending, which limits their length [12] and also sarcomere length. Thus, even at a molecular level, the motor which drives our hearts contains multiple built-in regulating functions, some of which act as an on-off switch, while others provide quantitative regulation of contraction and relaxation and yet others are involved in determining the passive properties of myocardium. And quite remarkably, components of the motor act as assemblers of the sarcomeres, which, organised in their trillions, generate the force which drives the heart.

Operation of the Heart Motor

Force is generated in the heart and skeletal muscle by active sliding of the thick filaments along the thin filaments. At the centre of this is the interaction between the myosin heads on thick filaments with specific binding sites on actin molecules on thin filaments. This requires an energy source, which in the heart is provided by ATP (adenosine triphosphate) hydrolysis. ATP is a molecule of adenine, a sugar, ribose and three phosphate groups. Hydrolysis simply means a chemical reaction in which a molecule of water is added and one of the phosphate groups is cleaved off, giving ADP (adenosine diphosphate) and a free phosphate group (P). A further crucial element of the reaction is that energy which had been bound up in the third phosphate group is released during the reaction and this is used to fuel not only muscle contraction but virtually all energy requirements of our cells.

This is not unlike the workings of an internal combustion engine. Petrol is mostly composed of octane. This is a chain of 8 carbon atoms linked to 18 hydrogen atoms (C_8H_{18}) and is decomposed in the presence of oxygen to carbon dioxide and water ($8CO_2 + 9H_2O$). As with ATP, breaking the carbon bonds of octane releases energy in the form of heat, while the explosive expansion of the CO_2 gas formed drives the piston which rotates the engine. The CO_2 and H_2O generated are released into the atmosphere and the heat is partly used to warm the car interior, but most is released to the atmosphere by the radiator. As with many chemical reactions, some energy input is needed to start it and in these engines, an electric spark does this by igniting the octane.

In our cells, similar mechanisms are at work but with modifications to make them much more efficient. As in the case of octane decomposition, some energy input is needed to initiate the hydrolysis of ATP, but this is reduced because the reaction is assisted by the enzyme, ATPase.

Enzymes act as catalysts; they facilitate chemical reactions by providing an intermediate step requiring a lower initiating energy. An analogy might be the distance between stepping stones across a stream. The greater the distance, the higher the jump needed to cross them; intermediate steps would require jumps of less height and less energy to get across the stream. In the case of octane, most of the heat energy generated is released into the atmosphere and not used. This is avoided in our cells because the release of energy is coupled to energy-requiring processes. An example of this is the phosphorylation of myosin regulatory light chains (MyRLC), an important mechanism that regulates contraction. As discussed above, this involves adding a phosphoryl group at the site of a serine amino acid by the enzyme myosin light chain kinase (MyRLCK). MyRLCK does this by cleaving ATP to ADP + P and using the energy released to bind the freed phosphate group to MyRLC. There are many different kinase enzymes in our cells which catalyse similar processes. They are, in effect, ATPases but named for how the freed phosphate group is used. And just as we have ATPase enzymes, we also have ATP synthase enzymes, which catalyse the synthesis of ATP by combining a phosphate group to ADP. This then forms a new bond that contains energy for future use, rather like a rechargeable battery. In the heart and all skeletal muscles, the energy released from ATP hydrolysis is used to generate force during an interaction between the myosin heads and actin, and because the energy released is coupled to its use, it is highly efficient. This process is usually referred to as "cross bridge interaction".

Cross bridge interaction

Myosin heads (Figure 3.4) are the primary sites at which energy is transduced to generate force. Each head contains a specific actin binding site and an ATP binding site which acts as an ATPase. In a cyclic process involving hydrolysis of ATP with release of energy, myosin heads bind to actin and undergo a shape change that pulls the thin filaments towards the centre of the sarcomere. The myosin head then detaches from actin and the cycle is repeated resulting in shortening of the sarcomere as illustrated in Figure 3.2. The operation of this remarkable engine requires several coordinated and interacting steps.

Activation

At rest, sarcomeres may be considered in an "off state". This is not quite the same as simply not engaged in contraction. It means an enforced off state and it is important because it ensures that no energy is needlessly wasted in unnecessary idling. Tropomyosin (Figure 3.4) achieves this because it obstructs access to myosin binding sites on actin during diastole, meaning the thin filament is inaccessible to myosin heads. Recent research has shown that thick filaments are also involved in sarcomere activation. Cryo-electron microscopy and X-ray analysis have demonstrated that myosin necks lie along the thick filament when inactive, suggesting an "off state" [13]. When active, myosin necks are more perpendicular to thick filaments and therefore better able to interact with actin. The active state of myosin heads is promoted by passive stretch of heart muscle. This relies on a mechano-sensing system mediated by titin interaction with thick filaments [14]. Myosin activation is also regulated by phosphorylation of MyRLCs and MyBP-C by controlling the angle between the thick filament backbone and myosin necks. Myosin binding to actin also activates a recruitment process which amplifies the active state. This is because actin–myosin engagement at one site promotes their engagement at nearby sites by drawing the filaments closer together. Thus, the on/off state of muscle is actively regulated by several factors. The on state is necessary for contraction and the ejection of blood; the off state is equally essential to allow relaxation and filling of the ventricles.

Actin myosin cyclic interaction

Cyclic interaction between myosin and actin is the basis for sliding of thin and thick filaments resulting in contraction and relaxation of muscles. The onset of contraction is triggered by an electrical event which leads to a series of changes within the cell resulting in activation of actin myosin cross bridge interaction.

- **Electrical triggering:** The initiation of muscle contraction is signalled by electric pulses. Skeletal muscle responds to stimulation by nerves which direct signals from the brain or reflexes located in the spinal cord. Heart cells are triggered to contract by an electrical signal which starts in the pacemaker (located in the right atrium) and spreads through the heart tissue to reach all parts in less than a quarter of a second. An action potential is induced in each cell by this process. This involves an abrupt change in the electrical potential across the cell membrane. Briefly, heart cells at rest are akin to tiny batteries; their interior is electrically negative compared to the exterior by around 72 mV. During an action potential this voltage gradient is abruptly and rapidly neutralised (within a few milliseconds) followed by a return to the resting state within around half a second (the duration varies with heart rate). Action potentials involve sudden movement of electrically charged ions, particularly sodium (Na^+) and potassium (K^+), across the cell membrane. The mechanism for this will be described in more detail in Chapter 4.

- **Calcium transient and troponin activation:** This voltage change also induces a rapid and transient increase in the concentration of ionised calcium (Ca^{2+}) within myocytes to 10 μm from a resting level of 0.1 μm. Known as the calcium transient, it is the result of an inward movement of Ca^{2+} across the cell membrane as well as its release from local storage sites (the sarcoplasmic reticulum) within the cell. As a result, troponin C located on thin filaments binds Ca^{2+}, which causes a change in its conformation. This is transmitted to its adjoining components, troponin I and T, and causes tropomyosin to be displaced from its resting position covering myosin-binding sites on actin, exposing them to myosin heads (Figure 3.4, bottom panel). Figure 3.6 shows a schematic illustration of myocyte activation and contraction together with mechanisms involved in its regulation by sympathetic stimulation.
- **Myosin-ATP-Actin interaction:** As discussed earlier, myosin heads have an "off state" in which they lie along the thick filament and do not interact with actin. The "on state" exists when the myosin neck is angled towards the thin filament, and interaction with actin becomes possible. Their interaction with actin may be described in four states, (1) detached, (2) weakly bound, (3) strongly bound and (4) stroke phase, as labelled in Figure 3.5.
    1. Myosin heads have a site that binds to actin and one which binds to ATP. Binding to ATP occurs at the end of a power stroke. This changes its configuration which forces the head to detach from actin.

2. Once detached, the head hydrolyses the ATP to ADP and a phosphate group (P) and undergoes a change in configuration which brings it close to actin again.
3. The myosin head binds weakly to actin and releases the P product of ATP hydrolysis.
4. This results in myosin binding strongly to actin and undergoing a configuration change that pulls the thin filament towards the centre of the sarcomere. This is usually referred to as the power stroke. The myosin head then releases the ADP product of ATP hydrolysis leaving the ATP site vacant. This returns the head to state (1) and the process can be repeated.

Regulation of the myosin–actin cycle

Repetition of steps 1–4 in multiple sarcomeres results in muscle shortening and the pump action of the heart. As discussed at the beginning of this chapter, contractions of the heart are carefully regulated to meet the demands of the body and these must inevitably operate at the level of the basic motor. This, in turn, means there have to be mechanisms by which the basic cycle described above can be modulated to implement them.

- **A gearbox for the heart:** We are familiar with the gearbox in our cars which matches the output of the engine to the demands of moving the car. A gearbox operating in the heart would have to be automatic and capable of detecting an increase in load. There are indeed such systems built into the actin–myosin motor. The variations in loads which the heart must respond to are of two kinds.
  - (i) An increase in filling volume, as occurs during exercise, stretches the chamber volume in diastole and requires an increase in ejection force to empty. As discussed above, this ability of the heart to respond to increased filling is known as the Frank–Starling mechanism and has been known since 1914.
  - (ii) An increase in blood pressure means the ventricle has to eject against a greater resistance and must respond by increasing the force of ejection to overcome it. The ability to do this is known as the Anrep effect.

- **Frank–Starling:** In skeletal muscle, sarcomere length has long been known to influence the force of contraction such that increasing length is associated with increased force generation [15]. This also occurs in heart muscle and explains how the heart is able to increase ejecting force in response to increased filling or preload. When initially observed, this was considered to be due to optimisation of actin and myosin filament overlap making more myosin heads available for cross bridge interaction. In 2016, Ait-Mou and colleagues [16], using X-ray diffraction techniques, were able to show that gently stretching heart muscle changes the angulation of myosin heads making more available to interact with actin. This effect is mediated through the stretching of titin and its interaction with thick filaments with which it is closely aligned (Figure 3.3, lower panel). This is a remarkable discovery and indicates that the molecular motor in the heart has a built-in sensing mechanism sensitive to stretch which prepares the sarcomere to contract with greater force in response.
- **The Anrep effect:** The heart responds to increases in afterload by increasing the force of contraction. It does so in two phases; phase 1 is an immediate increase in force generation; phase 2 is a further gradual increase in force over the succeeding 10–15 min. The initial response is mediated by the same mechano-sensing stretch mechanism which drives the Frank–Starling response, i.e., stretch induced increase in myosin heads available for cross bridge interaction. Phase 2 is due to an alteration in the process that triggers contraction. As discussed above, contraction of the heart is triggered by an electrical impulse propagated from the pacemaker that induces an abrupt voltage change in all contracting cells. This causes a transient rise in the concentration of ionised calcium (Ca^{2+}) within cells (called the calcium transient) which then activates the actin myosin motor by binding to troponin. The phase 2 increase in force generation in response to increased afterload is due to a gradual increase in this calcium transient. This results in more calcium binding to troponin and a stronger activation of the actin myosin motor. The mechanisms by which this increase in the Ca^{2+} transient occurs involves several intracellular systems that regulate Ca^{2+} ions in myocytes. This can also be the beginning of a process which ultimately leads to the growth of

heart muscle in cases where the increase in afterload is maintained. Thus, the Anrep phase 1 response can be seen as an initial response to increased afterload to offset the increased demand on the heart and occurs at a molecular level within the actin myosin motor. The gradual phase 2 response is the beginning of a longer-term adaptation to increased afterload as occurs in patients with persistent hypertension. The Frank–Starling and Anrep effects adjust the performance of the heart to varying workloads and in that sense are similar to the operation of an automatic gearbox in a car.

- **The force-frequency response:** Our heart rate is constantly changing in response to varying degrees of stimulation and inhibition. The heart has an intrinsic ability to modulate the force and pattern of contraction in response to changes in heart rate. This is an essential function because, without it, filling would be compromised at higher rates. As heart rate increases, the intervals between beats are reduced and so the contraction time has to shorten to allow adequate time for filling. The effect of changes in heart rate on contraction was first observed by Henry Bowditch in 1871 [17]. Bowditch, working with isolated frog hearts (which therefore could not be influenced by other stimuli), noticed that an abrupt increase in stimulation frequency caused an increase in force generation in successive beats until a new stable level was reached. Later researchers were able to show that in addition to increasing the peak force of contraction, the rate of force generation and the speed of relaxation also increased. This allows the heart to eject blood with greater force and more quickly at higher heart rates. The mechanism underlying the increase in force generation has long been thought to be due to an increase in the Ca^{2+} transient which activates contraction. This was confirmed in 1987 in experiments in which cardiac myocytes were injected with aequorin, a protein that emits light in proportion to Ca^{2+} concentration. The results showed a close correlation between an increase in the Ca^{2+} transients and the force of contraction in response to changes in stimulation frequency [18]. The mechanisms underlying this are still the focus of research but are likely to reflect greater availability of Ca^{2+} from storage sites within cells consequent on (i) increased accumulation due to the higher frequency of Ca^{2+} transients and (ii) alterations in the way Ca^{2+}

is removed and stored to terminate Ca^{2+} transients (for a detailed review, see Ref. [19]).

- **The frequency-relaxing response:** Relaxation of heart muscle at the end of contraction restores the ventricles to a resting state in diastole when filling occurs. This is not simply a passive process; it is actively regulated and one example of this is acceleration of relaxation at higher heart rates. This property known as lusitropy is an important contributor to the preservation of adequate diastolic time for ventricular filling. One mechanism for this relates to how myocytes manage Ca^{2+} (Figure 3.6). As we saw with the force frequency response, the Ca^{2+} transient increases at higher heart rates. In order to terminate contraction, this rise in Ca^{2+} concentration has to be reversed and this is achieved by (i) extrusion of Ca^{2+} by the cell membrane and (ii) reuptake of Ca^{2+} into storage vesicles called sarcoplasmic reticulum (SR) within the cell. This is an active process driven by SERCA (SR-Ca^{2+}-ATPase) an enzyme on the SR membrane which regulates Ca^{2+} by removing it from the cytoplasm and storing it in SR vesicles. SERCA activity itself is also tightly regulated, providing feedback control of Ca^{2+} concentration. Ca^{2+} is a major signalling molecule in all cells and is tightly regulated. Calmodulin (CAM) is a protein that acts as a Ca^{2+} sensor; it binds to Ca^{2+} and acts on many systems to help regulate it. In cardiac myocytes, CAM bound to Ca^{2+} increases SERCA activity and this increases reuptake of Ca^{2+} into SR and thus its removal from the cytoplasm, which in turn accelerates relaxation [20]. SERCA activity is also modulated by an inhibitory protein called phospholamban and there is some evidence that phospholamban is also modified at higher heart rates in a way that reduces its inhibitory effect on SERCA. This provides a second mechanism for increasing the removal of Ca^{2+} from the cytoplasm at higher heart rates. Thus, contraction in the heart is triggered by a rise in Ca^{2+}, and like all cells, cardiac myocytes contain powerful systems to regulate its level. This provides a mechanism to regulate the force and duration of contraction by modulating the degree and timing of Ca^{2+} release and removal into and from the cytoplasm.

Troponin I is part of the troponin complex which controls the on/off state of thin filaments (Figure 3.4). As discussed above, troponin I normally acts as an inhibitor, maintaining thin filaments in an off state. However, when Ca^{2+} binds to troponin C, this induces a conformation change in troponin I which moves tropomyosin away from the myosin binding sites on actin so that they become available for cross bridge interaction. Troponin I is phosphorylated by protein kinase A and adding a phosphoryl group increases its response to troponin C and results in an increase in contractile force. Phosphorylation of TnI has also been shown to augment the force frequency and relaxation responses [21].

Autonomic Regulation of the Heart

The most important external regulator of the heart is the autonomic nervous system. This consists of the sympathetic and parasympathetic system which is based in the hypothalamus, a small centre (often described as the size of an almond) at the base of the brain. It is connected to *afferent* nerves which carry information to it from various sensors around the body and to *efferent* nerves which deliver stimulating or inhibitory pulses to control the heart and many other functions of the body such as breathing, digestion, sexual arousal, urination and adjustment of the pupils to light. The autonomic system stimulates cardiac activity via the sympathetic nerves and inhibits it via parasympathetic nerves. The sympathetic nerves reach the heart from the hypothalamus by passing through the spinal cord, emerging between the upper five thoracic vertebrae and travelling along the large blood vessels. The parasympathetic nerves emerge directly from the medulla oblongata located at the base of the brain. They then travel in the vagus nerve through the neck into the chest, which gives off branches to the heart. The sympathetic nervous system stimulates the heart and is part of the "fight or flight" response which protects us from danger and has powerful effects on the heart. The sympathetic nerves connect to the pacemaker (sinoatrial node), the atria and the ventricles. The parasympathetic nerves innervate the pacemaker and have an inhibitory action, slowing heart rate. Autonomic control works at an unconscious level. It responds to our emotions, notably fear and alarm, and to conscious decisions that involve greater exertion. While it may also respond to the conscious recollection of past or anticipated threats, it does so automatically and is not directly within our conscious control.

Autonomic control of the molecular motor of the heart

The parasympathetic nerves do not innervate the ventricles and therefore have no direct effect on ventricular myocytes. They do, however, innervate the sinoatrial node and modulate the heart rate. As a result, they also influence cardiac contraction via the force/relaxation frequency response, albeit indirectly. Sympathetic nerves do innervate the ventricles including cardiac myocytes and have a direct effect of contraction via the molecular motors.

Sympathetic nerves within the heart

The presence of tiny nerves within the heart was first observed by light microscopy in 1866 by Gerlach [22]. Small efferent nerves are present throughout the ventricles and deliver signals from the sympathetic system, which control myocardial function. These sympathetic nerves make contact with myocytes at specialised junctions called synapses. The nerve endings contain noradrenaline (also called norepinephrine) stored in vesicles, while the myocyte side of the synapse has receptors to which noradrenaline binds. Researchers in the first half of the 20th century observed that stimulation of sympathetic nerves had a powerful effect on the heart, inducing a marked increase in heart rate and contractile force, and this was later shown to be due to the action of noradrenaline released from nerve endings in the heart [23]. Noradrenaline is an example of a neurotransmitter; in that it is released from the nerve ending in response to nerve pulses originating higher in the sympathetic system. The released noradrenaline binds to β-receptors and exerts a powerful effect on the activity of myocytes, indeed, its effect is so powerful that its activity has to be time-limited. This is achieved by a protein called norepinephrine transporter (NET) on the membrane of the nerve ending. NET is a specialised pump that re-captures around 90% of the released noradrenaline back into the nerve ending for storage and re-use later [24]. Interest in NET has grown rapidly since the discovery of norepinephrine reuptake inhibitors (NRIs). These have been developed to treat a number of conditions including depression by acting on synapses in the brain.

By inhibiting the reuptake of noradrenaline, its effects are enhanced. However, noradrenaline is a powerful cardiac stimulant and is harmful in excess. This can occur with tumours such as phaeochromocytoma which produce and release large amount of noradrenaline into the circulation and also from cocaine abuse.

Cocaine is a potent NRI which acts on tissues in the brain, heart and circulation. Its abuse potential relates to its action in the brain producing psychoactive effects, however it is also known to produce toxic effects on the heart and circulation due to overstimulation, especially with binge abuse. This can take the form of hypertension, stroke, abnormal heart rhythms (arrhythmias), heart attacks (myocardial infarction) and cardiomyopathy (myocardial damage resulting in loss of contractile force and dilatation of the left ventricle). Prolonged and repeated exposure can result in significant accumulated damage. The action of noradrenaline on contraction is mediated by its effects on the molecular motor through signalling processes which transmit its activation of β-receptors on the cell surface.

Adrenergic signalling

The actions of noradrenaline on the actin myosin motor in the heart result from activation of β-receptors (Figure 3.6). These are proteins which straddle the cell membrane of myocytes, forming three intracellular and three extra cellular loops, spanning it seven times. One of its intracellular loops is closely coupled to a G-protein (named simply because it binds to guanine nucleotides such as Guanine diphosphate (GDP) and Guanine triphosphate (GTP)). β-Receptors do not directly stimulate cells, but act indirectly through G-proteins, which then stimulate enzyme systems within the cell. This was an important discovery in understanding how signal transduction works in cells for which Alfred Gilman and Martin Rodbell received the Nobel prize in 1994. When noradrenaline binds to its receptor, the receptor undergoes a conformational change, which in turn induces a conformational change in its coupled G-protein, resulting in its activation. G-protein activation involves its dissociating the GDP groups which they normally bind when inactive and binding instead to GTP. In the case of β-receptors, GTP-activated G-protein then activates an adjacent membrane bound enzyme called adenyl cyclase, which leads to the formation of cyclic adenosine mono phosphate (cAMP). cAMP is a "second messenger" involved in transmitting signals originating outside cells which regulate many intracellular processes. Its discovery by Eric Sutherland was an important step in understanding how external stimuli such as hormones affect cells and for which he received the Nobel prize in 1971. The increase in cAMP in response to β-receptor stimulation leads to activation of the enzyme protein kinase A (PKA). PKA catalyses the phosphorylation of many different proteins in cells including those which regulate the function of the actin–myosin motor in the heart. In summary, sympathetic stimulation of the heart releases noradrenaline which activates β-receptors. This activates β-receptor coupled G-proteins, which in turn activate adenyl cyclase leading to increased formation of cAMP and cAMP activates PKA an enzyme which regulates many functions of actin–myosin interaction in cardiac myocytes.

Protein Kinase A Modulation of the Heart's Motor

Kinases (from the Greek word κιν-εῖν, to move + ase) are enzymes which transfer a phosphate group from ATP to proteins, a process called phosphorylation. It is important in biology because phosphorylation at specific sites can alter the shape and function of proteins. There are more than 500 protein kinases in humans of which PKA is just one. Why are there so many and what is special about PKA? In fact, the action of PKA is quite specific. This is possible because PKA phosphorylates proteins only at sites where the amino acid serine or threonine is exposed. This limits the range of proteins on which it can act. Its specificity is further focused by the position in the amino acid sequence at which serine/threonine is located. Once a phosphate group is attached to a serine or threonine amino acid, it induces a conformational change in the protein, which depends on the location of the serine/threonine in the protein and on its original shape. The proteins which PKA targets regulate a range of intracellular processes and their activity is determined by the extent to which they are phosphorylated. One example of this is the phosphorylation of myosin regulatory light chains. But there are several others which, as we shall see, orchestrate heart pump function.

PKA control of the Ca^{2+} transient

Contraction of the heart is triggered by a transient rise in intracellular Ca^{2+} (Figure 3.6). To recap briefly, this comes from two sources. A Ca^{2+} influx through voltage-gated Ca^{2+} channels located in the cell membrane is triggered by an action potential and this activates ryanodine receptors (RyR) on the sarcoplasmic reticulum (SR) resulting in a further release of Ca^{2+}. The resulting rise in intracellular Ca^{2+} then activates troponin and the actin–myosin interaction. The rise in Ca^{2+} is then rapidly terminated by its removal back into the lumen of the SR by SERCA (from SR-Ca^{2+}-ATPase). All the proteins involved in Ca^{2+} transient are amenable to phosphorylation by PKA, which modulates their activity. Phosphorylation of the Ca^{2+} channels on the cell surface and RyR receptors on the SR increases their permeability to Ca^{2+}, resulting in a larger Ca^{2+} transient and in turn stronger activation of contraction. This is a key factor in how sympathetic stimulation increases the contractility of heart muscle. However, noradrenaline does not just increase the force of contraction, it also reduces its duration. One of the ways this is achieved is by accelerating the removal of Ca^{2+} into the SR by SERCA. SERCA is normally bound to and inhibited by a protein called phospholamban (Figure 3.6). Phospholamban was discovered by Arnold Katz and co-workers and named from phosphate + lamban from the Greek "to receive" [25]. Phospholamban is also subject to phosphorylation by PKA, which causes it to dissociate from SERCA, releasing it from inhibition. As a result, SERCA removes Ca^{2+} more rapidly from the cytosol. In human heart muscle, SERCA is responsible for removing over 70% of the rise in Ca^{2+} and so its disinhibition reduces the duration of the Ca^{2+} transient and contraction. Thus, sympathetic stimulation of the heart activates β-receptors which then interact with a closely coupled G-protein resulting in activation of adenyl cyclase and production of cAMP. The increase in cAMP activates PKA, which then phosphorylates calcium channels in the sarcolemma, RyR receptors in the SR and phospholamban, resulting in a larger, but shorter Ca^{2+} transient and in turn greater and more rapid force development. This provides a very elegant system for controlling activation of the contractile machinery of heart cells. But as we shall see, PKA is also capable of regulating the behaviour of the molecular motor directly.

PKA modulation of actin–myosin cross bridge interaction

(a) Troponin I (TnI) is a subunit of the troponin complex on thin filaments and plays a crucial role in their activation in response to Ca^{2+} binding to troponin C. TnI inhibits thin filament activation by maintaining tropomyosin in a position which blocks myosin binding sites on actin (Figure 3.4). TnI has sites which are targets for PKA and when phosphorylated results in an increase in contractile force and accelerated relaxation [26] (Figure 3.6). The increase in contractile force is believed to be due to a change in conformation of TnI that increases access to myosin binding sites on actin, promoting cross bridge interaction [27]. Phosphorylation of TnI also reduces sensitivity of the contractile apparatus to Ca^{2+} [28]. This would explain earlier detachment of cross bridges as the Ca^{2+} transient declines and earlier relaxation. Thus, sympathetic stimulation modulates actin–myosin interaction via PKA induced phosphorylation of TnI resulting in increased force generation and accelerated relaxation.

(b) Myosin binding protein-C (MyBP-C) binds to myosin at the C zone of sarcomeres. Originally thought to maintain the structural integrity of thick filaments, it is now known to be a target for PKA, and phosphorylation plays an important role in regulating actin–myosin interaction. MyBP-C is bound to myosin and has a tethering effect on myosin necks which tends to limit their interaction with actin and this, in turn, inhibits contraction (Figure 3.6). Phosphorylation reduces this effect resulting in greater force generation. Phosphorylation of MyBP-C and TnI act in consort to increase contractile force, the former being more powerful [27]. MyBP-C also exerts an important effect on the duration of contraction and how it responds to an increase in afterload. When faced with increased afterload, the heart responds with greater force which is sustained for a longer period to allow the ventricle to empty. This increase in force, known as the Anrep effect, is mediated by a mechano-sensing mechanism and a gradual rise in the Ca^{2+} transient. MyBP-C also has a role in the duration of contraction in that in the unphosphorylated state its tethering effect on myosin heads inhibits binding to actin and also delays detachment which prolongs the duration of contraction. Phosphorylation has been shown to increase the mobility of myosin heads [29], allowing more rapid engagement and detachment from actin. Thus, MyBP-C may be seen as a mechanism for regulating the flexibility of myosin heads and the speed of actin–myosin interaction. Increased myosin head mobility in response to phosphorylation following sympathetic stimulation would increase the rate of engagement and detachment resulting in stronger contraction and more rapid relaxation. In contrast, MyBP-C in its unphosphorylated state stabilises myosin heads and slows the rate of contraction and relaxation.

(c) Myosin regulatory light chains (MyRLCs) are also targets for phosphorylation by a specific enzyme, myosin light chain kinase (MyRLCK), but not by PKA. MyRLC phosphorylation renders myosin necks more perpendicular to the thick filament backbone and therefore more accessible to actin. This facilitates binding to actin and greater force generation. Mutations in MyRLC are known to cause congenital cardiomyopathy [30], indicating their importance for normal myofilament function. MyRLCs are normally phosphorylated to a certain extent and this is necessary for normal contractile function; inadequate MyRLC phosphorylation results in heart failure, while increased phosphorylation increases contractile force. Baseline MyRLC phosphorylation is controlled by the activity of its phosphorylating enzyme MyRLCK and the dephosphorylating enzyme myosin regulatory light chain phosphatase (MyRLCP) and appears to be maintained at a fairly constant level [31]. The factors which regulate the level of MyRLC phosphorylation have yet to be clarified but appear to relate to the level of Ca^{2+} in the cytoplasm [31, 32]. If this proves the case, it would indicate a role in the force frequency response and elevated afterload since both are associated with elevation in Ca^{2+}.

(d) Titin is amenable to phosphorylation [33] by several enzymes of which PKA is one, each targeting specific sites on titin. At present, these are the subject of intense research focus, however a picture is emerging which suggests that phosphorylation alters the stiffness of titin and therefore is likely to be yet another important regulator of cardiac function. Of the many phospho-sites on titin, two regions have received particular attention. Both regions are in the sarcomere I-band region, one named N2Bus close to the Z disk and the PEVK region. As discussed above, titin has spring-like qualities and is the main determinant of the passive stiffness of sarcomeres and myocardium. PKA phosphorylation at the N2Bus region reduces the passive stiffness of sarcomeres. The PEVK region is not amenable to PKA but is phosphorylated by other enzymes, notably PKCα and CaMKIIδ and early research suggests that phosphorylation also increases passive stiffness. This is important for the behaviour of cardiac muscle because it alters sensitivity of the mechano-sensing system involved in the Frank–Starling response and the Anrep effect. A reduction in passive stiffness decreases the force response to stretch while an increase in passive stiffness increases it. Studies of myocardium from patients with heart failure suggest that, in general, phosphorylation of N2Bus sites tends to be reduced while that of PEVK sites is increased. This would be consistent with increased stiffness and reduced chamber compliance, which is a feature of heart failure. Reduced compliance impairs ventricular filling and is therefore an important potential target for intervention and will no doubt, be the focus of future research.

Conclusion

The picture which emerges from research into the way the heart pumps is one of extraordinary complexity, but also one of elegance and even beauty. The motor which drives the heart is built on actin and myosin molecules, which originally appeared around 1.8 billion years ago in an ancestral single-cell organism. We have about 300 myosin heads on each thick filament and roughly 10^{15} filaments in an adult heart. Their capacity to provide the pumping function of the heart derives from their number; although the power of each is tiny, in summation they become highly significant. Thus, in a resting adult, the heart generates about one watt of power and during exercise this can be increased about sixfold.

To achieve this, the activation of all these tiny motors has to be accurately coordinated so that the sarcomeres in each myocyte contract in unison. The contraction of the 2.6 billion myocytes in the heart must also be coordinated to ensure an efficient contraction of the ventricles with each heartbeat.

The heart pumps around two billion times in an average lifetime and responds to constantly varying stresses and demands. It is remarkably well adapted to do this through a series of external and internal mechanisms, which regulate its beating rate and power output.

The heart has been a major iconic symbol for at least 5,000 years, but only in the last century have we begun to understand the motor which drives it. Molecular biology and protein chemistry have revealed an extraordinary assembly of working parts; however, it also becomes clear that we still have much to learn. That is hardly surprising when we remember that evolution has been working on this for more than a billion years and humans with sufficient intelligence to consider it, for a very tiny part of that.

References

[1] Von Anrep G. On local vascular reactions and their interpretation. *Journal of Physiology*. 1912;45:318–327.

[2] Betts HC, Puttick MN, Clark JW, Williams TA, Donoghue PCJ, Pisani D. Integrated genomic and fossil evidence illuminates life's early evolution and eukaryote origin. *Nature Ecology & Evolution*. 2018;2(10):1556–1562. https://doi.org/10.1038/s41559-018-0644-x.

[3] Rottner K, Faix J, Bogdan S, Linder S, Kerkhoff E. Actin assembly mechanisms at a glance. *Journal of Cell Science.* 2017;130:3427–3435.

[4] Richards TA, Cavalier-Smith T. Myosin domain evolution and the primary divergence of eukaryotes. *Nature.* 2005;436:1113–1118.

[5] Foth BJ, Goedecke MC, Soldati D. New insights into myosin evolution and classification. *Proceedings of the National Academy of Sciences*. 2006; 103:3681–3686.

[6] Huxley AF, Niedergerke R. Structural changes in muscle during contraction; interference microscopy of living muscle fibres. *Nature*. 1954;173: 971–973.

[7] Huxley H, Hanson J. Changes in the cross-striations of muscle during contraction and stretch and their structural interpretation. *Nature*. 1954;173: 973–976.

[8] Irving M. Regulation of contraction by the thick filaments in skeletal muscle. *Biophysical Journal*. 2015;113:2579–2594.

[9] Lehman W, Craigt R, Vibertt P. Ca^{2+}-induced tropomyosin movement in *Limulus* thin filaments revealed by three-dimensional reconstruction. *Nature*. 1994;368:65–67.

[10] Hoorntje ET, van Spaendonck-Zwarts KY, *et al.* The first titin (c.59926 + 1G > A) founder mutation associated with dilated cardiomyopathy. *European Journal of Heart Failure*. 2018;20:803–806.

[11] Chopra A, Kutys ML, Zhang K, Polacheck WJ, Sheng CC, Luu RJ, Eyckmans J, Hinson JT, Seidman JG, Seidman CE, Chen CS. Force generation via β-cardiac myosin, titin, and α-actinin drives cardiac sarcomere assembly from cell-matrix adhesions. *Developmental Cell*. 2018;44:87–96.

[12] Kellermayer M, Sziklai D, Papp Z, Decker B, Lakatos E, Mártonfalvi Z. Topology of interaction between titin and myosin thick filaments. *Journal of Structural Biology* 2018;203:46–53.

[13] Irving M. Regulation of contraction by the thick filaments in skeletal muscle. *Biophysical Journal*. 2017;113:2579–2594.

[14] Marcucci L, Washio T, Yanagida T. Titin-mediated thick filament activation, through a mechanosensing mechanism, introduces sarcomere-length dependencies in mathematical models of rat trabecula and whole ventricle. *Scientific Reports*. 2017;7:1–10.

[15] Gordon AM, Huxley AF, Julian FJ. The variation in isometric tension with sarcomere length in vertebrate muscle fibres. *Journal of Physiology*. 1966;184:170–192.

[16] Ait-Mou Y, Hsu K, Farman GP, Kumar M, Greaser L, Irving TC, de Tombea PB. Titin strain contributes to the Frank–Starling law of the heart by structural rearrangements of both thin- and thick-filament proteins. *Proceedings of the National Academy of Sciences of the United States of America*. 2016;113:2306–2311.

[17] Bowditch HP. Über die Eigentqmlichkeiten der Reizbarkeit welche die Muskelfasern des Herzens zeigen. *Ber. Verh. Saechs. Akad. Wiss.* 1871;23: 652–689.

[18] Allen DG, Blinks JR. Calcium transients in aequorin-injected frog cardiac muscle. *Nature*. 1978;273:509–513.

[19] Endoh M. Force–frequency relationship in intact mammalian ventricular myocardium: Physiological and pathophysiological relevance. *European Journal of Pharmacology*. 2004;500:73–86.

[20] Pieske B, Maier LS, Bers DM, Hasenfuss G. Ca^{2+} handling and sarcoplasmic reticulum Ca^{2+} content in isolated failing and nonfailing human myocardium. *Circulation Research.* 1999;85:38–46.

[21] Ramirez-Correa GA, Cortassa S, Stanley B, Dong Gao W, Murphy AM. Calcium sensitivity, force frequency relationship and cardiac troponin I: Critical role of PKA and PKC phosphorylation sites. *Journal of Molecular and Cellular Cardiology.* 2010;48:943–953.

[22] Miller MR, Kasahara M. Studies on the nerve endings in the heart. *American Journal of Anatomy.* 1964;115:217–234.

[23] Braunwald E, Harrison DC, Chidsey CA. The heart as an endocrine organ. *American Journal of Medicine.* 1964;36:1–4.

[24] von Euler US, Lishajko F. Reuptake and net uptake of noradrenaline in adrenergic nerve granules with a note on the affinity for 1- and d-isomers. *Acta Physiologica Scandinavica.* 1967;71:151–162.

[25] Kirchberber MA, Tada M, Katz AM. Phospholamban: A regulatory protein of the cardiac sarcoplasmic reticulum. *Recent Advances in Studies on Cardiac Structure and Metabolism.*1975;5:103–115.

[26] Kentish JC, McCloskey DT, Layland J, Palmer S, Leiden JM, Martin AF, Solaro RJ. Phosphorylation of troponin I by protein kinase A accelerates relaxation and crossbridge cycle kinetics in mouse ventricular muscle. *Circulation Research.* 2001;88:1059–1065.

[27] Kumar M, Govindan S, Zhang M,. Khairallah RJ, Martin JL, Sadayappan S, de Tombe PP. Cardiac myosin-binding protein C and troponin-I phosphorylation independently modulate myofilament length-dependent activation. *Journal of Biological Chemistry.* 2015;290:29241–29249.

[28] Takimoto E, Soergel DG, Janssen PML, Stull LB, Kass DA, Murphy AM. Frequency- and afterload-dependent cardiac modulation in vivo by troponin I with constitutively active protein kinase A phosphorylation sites. *Circulation Research.* 2004;94:496–504.

[29] Moss RL, Fitzsimons DP, Carter Ralphe J. Cardiac MyBP-C regulates the rate and force of contraction in mammalian myocardium. *Circulation Research.* 2015;116:183–192.

[30] Guhathakurta P, Prochniewicz E, Roopnarine O, Rohde JA, Thomas DD. A cardiomyopathy mutation in the myosin essential light chain alters actomyosin structure. *Biophysical Journal.* 2017;113:91–100.

[31] Chang AN, Battiprolu PK, Cowley PM, Chen G, Gerard RD, Pinto JR, Hill JA, Baker AJ, Kamm KE, Stull JT. Constitutive phosphorylation of cardiac

myosin regulatory light chain in vivo. *Journal of Biological Chemistry*. 2015;290:10703–10716.

[32] Kampourakisa T, Suna Y-B, Irving M. Myosin light chain phosphorylation enhances contraction of heart muscle via structural changes in both thick and thin filaments. *Proceedings of the National Academy of Sciences of the United States of America*. 2016;113(21):E3039–E3047.

[33] Hamdani N, Herwig M, Linke WA. Tampering with springs: Phosphorylation of titin affecting the mechanical function of cardiomyocytes. *Biophysics Reviews*. 2017;9:225–237.

[34] Dabiri GA, Turnacioglu KK, Sanger JM, Sanger JW. Myofibrillogenesis visualized in living embryonic cardiomyocytes. *Proceedings of the National Academy of Sciences of the United States of America*. 1997;94:9493–9498.

4

An Intelligent Pump

Efficient pumping by the heart depends on coordinating its contraction so that its 2–3 billion myocytes and the molecular motors in each myocyte work together. This implies that there has to be a communication system to synchronise their behaviour. Chapter 3 showed that modulation of the intracellular concentration of Ca^{2+} acts as a signal to orchestrate contraction of all the sarcomere filaments within each myocyte. However, there must also be a mechanism to ensure that contraction of each myocyte is correctly coordinated to achieve efficient ejection of blood. This is achieved by a remarkable communication system which directs and controls the contraction and relaxation of every myocyte with each heartbeat.

Bio-communication

Communication is central to all life forms. As humans we use languages to communicate complex emotional and intellectual interchanges that are crucial for our lives as individuals and as communities. However, communication is also widespread in nature; it plays important roles in everything from the simplest to the most complex organisms; viruses, bacteria, fungi, yeasts, plants and animals all use systems for communicating. Bacteria may produce chemicals as a signal to indicate colony size (known as quorum sensing) and respond to them by modulating gene expression to control motility, virulence and other attributes. Plants attract pollinating insects using visual or olfactory (scents) signals and communicate with each other through their root systems. Internal communication also takes place within plants and animals. This occurs within and between individual cells as well as between their disparate organs. Plant roots produce chemicals which can travel to nearby cells to control their growth; other chemicals can travel up in the stem channels to control behaviour such as flower initiation or plant growth in response to soil health.

Modes of communication

Animals, including humans, produce chemicals called hormones in specialised organs that control the behaviour of other organs and tissues. They rely on the circulating bloodstream to deliver them to their targets. These may be relatively simple molecules (such as adrenaline), small or large proteins (like oxytocin and growth hormone, respectively) or steroids (such as testosterone and oestrogen). Each of them targets specific cells by occupying unique receptors that activate a series of signalling chemicals (called second messengers) within the target cell to communicate and initiate the intended response. The time course of responses to hormones varies depending on the circulation time and on cells' responses to them. Some act rapidly, as in the case of adrenaline secreted from the adrenal medulla in a fight or flight response. Others act over a period of minutes to hours (for example, insulin). Yet, others act as long-term maintenance systems as with thyroid and growth hormones. Rapid communication in animals is achieved by communication through nerves. Nerves consist of long extensions (called axons) of nerve cells (called neurones), most of which are located in the spinal cord and the brain. They transmit electrochemical impulses called action potentials along their length. In the case of efferent nerves, the impulses travel from neurones to muscles to control motion and other functions, while afferent nerves conduct sensory signals from peripheral tissues to the brain. Compared to the velocity of blood flow (5–20 cm/s in arteries), nerve pulses are very fast, some large axons conduct at 120 m/s. Bearing in mind the heart rate and the need to transmit a signal to each of its 2–6 billion myocytes with every beat, an effective system would also need to be rapid. In fact, the system which does this has many similarities with nerve conduction. Cardiac myocytes and neurones both have an electrical gradient across the cell membrane (called the resting membrane potential), they both propagate signals based on action potentials, and their action potentials share a common evolutionary origin.

Resting membrane potential

Resting potential simply means that cells have an electrical charge difference between the inside and outside of the cell membrane. The electric circuits we experience daily in our lives involve a flow (current) of electrons through copper wires from power stations to the switches and power sockets of our homes. Electrons are negatively charged subatomic particles, which, if removed from the atoms they orbit, can travel through conducting materials at close to the speed of light towards atoms lacking an electron. The electric potentials referred to in biology differ in that they concern electrically charged ions. Ions are atoms or molecules which have a positive or negative electrical charge. Atoms contain positively charged protons and negatively charged electrons in equal number and are therefore neutral. The loss of an electron from an atom leaves the remainder imbalanced with a positive charge. These are called cations, of which sodium (Na^+) and potassium (K^+) are examples. Atoms that gain an electron are negatively charged ions called anions, of which chlorine (Cl^-) is an example. In addition, many molecules are negatively charged and are also called anions. The development of resting membrane potentials in cells is related to differences in the concentrations of ions and ionic charges across cell membranes. This has its origin in the way cells evolved membranes to surround and contain their functions.

Cells contain essential atoms and molecules of many kinds which must be preserved undisturbed to function correctly. This is achieved by the cell membrane which separates its contents from the extracellular environment. In animals, cell membranes are relatively permeable to small ions such as Na^+ and K^+ and to water, but not to larger molecular ions, Figure 4.1. This leads to an osmotic imbalance between the inside and outside of the cell. Osmolarity is simply the sum of all the ions and molecules dissolved in a liquid and this generates a pressure (called osmotic pressure) which seeks to distribute solutes equally throughout. In the case of cells with permeable membranes, the inability of the large protein ions

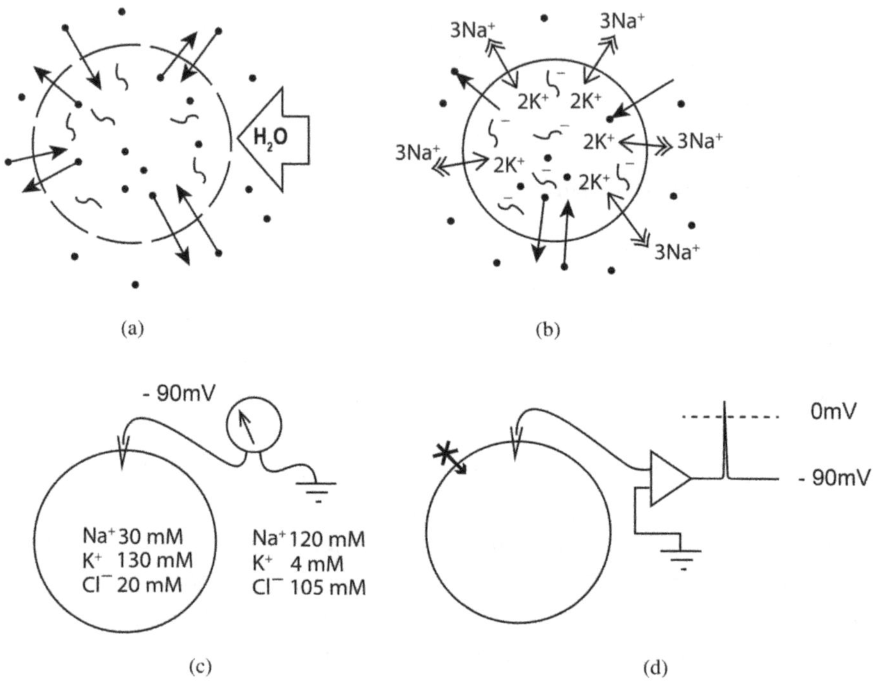

Figure 4.1: Transmembrane ionic and electrical gradients. (a) shows a hypothetical cell containing atomic ions (black dots) and larger protein ions (squiggly lines) within a membrane that is permeable to small but not larger ions. Osmotic pressure due to the impermeable large anions draws water into the cell and ruptures it. (b) Sodium pumps (double-headed arrows) extrude three Na^+ ions from the cell in exchange for admitting two K^+ ions. The net loss of one ion offsets the osmotic imbalance due to the impermeable anions and serves to stabilise the cell volume. It also depletes the cell of one positive charge. (c) The net loss of a positive charge (one Na^+ ion) with each cycle of the sodium pump removes positive charge from the cell. This creates a charge difference between the inside and outside of the cell, which in turn effects the diffusion of other ions, repelling inward diffusion of anions and attracting cations. In time, an equilibrium is reached, in which transmembrane ion and voltage gradients are stable. The values here are typical for cardiac myocytes. Not shown here are other cations and anions within the cell, which together with Na^+, K^+ and Cl^- result in charge neutrality inside the cell. (d) In response to a stimulus (asterisk), specialised cells undergo an abrupt change in membrane voltage, rapidly depolarising from a resting negative potential to positive values and just as rapidly returning to normal resting value. These are called action potentials and are characteristic of excitable cells such as neurones and muscle cells. They are caused by rapid, transient movements of ions across the cell membrane when stimulated.

to cross the membrane generates an osmotic pressure which would draw water into the cell in an effort to equalise their concentration on both sides of the membrane. This would result in rupture of the cell membrane (Figure 4.1(a)).

Ion pumps

August Krogh, in his Croonian lecture to the Royal Society in 1946 [1], drew attention to the fact that all such cells are in constant danger of being flooded and rupturing through this effect. He observed that plant cells overcome this by developing a thick cell wall that can withstand the osmotic pressure. However, animal cells have cell membranes which in physical terms are fairly flimsy and, without some other mechanism, would flood and rupture quite easily. One way to prevent this would be to exclude some solutes from the cell to offset the higher concentration of impermeable solutes inside the cell and thereby lower the osmotic pressure. Just three years after Krogh's lecture, fellow Dane, Jens Christian Skou, discovered that membranes obtained from crab nerves contain an enzyme capable of extruding Na^+ ions from the cell interior while hydrolysing ATP as an energy source to power the process (energy is needed because the extrusion takes place against a concentration gradient). He also showed that its activity is responsive to changes in the concentration of Na^+ and K^+ in a manner that maintains them within narrow concentration ranges. He suggested that it may be a mechanism for active extrusion of Na^+ [2]. This was an important discovery for which Skou shared the Nobel Prize in 1997. It is now known as the sodium pump (Na^+ ATPase) and is one of a large family of ATPase enzymes that span cell membranes and regulate the active transport of cations. Na^+ ATPase transports three Na^+ ions out of the cell and two K^+ ions into it with the hydrolysis of one ATP molecule. Briefly,

- Na^+ ATPase binds to ATP and then binds three intracellular Na^+ ions.
- The ATP is hydrolysed, releasing ADP. The retained phosphate phosphorylates the enzyme.
- This induces a conformational change in the enzyme, which moves the Na^+ ions to the cell exterior and releases them.

- Na^+ ATPase then binds two extracellular K^+ ions, which dephosphorylates the enzyme and reverses the conformational change, moving the K^+ ions to the cell interior and releasing them.

***Sodium pumps*:** Na^+ ATPase is present in all animal cells and is a key factor in maintaining ion concentrations and cell volume. Variants of the enzyme are also used by different cells to transport other cations, for example H^+ ions to produce acidity in the stomach and Ca^{2+} ions into the sarcoplasmic reticulum in cardiac myocytes (SERCA, discussed in Chapter 3). Phylogenetic studies indicate that Na^+ ATPase evolved very early in living organisms. They are present in prokaryotes (single cell organisms which lack a nucleus), of which the earliest fossil record is 3.5 billion years old [3]. This is not surprising however, when we recall their fundamental function in preserving the integrity of cell volume, osmolarity and ionic composition. It seems likely that the emergence of the sodium pump may have been a critical step in the evolution of cell membranes and was already present in the last common universal cell ancestor from which all living organisms evolved.

Osmotic control and electrochemical gradients

Thus, sodium pumps extrude Na^+ ions from cells as a way to counter the osmotic effect of non-permeable anions and prevent osmotic rupture (Figure 4.1(b)). As it does so, it transports two K^+ ions into the cell for every three Na^+ ions it extrudes. The net loss of one ion with each cycle of the sodium pump serves to offset the osmotic imbalance and stabilise cell volume. This changes the ionic composition of cells and results in significant ion concentration gradients across the cell membrane (Figure 4.1(c)). The concentration of Na^+ in cardiac myocytes is 30 mM compared to 120 mM outside. A reverse gradient occurs in the case of K^+ with 130 mM inside compared to 4 mM outside. The activity of the sodium pump also changes the charge status across the cell membrane. Each cycle of the pump extrudes three Na^+ ions out of the cell in exchange for admitting two K^+ ions, a net loss of one positive charge. The result is that the interior of the cells is negatively charged compared to the outside, or in other words, there is an electrical potential across the cell membrane (Figure 4.1(c)).

It should also be recalled that cells contain other ions, for example, chloride (Cl^-) and calcium (Ca^{2+}) ions which can readily diffuse through cell membranes). These would normally distribute equally on both sides of cell membranes, but because they are charged, they are affected by the action of the sodium pump. This is because the negative charge inside cells repels negatively charged ions such as Cl^- and attracts positively charged ions such as Ca^{2+}. Ultimately an equilibrium state is reached in which the force due to the charge gradient is balanced by reverse effects of the concentration gradients generated and this applies to all the permeant ions present in cells. This results in a steady state in which cells have concentration gradients of Na^+ and K^+ and other ions across the cell membrane and also a charge difference, the interior of cells being negative compared to the exterior (Figure 4.1(c)). The latter is referred to as the membrane potential and is measured in Volts. All animal cells have such a transmembrane potential, although its strength varies with different cell types. For example, skeletal muscle cells have a membrane potential of –95 mV, cardiac ventricular cells of around –90 mV, neurones of –60 mV to –70 mV, red blood cells of –9 mV.

Thus, the origin of resting membrane potential in cells dates back to the earliest unicelled organisms. The evolution of action potentials is similarly ancient and linked to the way cells came to regulate Ca^{2+} ions and use them as signals.

Action potentials

The transmembrane potential in some cells has evolved to confer the property of excitability. Excitable cells have the ability to undergo a sudden change in membrane potential in response to certain stimuli and some have evolved this in specialised ways. For example, neurones respond to specific stimuli by undergoing a sudden change in membrane potential and use it to signal other cells. These abrupt changes in transmembrane potential are called action potentials and are characteristic of all excitable cells. Typically, the negative resting transmembrane potential abruptly depolarises to a positive value and then just as abruptly reverts to its stable resting state (Figure 4.1(d)). Nerves are composed of axons, which are extensions of neurones in the brain or spinal cord and axons propagate action potentials generated in their parent neurones along their length.

Skeletal muscle cells are controlled by the impulses propagated by such nerves. On arrival at a neuromuscular junction, an action potential causes release of a neurotransmitter called acetylcholine. This causes an action potential to occur in the muscle cell which then activates its contractile machinery. The heart rate and rhythm are also controlled by action potentials as is the process of contraction and relaxation. A question that has intrigued many researchers is how action potentials evolved and how did they come to have such important roles in biology.

Regulation of calcium ions

Our bodies contain many ions other than Na^+ and K^+ and these are also maintained in balance across the cell membrane. Two others to mention here are chloride ions (Cl^-) and calcium ions (Ca^{2+}). Chlorine atoms in solution tend to gain an electron which gives them a negative charge, forming chloride ions (Cl^-). Cl^- ions are abundant in biology and are the main atomic anion in cells. They are freely diffusible across cell membranes but being negatively charged are repelled by the negative potential in cells, which causes a net outward diffusion. This results in a higher extracellular concentration, and as it develops, counters further extrusion. In balance, the concentrations are 105 mM outside and 20 mM inside cells.

Calcium is of special importance for two main reasons which, as we shall see later, have an evolutionary link. Firstly, calcium ions (Ca^{2+}) at a high concentration are highly toxic to the interior of cells. This is because they quickly react with phosphate groups to form insoluble calcium phosphate precipitates. All our energy producing chemistry is based on phosphates in ATP and phosphates are a by-product of protein synthesis. Exposure to high concentrations of calcium would disrupt both of these processes. Secondly, cells have evolved to utilise Ca^{2+} as one of the most important second messengers (ions and molecules that transmit signals to the interior of cells from outside) in biology. As we saw in Chapter 3, a brief rise in intracellular Ca^{2+} initiates contraction in heart and skeletal muscle; it also signals the secretion of many hormones. As a consequence, the concentration of Ca^{2+} is carefully controlled in cells and maintained at a very low concentration. In humans, this is achieved by storing 98% of total body calcium in bones. In plasma, just over half the calcium present is in a form that can diffuse through cell membranes (2.3 mM); the remainder is bound to proteins and non-diffusible. In contrast, Ca^{2+} concentration inside cardiac myocytes at rest is 0.1 mM, meaning that the transmembrane gradient is far greater than for any other ion. To achieve and maintain such a low concentration, heart cells have three Ca^{2+} pumps. (i) A variant of the Na^+ ATPase pump has evolved which pumps Ca^{2+} rather than Na^+ out of cells. This has a high affinity for Ca^{2+}, meaning it avidly binds Ca^{2+} even at low concentrations and extrudes it. It uses the energy from hydrolysing one ATP molecule to extrude one Ca^{2+} ion. This is a relatively slow process, but ideal for maintaining Ca^{2+} in the resting state when there is little Ca^{2+} flux. (ii) A second pump attached to the cell membrane works by using the high extracellular concentration of Na^+ as an energy source to extrude Ca^{2+} ions. The pump creates a pore that allows three Na^+ to enter the cell, which they do readily due to the high gradient. Then, rather like a water wheel powered by a column of water, the pump then binds a Ca^{2+} ion inside the cell, which it releases outside the cell. Called the Na^+/Ca^{2+} exchange pump, it is highly efficient in removing Ca^{2+} from the cell due to the strong concentration gradient of Na^+, having an optimal capacity to remove 2,500–5,000 Ca^{2+} ions per second [4]. (iii) Cardiac myocytes have a third Ca^{2+} pump which we encountered in Chapter 3, sarcoplasmic reticulum Ca^{2+} ATPase (SERCA). Briefly, SERCA is bound to the membrane of sarcoplasmic reticulum (SR). The latter acts as a store of Ca^{2+} in cardiac myocytes and SERCA captures Ca^{2+} ions from the cytosol and pumps them into the SR.

Ca^{2+} and action potentials signals

Action potentials and Ca^{2+} are the most widely used signalling systems in our bodies and their uses in this way are linked to evolutionary developments very early in the origins of life.

Origin of Ca^{2+} signalling

It may seem paradoxical that Ca^{2+} ions should be both highly toxic to cells and an important signalling ion. However, for early unicelled organisms, a rise in intracellular Ca^{2+} ions was both a threat and the first indication that membrane rupture had occurred. These cells were vulnerable to membrane rupture because a rapid influx of Ca^{2+} could cause precipitation of phosphate compounds in the cell. An influx of Ca^{2+} would also have caused membrane depolarisation due to the positive charge on Ca^{2+} ions and this is also harmful because the structures of many membrane bound proteins are voltage sensitive [5]. Thus, a rise in intracellular Ca^{2+} and membrane depolarisation were early indicators of membrane damage and evolved to become signals that initiated emergency responses to such damage. Later, both evolved more widely as signalling mechanisms for regulating many cell processes.

Ca^{2+} regulates contraction

Contraction of actomyosin in heart and skeletal muscle is triggered by a transient rise in Ca^{2+}. Actin and myosin are ancient proteins which evolved in single-celled organisms [6]. These would not have had muscle, and so why did such proteins evolve and what was their function? The answer to this relates to mechanisms that evolved for healing membrane wounds. All eukaryote cells have healing mechanisms to repair membrane wounds, Figure 4.2, and these are activated by a rise in intracellular Ca^{2+} [7]. These consist of:

1. Contraction of a "purse-string" of actomyosin around the wound in response to the local rise in Ca^{2+}, which reduces its size rapidly. Eukaryotes evolved a layer of actomyosin adjacent to the cell membrane which became available for this purpose. It is interesting to note here that even in growing cardiac myocytes the assembly of actin and myosin into sarcomeres also takes place close to the sarcolemma (see Figure 3.7 and related text in Chapter 3).
2. The local rise in intracellular Ca^{2+} also induces exocytosis which directly repairs the membrane wound. Exocytosis is a widely used process for moving molecules from inside cells to the outside. Briefly, molecules enclosed within a membrane called a vesicle are transported to the cell membrane. The vesicle then docks and fuses with the cell membrane which results in its interior facing outside the cell and its contents being released to the extracellular space, Figure 4.2.

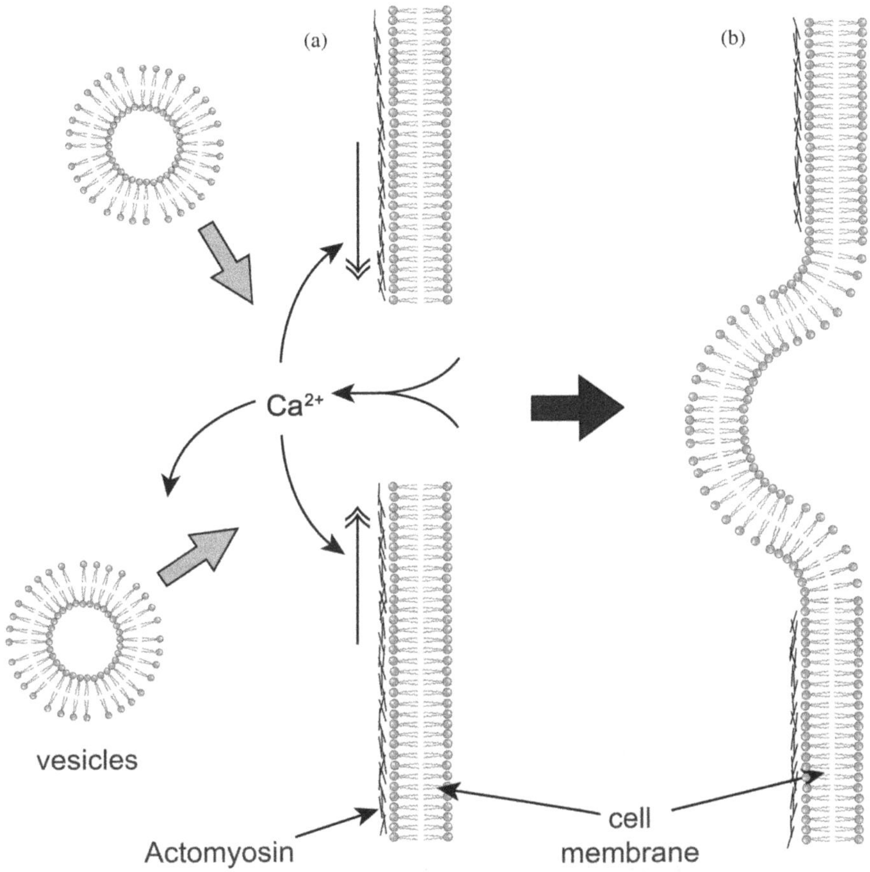

Figure 4.2: Schematic diagram of cell membrane wound healing. (a) A membrane rupture results in a rapid influx of Ca^{2+} ions due to its high concentration gradient. This activates contraction of actomyosin adjacent to the wound which acts like a purse string contraction reducing the hole size. In addition, Ca^{2+} activates the formation of vesicles. Vesicles are spheres formed from bilipid membranes similar to these in the cell membrane. (b) The vesicles formed then dock with the cell membrane on both sides of the wound in a process known as exocytosis, which seals it.

Exocytosis is a widely used physiological mechanism to secrete hormones and to release neurotransmitters like noradrenaline; this form of exocytosis is regulated by Ca^{2+} signalling. The same process occurs in membrane wound healing. The rise in Ca^{2+} induces exocytosis and the vesicles formed fuse with the wound edges producing a plug that heals the hole. Thus, Ca^{2+} influx triggers contraction of local actomyosin and exocytosis to repair cell membrane wounds. Later these evolved as specialised functions. Actomyosin was highly conserved during evolution to become the molecular motor in all muscles including the heart; exocytosis is widely used as hormone and neurotransmitter release mechanisms and both are triggered by Ca^{2+}.

The origin of action potentials

An influx of Ca^{2+} following cell wall damage causes depolarisation of cells due to its positive charge. This is a gradual change depending on the size of the wound. Action potentials differ from this in being characteristically "all or none" events; once an excitable cell depolarises to a threshold voltage, a rapid amplification of the initial depolarisation occurs for a short period, after which it quickly returns to the resting potential. Importantly too, action potentials have the ability to propagate along the entire cell membrane in excitable cells. Thus, the axons in our nerves propagate action potentials originating in neurones in the brain or spinal cord as signals to muscles and other tissues. Similarly, action potentials initiated in cardiac myocytes spread to the entire cell surface to ensure contraction is activated uniformly. The evolution of voltage gated ion channels was a crucial step in how this emerged.

Voltage-gated ion channels

The active amplification of depolarisation during action potentials requires the opening of membrane pores (called voltage-gated ion channels) which are permeable to cations (typically Na^+ or Ca^{2+}) in response to a threshold level of depolarisation. The concentration gradients of these ions result in rapid influxes through open pores and the positive charge they carry rapidly depolarises the cell. The resting potential is then quickly and actively repolarised (described in what follows) to terminate the action potential. Thus, voltage-gated ion channels are closed at rest, but open briefly in response to a small positive change in membrane potential resulting in a rapid influx of positive charge which causes an action potential. All excitable cells have these voltage-gated channels, and they were a necessary step for the evolution of action potentials.

The origin of voltage-gated ion channels

Traditionally, phylogenetic studies were based on anatomical comparisons. However, the introduction of gene sequencing around 30 years ago was a major advance which allowed scientists to compare gene sequences across many species. For example, if the same gene sequence is found in plants and animals, it follows that the gene was likely to have been present in their common ancestor and can be dated to before their separation. Such studies indicate that voltage-gated ion channels are ancient and are widely distributed in living organisms. The earliest and founder member of what became a large family of ion channels was a voltage-gated potassium channel (Kv), which appeared in bacteria around three billion years ago [8], Figure 4.3(a) (i). This consists of a membrane bound protein with six transmembrane spanning segments. The fourth transmembrane segment (S4, shown in red in Figure 4.3(a)) contains several positively charged residues, the gene code for which is highly conserved in all other voltage-gated ion channels. This segment moves in response to changes in the membrane potential causing a conformational change which alters the shape of the protein and is the voltage sensing element of all voltage-gated ion channels. Kv channels respond to depolarisation by opening a pore which allows an efflux of K^+ ions from the cell (down its concentration gradient) [9]. This remains a key element in restoring resting potential following depolarisation in many excitable cells. The Kv channel is referred to as founding member of voltage-gated channels because the later evolution of voltage-gated channels specific for Na^+ (Nav) and Ca^{2+} (Cav) were formed from units of the Kv channel. At some point, duplication of the gene that encodes the Kv channel occurred, forming a new gene that produced an ion channel consisting of two linked units (called a dimer) of the Kv channel and which evolved selectivity for Ca^{2+} ions, Figure 4.3(a) (ii). Later still, this dimer gene underwent a further duplication to form a four domain channel, which retained the Ca^{2+} selectivity of the dimer and later evolved further to form a channel selective for Na^+, Figure 4.3(a) (iii). The evidence for this progression is based on (a) the fact the gene duplication is well recognised in genetic evolution, (b) the high degree of homology between each domain in two and four domain channels and the single domain of Kv channel and (c) the existence of intermediate two pore (TPC) channels in extant lysosomes and endosomes (intracellular vesicles) [10].

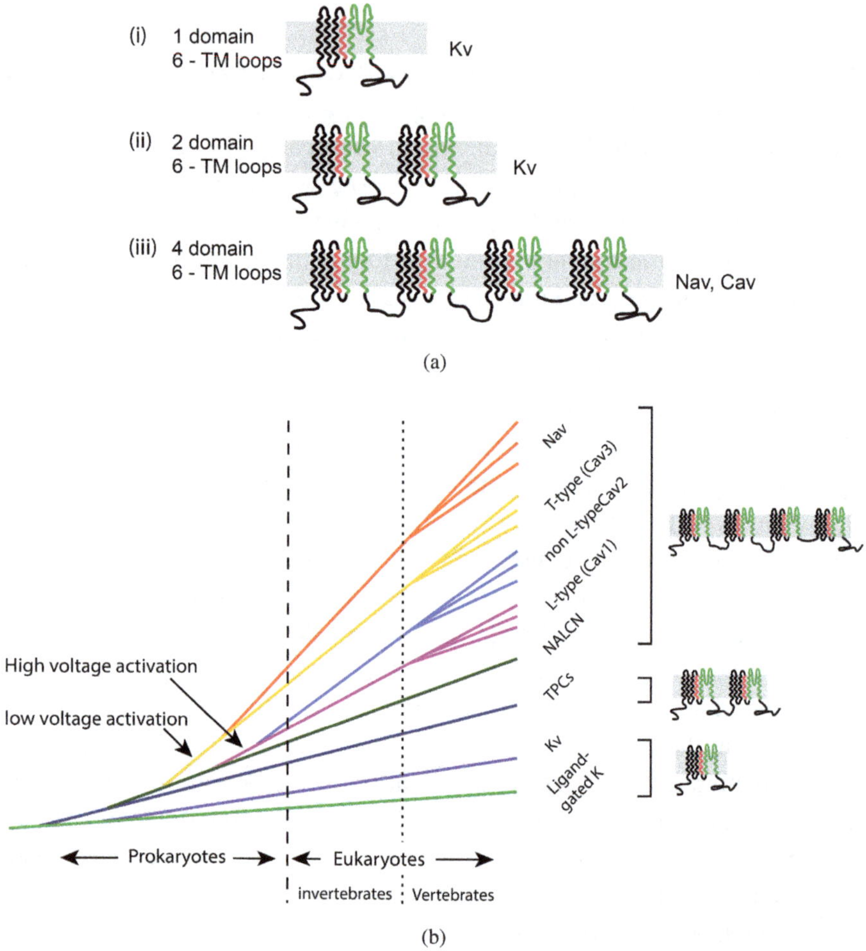

Figure 4.3: (a) Schematic diagram of voltage gated ion channels. (i) The founder member was a single-domain protein with six transmembrane segments. The fourth segment (S4, shown in red) is voltage sensitive and opens a pore segment (shown in green) selective for potassium ions in response to a change in membrane potential. In this case, the pore opens in response to a negative potential change (that is, the cell interior becoming more negative) and the resulting K^+ influx reverses that. (ii) A gene duplication resulted in a two-domain (dimer) channel which is still present in lysosomes. (iii) A further gene duplication of the dimer resulted in a four-domain voltage-gated channel that later evolved selectivity for Na^+ and Ca^{2+} ions. These four-domain channels are responsible for action potentials in nerves and muscles including cardiac muscles. In the case of Nav channels, activation of the S4 segments of domains 1 and 3 is required for channel opening; activation of the S4 segment of domain 4 is linked to that of domains 1 to 3 and this initiates pore closure. (b) Phylogenetic

Voltage-gated ion channels in heart cells

Thus, voltage-gated ion channels are ancient. The protein structure of the single domain founder Kv, which was already present in bacteria three billion years ago, evolved to form two and four domain channels with a variety of ion selectivities, Figure 4.3(b). Other variants which have evolved include the non-selective sodium leak channel (NALCN), which is important in brain cell function. In addition, some ion channels evolved with different activating voltage thresholds. Thus, low voltage activating (LVA) channels are activated by relatively small degrees of membrane depolarisation and the Nav channels responsible for most action potentials are examples of these. Others are activated at much higher levels of depolarisation such as Cav channels which are activated at voltages above –30 mV. Evolution of ion selectivity would only have required small changes because ion selectivity filters for Na^+ and Ca^{2+} are dependent on the presence of a single amino acid in each of the four domains [11]. A further modification occurred in Nav channels which linked their activation and inactivation, and this became important for the way action potentials are initiated and terminated. All ion channel domains consist of six transmembrane segments, of which segment 4 (S4) acts as the voltage sensor. This segment moves in response to membrane depolarisation, initiating the change in pore permeability. Four domain ion channels then have four voltage sensing S4 segments. All four must be activated for K^+ channel pores to open. In the case of Nav channels however, the S4 segments in domains I, II and III respond to small changes in membrane depolarisation resulting in pore opening and provide the mechanism for the influx of Na^+ ions (called the sodium current) which initiates the onset of the "all or

Figure 4.3: *(see figure on facing page)* tree model of voltage-gated ion channels. Genetic evidence indicates that these channels evolved from a founder 6-transmembrane protein present in an early prokaryote. The original is believed to have been gated by a signalling molecule and later evolved into a K^+ gated channel. Then two successive gene duplications gave rise to two- and four-domain 6-transmembrane channels. The original four-domain channel may not have been ion selective but later evolved further to form low and high voltage channels (discussed in more detail in the text) from which sodium and calcium channels evolved. (After Anderson *et al.* [8])

none" phase of action potentials. The activation of the S4 segment in domain IV is coupled to activation of those in I–III and once it becomes active, it initiates inactivation of pore opening. Thus, S4 in domain IV provides a mechanism by which inactivation of Na^+ influx is linked to its onset within the molecular structure of the channel [12]. This is reflected in the brief duration of the sodium current, typically just a few milliseconds. It also corresponds to the duration of action potentials in neurones and skeletal muscle. In cardiac cells, action potential duration is much longer, typically a few hundred milliseconds, due to the engagement of secondary currents (these are discussed in what follows).

How did action potentials evolve?

The threat of injury would have been an ongoing evolutionary challenge for single-celled organisms. Some eukaryotes developed cilia and flagella as narrow extensions of the cell which had sensory functions and some also had a motile function, which could propel the cell. Thibaut Brunet and Detlev Arendt [13] have proposed that action potentials evolved as a signal to engage flagella motility. Green alga (Chlamydomonas) have a primitive eye that can sense light, and action potentials are known to occur in its flagellum following photo-stimulation. Remarkable features of this are (i) the eye of green alga is remote from its flagella, (ii) photo-stimulation of the eye causes membrane depolarisation due to an inward Ca^{2+} current, (iii) this acts as a signal to activate an "all or none" depolarisation in the membrane of the flagella, which does not involve the rest of the cell [14]. This behaviour demonstrates that this single-celled organism has voltage-gated ion channels that can generate action potentials to control flagella motility and that it uses a depolarisation signal originating from a remote site in the cell to activate them. It is a remarkable harbinger of the way the nervous system and its control of motion developed. The hypothesis that action potentials evolved with flagella motility is supported by the fact that the loss of flagella in later evolution among some eukaryotes closely paralleled the loss of their Cav and Nav channels, suggesting they evolved together. Thus, the action potentials which act as signals in our nerves and muscles may have evolved initially in single-celled organisms and are likely to have been present in the last eukaryote common ancestor.

Action potentials in the heart

The generation of action potentials by heart cells is central to the regulation of the heart rhythm and in organising coordinated contraction of its chambers. To do this, cardiac cells in different parts of the heart have evolved specialised action potentials which control separate functions. For simplicity these will be described under four headings; (i) pacemaker cells which control the heart rhythm, (ii) atrial myocytes, which are the contracting cells of the atria, (iii) specialised conducting cells that control the speed of propagation through the heart and (iv) ventricular myocytes which are the contracting cells of the ventricles. Before describing these in more detail, a brief description of some aspects of cardiac anatomy will help to put them in context.

Electrical Anatomy of the Heart

In 1852, Herman Stannius demonstrated that in the frog the heartbeat is initiated in the atria and conducted to the ventricles. However, throughout the 19th century it was unknown whether this occurred through nerves in the heart or by direct conduction through heart cells. This uncertainty remained even as recently as 1951 [15]. Nevertheless, the work of Stannius and others did show that the region of the atrio ventricular junction was an important gateway for propagation to the ventricles. In a landmark study in 1907, Sunao Tawara discovered the atrio ventricular node now known to be a specialised conducting pathway between the atria and ventricles. In the same year, the site of the heart pacemaker, now known as the sinus node, was discovered [16]. These studies relied entirely on anatomical and histological methods. In 1910, it was realised that the conducting cells which continue from the atrio ventricular node into the ventricles are surrounded by fibrous tissue, and, since this does not conduct, it meant that the atria and ventricles are insulated except for atrio ventricular node cells. These were important observations made decades before the electrical properties of cells were understood and form the basis of much of our current understanding of the anatomy of the electrical system in the heart. A schematic illustration of this is shown in Figure 4.4.

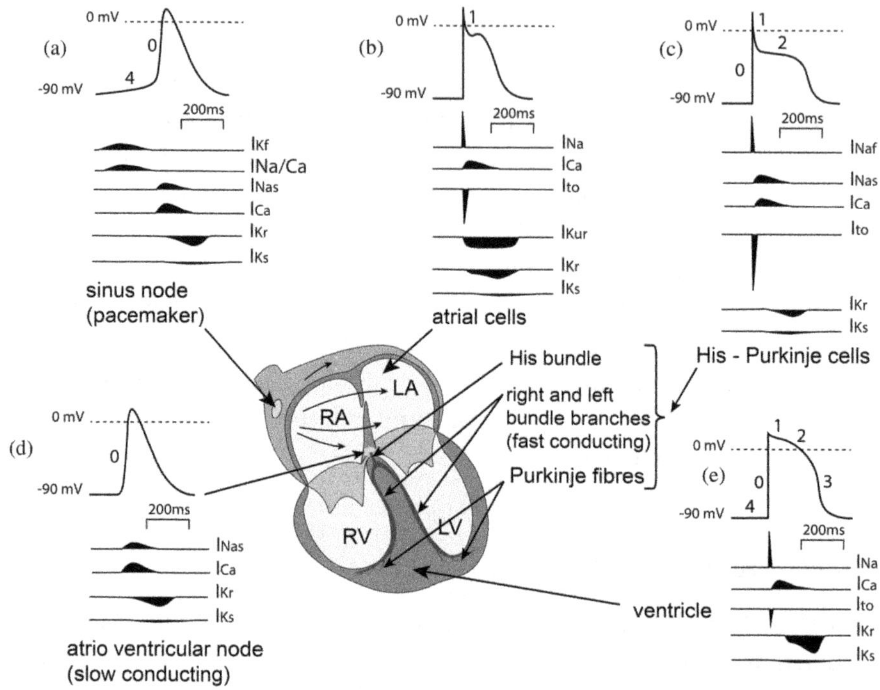

Figure 4.4: Diagram to illustrate the electrical anatomy of the heart and the action potentials corresponding to each specialised region. The sinus node is the pacemaker of the heart and is located in the right atrium close to the junction with the superior vena cava. From there the impulse generated is conducted through the atria and to the atrio ventricular node (AVN). The AVN is the only pathway by which the impulse can reach the ventricles and it is specialised to conduct the impulse slowly to the ventricles to allow coordination of atrial and ventricular contraction. From the AVN, the impulse reaches the His bundle through which it is conducted rapidly to the ventricles and thereby facilitates synchronised contraction. Action potentials are changes in transmembrane voltage that occur in excitable cells in response to stimulation. In heart cells they virtually always occur in response to activation by the action potentials which occur in adjacent cells. Those generated in each region of the heart are shown in (a)–(e). In each case they are caused by movements of positive and negative ions across the cell membrane. Cardiac action potentials have traditionally been described as having phases (0) to (4). These are illustrated here in the example of a ventricular action potential and are used in what follows to describe events at different stages. In ventricular cells (e), action potentials involve a rapid depolarisation of the cell (0), followed initially by a brief transient early repolarisation (1), then a much slower (2) and more rapid (3) repolarisation. Membrane potential during diastole (4) is stable. Atrial cells (b) have a similarly rapid depolarisation, but a much more marked transient early repolarisation (1). Repolarisation phases (2) and (3) are also more marked and the overall duration of atrial action potentials is shorter than in ventricular cell. Action potentials in His–Purkinje cells

Sinus node and atrial activation

The sinus node is the heart pacemaker and is located in the right atrium close to the junction with the superior vena cava (which drains blood from the upper body). The impulse generated there then spreads through the atria from cell to cell, somewhat akin to the waves from a pebble dropped in a pool of water. As it does so, it stimulates the atria to contract, which primes the ventricles at the end of their filling period and just before they contract. This wave of excitation will eventually reach the atrio ventricular node and activate the cells there.

Atrio ventricular node

The atrio ventricular node (AV node) is located at the junction of the atria and ventricles just above the ventricular septum. Atrial cells are in direct contact with those in the upper part of the AV node and when the wave of atrial activation reaches them, they immediately activate these cells in the AV node. AV node cells have evolved a specialised conduction property; propagation of action potentials from cell to cell here is much slower

Figure 4.4: (*see figure on facing page*) (c) have more rapid depolarisation (0) than atrial or ventricular cells and this contributes to faster conduction of the cardiac impulse. The early transient repolarisation phase (1) is very marked and the plateau phase (2) occurs at a more negative potential. Thereafter, repolarisation is slower in these cells and their action potentials are longer. Action potentials in AVN cells (d) have a much slower rate of depolarisation (0) than atrial or ventricular cells and this contributes to their slow rate of conduction. The early transient repolarisation is absent in these cells and action potential duration is shorter than ventricular cells. Action potentials in the sinus node (a) are unique in that the cell membrane gradually depolarises spontaneously during diastole (4) until the firing threshold is reached and an action potential follows. In this way, sinus node cells are inherently rhythmic and are able to function as pacemaker cells. The varying features of cardiac action potentials reflect differences in voltage gated ion channels expressed in each cell type and the ionic currents they mediate. These are described in detail in the text.

(discussed in more detail in the following) than in other parts of the heart. This is related to the need to coordinate atrial contraction and ventricular filling, discussed in more detail later.

The bundle of His

In 1893, Wilhelm His discovered a bundle of cells which were continuous with the AV node above and penetrated a plane of fibrous tissue below to continue into the ventricles [17]. This was important because the fibrous plane it penetrates is the basis for electrical insulation between the atria and ventricles. In other words, the His bundle, as a continuation downstream of the AV node, is the only conducting tissue which connects the atria and ventricles. It might seem paradoxical that the activation of the ventricles should be limited to such a small conducting pathway. However, it does have advantages in that restricting atrio ventricular conduction to the His bundle, together with the slow conduction through the AV node means that the ventricles are protected from abnormally rapid rhythms in the atria. For example, atrial fibrillation is a common arrhythmia that originates in the atria and may cause them to contract very rapidly (250–350 times per minute). This rate of atrial contractions reduces the efficiency of the heart by impairing ventricular filling; however, if such a rate were conducted to the ventricles, it would be life threatening because ventricular contraction cannot function at such a rate. Typically in atrial fibrillation, only a fraction of the atrial activations is conducted to the ventricles due to its insulation from the atria and the slow conduction through the AV node.

Bundle branches

The His bundle divides at the top of the ventricular septum into right and left branches. The left branch divides into two and in some hearts three interlinked fascicles (this detail is omitted in Figure 4.4 for simplicity) which continue to the apex of the ventricle. The right branch continues as a single structure into the right ventricle where it ramifies widely. Eventually, the terminal fascicles of the bundle branches end in a network of Purkinje fibres named after Jan Purkinje, a Czech physiologist who discovered them in 1839 [18].

Cell-to-cell conduction

An important aspect of the electrical anatomy of the heart is that adjacent cells are connected and this allows conduction of the electrical impulse from cell to cell through the ventricles following its rapid distribution through the His Purkinje fibres. We saw in Chapter 3 that strong physical connections between myocytes ensure that their contraction is coordinated. Myocytes also have specialised electrical connections (called gap junctions). These are areas where the cell membranes (sarcolemmas) of two adjacent cells are fused to form a single thin membrane, as illustrated in Figure 4.5.

In this example, a gap junction is shown forming a continuation of the sarcolemmas of two cells at one end and the other is continuous with the intercalated disk which fuses sarcomeres at the end of each cell. Gap junctions are composed of proteins called connexins arranged in groups of six

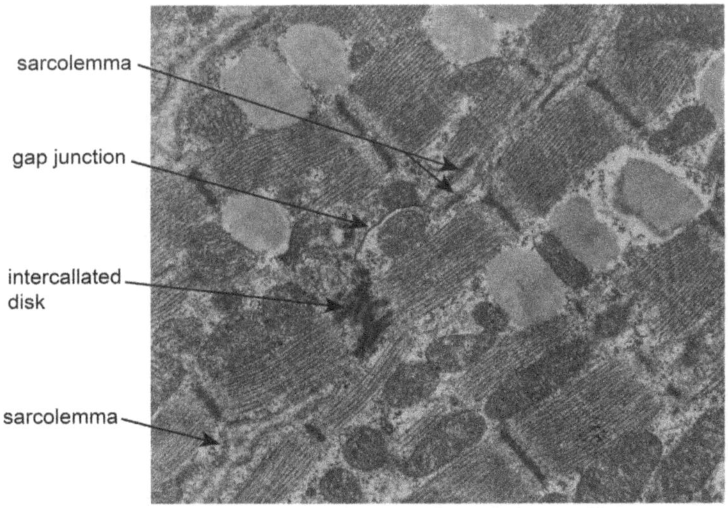

Figure 4.5: Electronmicrograph of myocardium. Cardiac muscle cells conduct electrical impulses from cell to cell. They do this via gap junctions which are areas where the sarcolemmas of two adjacent cells form a single narrow junction composed of connexins. These are specialised pores through which ion fluxes from an excited cell cause depolarisation in an adjacent one which induces an action potential. In this figure, a gap junction is shown, with one end originating from the sarcolemma of two adjacent cells and the other is continuous with an intercalated disk.

in a circular cluster, called a connexon, such that the centre can form a pore. Connexons in the membrane of two adjacent cells then align and dock to form a single gap junction. Gap junctions permit ion fluxes between adjacent cells. When one cell becomes depolarised while an adjacent cell is not, a potential across the gap junctions which unites them is created. This causes a flux of ions through the gap junctions and results in some depolarisation of the formerly quiescent cell. Once that cell reaches its threshold potential, it too will fire an action potential and similarly activate its neighbours. There are several connexin isoforms, named in relation to their molecular weight. For example, connexin 43 (molecular weight 43 KDa) is a common form in ventricular cardiac myocytes, connexin 30.2 is found in sinoatrial and atrioventricular node cells, while connexin 40 is abundant in His bundle and Purkinje cells [19]. The distribution of connexin isoforms plays an important role in cardiac conduction because of their different conductances, connexin 40 being fastest, connexin 43, intermediate and connexin 30.2, slowest. Thus, conduction in the heart occurs on the basis of action potential propagation from cell to cell. When one cell depolarises during an action potential, it induces an action potential in adjacent cells which in turn activate their adjacent cells. In this way, action potentials initiated in the pacemaker are propagated throughout the heart with each heartbeat. Initially, the impulse travels along atrial myocytes to activate atrial contraction; it then passes through the atrioventricular node to reach the specialised conducting cells of the His–Purkinje system which conveys it to the ventricular cells which propagate it from cell to cell.

Velocity of propagation

Clearly, impulse propagation through the heart has to be rapid to control the contraction of each heartbeat. However, it is also organised in a special way to ensure that the pumping of each chamber is correctly synchronised and that the heart rhythm remains stable. The velocity of conduction contributes to this. Beginning in the atria, action potentials are conducted from cell to cell at a speed of around 50 cm/s. It reaches the atrioventricular node at around 70 milliseconds after the pacemaker impulse, where the speed slows markedly to just 5 cm/s. This can be attributed to the presence of gap junctions composed of connexin 30.2, which has a slow ionic conductance [19]. Slow AV node conduction has the benefit of allowing time for atrial contraction to prime the ventricles before the onset of ventricular contraction. The impulse then reaches the His bundle at around 160 milliseconds after the pacemaker impulse, where the speed of conduction increases markedly to around 200 cm/s and to 400 cm/s through the Purkinje fibres. The high speed of conduction in Purkinje fibres is due to abundance of gap junctions composed of connexin 40 [19]. From there, ventricular conduction is from cell to cell at a speed of around 50 cm/s and again this relates to the presence of a mix of gap junctions containing connexin 43 and 40. Thus, conduction velocity is coordinated so that slow conduction through the atrioventricular node allows time for atrial contraction to fill the ventricles in advance of their contraction. Once the impulse reaches His–Purkinje system, conduction speeds up dramatically, which ensures that activation of all parts of the ventricles is synchronised. This variation in the speed of conduction in different tissues of the heart is partly explained by differences in the connexins present in gap junctions. However, it is also determined by differences in the action potentials generated in each region of the conducting system.

Specialisation of cardiac action potentials

Here we will consider action potentials that occur in each part of the heart. For simplicity, we will begin at the end with ventricular action potentials. Action potentials that occur in each section of the heart are illustrated diagrammatically in Figure 4.4 which retains the traditional four-stage terminology for describing them.

Understanding that electrical signals occur in nerves and other tissues began in the late 18th century and hypotheses that ion fluxes may underpin the voltage changes in cells emerged in the nineteenth. Then in the mid-20th century these ideas were confirmed in experiments by Alan Hodgkin and Andrew Huxley, which showed that changes in membrane permeability for sodium underpinned action potential depolarisation in nerves [20]. They made the first recordings of action potentials by placing small electrodes inside nerve fibres (axons). They also developed a membrane voltage clamp technique which involves holding the membrane voltage constant during an action potential by applying an electric current across the membrane to counter the charge movement due to the ion fluxes of the action potential. Using this technique, they confirmed that action potentials depended on sodium and potassium ion movements across the membrane. In simple terms, depolarisation resulted from an inward flow of sodium ions (carrying positive charge into the cell) and repolarisation from an outward flow of potassium ions (carrying positive charge out of the cell). Later in the 20th and 21st centuries, much more was learned about the molecular structure and function of the voltage gated ion channels which control these ion movements. Cardiac action potentials are mediated by the activation of several of them in a tightly orchestrated manner. Here we are particularly concerned with voltage gated channels selective for sodium (Nav), calcium (Cav) and potassium (Kv) ions, each playing a crucial role in the speed and timing of action potentials and also in initiating the process of contraction. All cardiac action potentials involve membrane depolarisation and repolarisation and their mechanisms have many similarities. However, they differ in speed and timing in different cells. Here, ventricular action potentials are described in some detail. Those which occur in other regions of the heart are described in comparison to ventricular cells and the differences used to explain their unique functions.

Ventricular action potentials

Figure 4.4(e) illustrates the voltage changes which occur during a ventricular action potential. Typically, ventricular action potentials commence with a rapid rise in transmembrane voltage (depolarisation) from a stable resting potential of around –90 mV to +20 mV in just a few milliseconds and this is followed by restoration (repolarisation) to the resting potential. Repolarisation is initially rapid, followed by a more gradual phase (the latter is often described as the plateau phase) and then a further rapid return to the resting potential. The ionic flows (called currents) which generate these voltage changes are indicated below the voltage trace in Figure 4.4(e). In this and all other current traces, inward currents are represented as upward and outward currents as downward.

Depolarisation

The initial phase of depolarisation is due to an inflow of Na^+ ions, called the sodium current (INa) into the cell, Figure 4.4(e), (0). Since it was initially discovered by Hodgkin and Huxley in the 1950s, it is now known to be due to the opening of a voltage gated sodium ion channel (Nav) as described above. The voltage sensitive transmembrane domain (S4) of this channel responds to small voltage changes in membrane. Such levels of depolarisation are readily produced by ion leakage through gap junctions from an adjacent depolarised cell. Once the threshold activation voltage has been reached, the channel opens in an "all or none" manner and the inflow of Na^+ depolarises the cell. This occurs very rapidly; the transmembrane potential typically rises from around –90 mV to +20 mV in just a few milliseconds, after which the process of repolarising the cell then commences.

Early repolarisation and the early transient outward K^+ current (Ito)

The first stage in repolarisation is termination of the inward sodium current. As discussed above, activation of domain IV of the Nav channel is linked to activation of domains I and III and once it is activated, it deactivates the pore, terminating the sodium current in just a few milliseconds. However, cardiac action potentials have a long time course (around 100 times longer than nerves or skeletal muscle) and this is due to the engagement of additional ion channels.

The first repolarising event following depolarisation is a transient outward potassium current, Ito. This is mediated through a voltage gated ion channel selective for potassium ions. It is a member of the four domain 6-transmembrane ion channel family and is rapidly activated at the voltages of depolarised membranes and results in an outward K^+ current (down its concentration gradient). The loss of positively charged ions from the cell begins to reverse the depolarisation causing early repolarisation, Figure 4.4(e) (1). However, the Ito current is short lived and so its repolarising effect is also transient, effecting only the early repolarisation phase of action potential. The magnitude of Ito varies in different parts of the heart reflecting the density of the channel expressed in each cell type. For example, Ito current is larger in atrial than ventricular cells and consequently has a bigger effect on early repolarisation in atrial (Figure 4.4(b) (1)) than ventricular cells (Figure 4.4(e) (1)). So Ito has a significant effect on the action potential during early repolarisation, however, it also has important consequences on later events during the action potential and on the way the action potential initiates contraction. This has to do with its influence on the next phase of the action potential, which is mediated by an inward calcium current.

Inward calcium current Ica

Depolarisation changes the transmembrane potential from around –90 mV to +20 mV. Once the potential becomes more positive than –30 mV, voltage gated calcium channels (Cav) are activated. These are known as L-type Cav (L stands for long-lasting because the channel remains open for longer than other depolarising currents). This causes an inward calcium current, Ica which contributes to the long duration of cardiac action potentials. It is also crucial for activating the contraction process. As discussed in Chapter 3, this in-rush of Ca^{2+} ions activates the release of more Ca^{2+} from the sarcoplasmic reticulum which increases the intracellular concentration and activates the actin myosin contractile motors. Therefore, Ica is the key to coupling the action potential to the onset of contraction, a process often referred to as cardiac excitation contraction coupling (ECC). However, L-Cav channels are voltage sensitive and it is for this reason that the early transient outward current is important; in cells where Ito is large, early repolarisation is more marked and this serves to curtail L-Cav opening [21]. The voltage sensitivity of L-Cav channels also means that they are deactivated further as the cell repolarises.

Late repolarisation and outward potassium currents

Ventricular action potentials last for about 0.4 seconds at a heart rate of 70 beats per minute, Figure 4.4(e). Restoration to the resting potential is due to a decline in Ica and two additional potassium currents, IKr and IKs. These are also voltage gated channels with 4-domain 6-transmembrane segments, but in this case selective for potassium ions. They are called "delayed rectifier channels" and were first discovered by Nobel and Tsien in 1963 [22]. Rectification means they pass current more readily inwards than outwards. This is important in that the transmembrane K^+ concentration gradient favours outward movement of K^+, so inward rectification blunts that driving force. At potentials near resting values, this plays an important role in stabilising resting membrane potential. However, at the positive membrane potentials which occur during action potentials IK channels open, permitting an efflux of K^+ ions out of the cell and providing the main mechanism for repolarisation. Like other ion channels, IK channels are voltage gated, but pore opening occurs after a significant delay. In the experimental conditions of their original work, Noble and Tsien reported time constants (time to reach ~63% activation) of 0.5 and 4 seconds for IKr and IKs, respectively. These intervals mean that IKr (the "r" stands for rapid) is most important for repolarisation under normal conditions. Like other voltage gated ion channels, IKs (here the "s" stands for slow) is activated by membrane voltage changes effecting its S4 transmembrane loop, but in this case activation involves two transition states of the S4 loop, which accounts for the delay in pore opening [23].

Recent research has shown that although IKs plays a relatively minor role in repolarisation at resting heart rates, it becomes more prominent at fast heart rates. This is because IKs inactivation is also slow. As a result, IKs remains in a partly activated state at fast heart rates from which it activates more quickly. Consequently, IKs pores open earlier following depolarisation at faster heart rates and therefore contribute more outward K^+ current, which accelerates repolarisation as the heart rate increases. This is an important mechanism for reducing action potential duration as the heart rate increases. As discussed in Chapter 3, heart muscle contracts and relaxes more rapidly at higher heart rates in order to maintain adequate time for ventricular filling during diastole, and reduction in action potential duration at faster heart rates contributes to this. As we also saw in Chapter 3, β-adrenergic stimulation increases intracellular levels of C-AMP and in turn activates phosphokinase A (PKA). In response to β-adrenergic stimulation, PKA causes phosphorylation of IKs which shifts its voltage sensitivity to more negative membrane potentials. As a result, it is activated earlier following depolarisation with a greater effect on repolarisation. And so, β-adrenergic stimulation, which has long been known to reduce action potential duration, does so in part by increased activation of IKs channels. Thus, IKs provides an important second mechanism for repolarisation. This can have considerable clinical importance in some patients where IKs is impaired by a genetic mutation or the effect of drugs. This may not be detectable under normal resting conditions when IKs contributes little to repolarisation. However, if repolarisation by Ikr also becomes impaired for any reason, then these patients can experience action potential prolongation and be at risk of rhythm disturbances.

The importance of normal repolarisation of action potentials is due to the fact that cardiac cells are unresponsive to further stimuli once an action potential has commenced and for its duration. This property is known as refractoriness and is a feature of all cardiac cells. It has an important role in preserving a normal rhythm and is discussed in more detail in the following.

Atrial action potentials

Figure 4.4(b) illustrates typical features of action potentials and the ionic currents which generate them in atrial cells. These have many similarities with ventricular action potentials, (Figures 4.4(e) and 4.4(b)) but also some important differences. Depolarisation is caused by a rapid inward sodium current mediated by voltage gated Nav in channels as with ventricular cells; however, other aspects of the action potential differ. The early repolarisation phase is more marked (Figures 4.4(e) (1) and 4.4(b) (1)), the plateau phase occurs at a more negative potential, and the action potential duration is shorter. These reflect differences in the ionic currents which generate action potentials in atrial cells and in turn the number of ion channels expressed in them. (i) The early transient outward potassium current, Ito, is larger in atrial than ventricular cells [24], causing a more marked early repolarisation. (ii) Atrial cells express an additional voltage gated potassium channel called the ultra-rapid delayed rectifier potassium channel IKur, which is activated early and sustained during repolarisation. (iii) In addition, IKr is also activated earlier during repolarisation. The increased and earlier K^+ outward current due to IKur and IKr results in a lower plateau voltage and a shorter action potential duration.

Purkinje fibre action potentials

Action potentials in His bundle and Purkinje fibre cells also have unique features, Figure 4.4(c). They undergo more rapid early depolarisation than other cells, the early repolarisation phase, Figure 4.4(c) (1) is more marked and their plateau phase Figure 4.4(c) (2) occurs at a lower voltage. They are also longer in duration than in ventricular or atrial cells. As would be expected, these are due to differences in ion channel function. Depolarisation is caused by the fast Ina current as in ventricular cells. The early transient outward K^+ current, Ito, is much larger in Purkinje than other cardiac cells accounting for the marked early repolarisation. Inward and outward currents during the plateau phase also differ in Purkinje cells accounting for its longer duration and lower voltage. (i) Purkinje cells have two sodium channels, labelled Inaf and Inas in Figure 4.4(c). Together they produce more rapid depolarisation. However, Inas inactivates slowly, resulting in some Na^+ current persisting during the plateau phase. (ii) In addition, outward K^+ currents are much weaker in Purkinje than other cells. This together with a more prolonged inward Na^+ current accounts for the longer duration of Purkinje action potentials. Thus, Purkinje cells depolarise more rapidly, have a lower plateau voltage and repolarisation occurs later than in other cells in the heart. These differences relate to their unique function. Purkinje fibres are specialised conducting cells; they do not contract or do so very little. Instead, they conduct the cardiac impulse faster than other cells, which is important in achieving synchronised ventricular activation and contraction. Their high speed of conduction is partly due to an abundance of the high conductance connexin 40 containing gap junctions [19]. This results in larger ion fluxes to adjacent cells, which depolarises them more rapidly. In addition, the rate of depolarisation is faster in Purkinje cells (around 400–500 V/s) than in other cells and this also contributes to faster depolarisation of adjacent cells to which they are connected [25]. As a result, Purkinje cells conduct the cardiac impulse at around 2–4 m/s compared with 0.5 m/s in ventricular or atrial tissue.

Atrioventricular node cells

The atrioventricular (AV) node is a small structure less than 2 mm in width and 4–5 mm long [26]. It performs a specialised function in the heart by conducting the cardiac impulse slowly from the atria to the ventricles, which as discussed above serves to delay ventricular contraction until atrial contraction has primed them. This is partly achieved by the fact that they have fewer gap junctions and those present contain connexin 30.2, which has a lower conductance than connexin 43 and 40 [19]. Action potentials in AV node cells also have a slow rate of depolarisation (Figure 4.4(d) (phase 0)), around 50–75 V/s compared with 400–500 V/s in Purkinje cells, and this also contributes to their slow conduction. Furthermore, AV node action potentials do not have the early repolarisation and plateau phase seen in atrial, ventricular and Purkinje cells. These differences reflect differences in the voltage gated ion channels expressed in AV node cells and in turn the ionic currents which operate in them, Figure 4.4(d). Thus, fast Nav channels are not expressed in AV node cells, so the rapid sodium inward current seen in atrial and ventricular cells is absent. Instead, depolarisation is caused by a slow inward current. This is largely carried by Ca^{2+} ions, but Na^{+} ions also contribute to it [27]. The early transient outward K^{+} current is absent in AV node cells and accounts for the lack of an early repolarisation phase. Repolarisation is achieved by two IK channels, IKr being dominant. Thus, the AV node consists of small non-contracting cells that are well suited to its specialised function. Their slow rate of depolarisation helps to ensure they conduct slowly, and this helps to coordinate atrioventricular contraction. In addition, AV node cells, as other cardiac cells, are refractory to further stimulation for the duration of their action potentials and this effectively prevents the conduction of very rapid arrhythmias (such as atrial fibrillation) that may occur in the atria. AV node cells do not contract and so have no need for the ionic and action potential features associated with that. An important clinical aspect of AV node cells is their ability to function as pacemaker cells in the event of sinus node failure. We will return to this aspect in discussing the sinus node.

Sinus node

The sinus node functions as the normal heart pacemaker, Figure 4.4(a). As discussed above, cardiac cells normally have a stable membrane potential during diastole, only firing when depolarised to the threshold voltage by an adjacent cell. Sinus node cells differ in this respect in that they gradually depolarise spontaneously during diastole, Figure 4.4(a) (phase 4). Then, on reaching their voltage threshold, an action potential follows. In this way, they develop action potentials in a regular manner, which then conduct to the atria and onwards through the heart. The rate of diastolic depolarisation determines how quickly the threshold voltage is reached and thus the heart rate. This is influenced by the autonomic nervous system which controls the heart rate. Apart from its diastolic depolarisation, sinus node action potentials are similar to those in the AV node, with a slow rate of depolarisation, Figure 4.4(d) (phase 0) and no early repolarisation or plateau phase, Figure 4.4(a). Thus, depolarisation is mediated by a slow inward current carried by Ca^+ ions. The early transient outward current is absent, and repolarisation is mediated by the rapid (IKr) and slow (IKs) outward potassium currents, the former being dominant under resting conditions as in other cells.

Maintaining a stable membrane potential during diastole in most cardiac cells (sinus node cells being the exception) is crucial to avoid competing pacemakers and rhythm disturbances. In non-sinus node cells, this is achieved by an inward rectifying K^+ current (inward rectification means the pore is configured to allow ions to pass inward more easily than outward). In simple terms, if the membrane potential becomes more negative than the correct resting value (called hyperpolarisation), then these channels open, allowing K^+ ions to enter the cell. This has a depolarising effect, bringing the membrane potential back to its resting value. However, the property of inward rectification means that these channels only open in response to hyperpolarisation, so if the membrane potential depolarises (becomes more positive than the correct resting potential), they remain closed. This ensures that the membrane potential is maintained close to correct resting value unless actively depolarised by an action potential.

Diastolic depolarisation

That process is altered in sinus node cells in that their membrane potential undergoes a gradual depolarisation. Early studies suggested that this resulted from a decay of the outward going K^+ current IKr, which repolarises cells at the end of the action potential. The idea here was that complete repolarisation failed to occur and gradual depolarisation resulted [28].

The role of IKf in diastolic depolarisation

In 1976, Noma and Irisawa discovered a new inward going current which was activated at diastolic membrane voltages in sinus node cells [29] and which is carried by Na^+ and Ca^{2+} ions. Later given the name IKf, it is mediated by a voltage gated ion channel known as HCN4 (hyperpolarisation-activated, cyclic nucleotide-gated) which is activated in the diastolic voltage range of pacemaker cells, between –45 mV and –15 mV. It was also found to be increased by adrenaline and therefore would fit with the role of sympathetic nerve stimulation in increasing the heart rate. The mechanism for its response to adrenaline is due to the fact that IKf channels are sensitive to levels of C-AMP. C-AMP binds to the channel and modifies its voltage sensitivity to more positive values, resulting in increased current flow and a steeper rate of cell depolarisation. As discussed in Chapter 3, sympathetic stimulation increases C-AMP synthesis by activating adenyl cyclase. Parasympathetic stimulation is well known to slow the heart rate. This has also been attributed to an effect on IKf, in that parasympathetic stimulation acts on tissues by releasing a neurotransmitter called acetylcholine, which inhibits C-AMP production. This has a negative effect on IKr channels by shifting their activation range to more negative potentials [30] and results in slowing the rate of diastolic depolarisation. Thus, diastolic activation of IKf has emerged as an important mechanism of sinus node pacemaker function.

The role of periodic SR Ca^{2+} in sinus node cells

As discussed in Chapter 3, the release of Ca^{2+} ions from the sarcoplasmic reticulum (SR) plays an important role in activating contraction in ventricular cells. It was also discussed how this transient rise in Ca^+ is terminated by reuptake into the SR and by extrusion of Ca^+ by the Na^+/Ca^{2+} exchange pump in the cell membrane. Pacemaker cells do not contract; however they do contain SR which contains an abundant store of Ca^{2+} ions. In recent years it has been discovered that Ca^+ ions are periodically released from the SR in pacemaker cells during diastole and this also plays an important role in sinus node cell automaticity [31]. The release of Ca^{2+} from the SR increases Ca^{2+} concentration in the cytosol and this activates the Na^+/Ca^{2+} exchange pump on the sarcolemma. This pump extrudes one Ca^+ from the cell in exchange for admitting three Na^+ ions, resulting in a net positive charge entering the cell and thereby contributes to depolarisation. The total amount Ca^{2+} released from the SR determines the magnitude of this effect and in this way contributes to modulating the diastolic depolarisation and heart rate. Ca^{2+} release from the SR is also increased by sympathetic stimulation and therefore provides an additional mechanism for increased diastolic depolarisation and heart rate [31].

Thus, pacemaker cells depolarise rhythmically and spontaneously. At least two mechanisms contribute to this. (i) An inward current, IKf, carried by Na^+ and Ca^{2+} ions depolarises sinus node cells during diastole until the firing threshold is reached and an action potential follows. (ii) Local release of Ca^{2+} from the SR also contributes to diastolic depolarisation through its effect on the Na^+/Ca^{2+} pump. Both mechanisms are responsive to sympathetic stimulation and are therefore involved in regulating the heart rate. Why would there be more than one mechanism and is one dominant? It may be that the flexibility and backup of two mechanisms provided an evolutionary advantage for such a vital function.

In the event of the sinus node failing to initiate the heart rhythm, another pacemaker usually emerges to undertake that function, most often in the AV node. Some cells within the AV node express the HCN4 ion channel responsible for the IKf current that underpins diastolic depolarisation in sinus node pacemaker cells and undergo diastolic depolarisation [32]. This allows them to function as pacemaker cells; however, their depolarisation rate is slower than sinus node cells and their spontaneous cycle length is longer. Consequently, heart rates initiated by the AV node are slower than the sinus node.

Maintaining a Normal Heart Rhythm

We have seen how the heart's pacemaker initiates each heartbeat and how its conduction through the heart activates and coordinates contraction of the atria and ventricles. Changes in heart rate provide an essential means to modulate the cardiac output under the control of the autonomic nervous system. The heart's rhythm is initiated by the pacemaker in the sinus node; however, it also depends on all other parts of the electrical system in the heart functioning normally. Disturbances in the heart rhythm (arrhythmias, also called dysrhythmias) are not uncommon and vary in severity from trivial to life threatening. All result from disturbances in some aspect of cardiac electrophysiology. What then are the electrical properties of the heart which maintain the normal heart rhythm and prevent such disturbances?

A dominant pacemaker

Ultimately, every heartbeat originates from a single cell in the sinus node acting as the pacemaker. This raises questions such as "are there others?" and "what is their function?" Clearly, having just one cell capable of pacing the heart would seem vulnerable. In fact, there are many. The rhythm of the heart is normally controlled by a cell located in the central part of the sinus node at rest. In reality, many other cells also undergo diastolic depolarisation leading to spontaneous action potentials; however they will fail to "capture" the heart rhythm unless they reach their threshold potential first. This is because whichever cell depolarises first will activate all others before they reach their firing threshold. In other words, the first cell to depolarise spontaneously and initiate conduction will always dominate the heart rhythm because it depolarises other potential pacemakers before they reach their firing threshold and this has the effect of resetting their pacing cycle.

Managing multiple pacemaker cells

The location of the leading pacemaker in the sinus node can change. For example, parasympathetic stimulation, which slows the heart rate, also results in cells in a lower part of the sinus node taking over the leading pacemaker role [33]. The availability of multiple potential pacemaker cells provides an important reserve if the impulse generated by a faster pacemaker cell fails to conduct outwards. This can occur due to scarring of the sinus node causing a condition known as sinoatrial exit block [34]. In these patients, the sinus node appears not to be working on a standard electrocardiogram (ECG), however, recordings made close to the node indicate that it does in fact generate impulses but these fail to conduct out to the atria. In such cases the heart rhythm depends on some other cell in the heart functioning as a pacemaker, most often located in the atrioventricular node. These have a slower inherent rate than sinus node cells resulting in a slower heart rate. This also results in reverse conduction through the atria, which alters the configuration of the atrial recording on the ECG (known as the P wave). Conduction to the ventricles remains normal, as is ventricular contraction. However, the normal coordination of atrial and ventricular contraction will be disturbed to some degree. This is because the atria and ventricles contract simultaneously rather than sequentially resulting in the loss of the normal atrial filling of the ventricles at the end of diastole. Thus, there are many cells in the heart with the capacity for pacemaker function. Normally these have slower firing rates than the sinus node and therefore never capture the heart rhythm.

Abnormal pacemaker function

Cardiac cells, with the exception of pacemaker cells, remain quiescent as long as their membrane potential is maintained below their threshold value and this is actively regulated to prevent abnormal firing. If a cell becomes stressed or injured, its ability to maintain a normal membrane potential may fail, resulting in a gradual depolarisation. If this reaches the firing threshold, an action potential will result and may be conducted through the entire heart as an "extra beat". These are also known as ectopic beats or extrasystoles and are distinguishable on the ECG due to the abnormal direction of impulse propagation. Ventricular ectopics may arise at some distance from the His–Purkinje system and fail to access its fast conducting pathway, relying instead on ventricular cell-to-cell propagation. In that case, conduction is slower and ventricular activation takes longer (more than 120 milliseconds).

In addition, because they occur early in the cardiac cycle, ventricular filling is usually incomplete and consequently the volume of blood they eject is less than normal. For this reason, they are often not palpable at the pulse. Most ectopics do conduct into the sinus node and reset its pacing cycle so that the post ectopic interval is similar to the cycle length during sinus rhythm. This results in post ectopic beats having a larger than usual stroke volume. This occurs because the ejection of blood by ectopic beats is deficient due to incomplete ventricular filling. The following beat then does catch-up; the ventricles receive and eject a larger than usual blood volume. Post-ectopic beats are also stronger than normal because heart muscle is sensitive to stretch and responds with increasing force of contraction. It is for this reason that our awareness of "extra beats" is most often due to the increased force of post-ectopic contraction.

The clinical significance of ectopics depends on the condition which causes them. In patients who experience a heart attack, blood flow to a region of heart muscle is blocked by clot formation in a coronary artery. This deprives that muscle of oxygen and other nutrients, which impairs its electrophysiological function and may cause many such ectopic beats. On the other hand, many people experience occasional ectopic beats, reflecting an occasional disturbance in one of the heart's 2–3 billion cells. Provided the structure and function of their hearts is otherwise normal, these are virtually always harmless. The heart continues unperturbed by such beats because they are single events with no significant lasting effects.

Abnormal pacemaker function may become permanent if conduction from the sinus node or the atrioventricular node is blocked. In the case of AV node conduction block due to scarring, the heartbeat then depends on a pacemaker emerging from a point below the block. This thankfully does occur, but usually with quite a slow rate of around 40 beats per minute. This is too slow to allow normal function for most people. Typically, such patients receive an electronic pacemaker. This monitors heart activity and if the rate falls below a pre-specified rate, introduces an electrical signal to the right ventricle, which is conducted through heart muscle to both ventricles, to ensure satisfactory heart rate. Electronic pacemakers can also monitor, correct and coordinate activity between atrial and ventricular activation.

Two cardiac electrophysiological functions of the heart help to maintain its normal rhythm. These are the properties of (i) refractoriness and (ii) ordered one-way conduction.

Refractoriness

Membrane depolarisation during the onset of an action potential is due to opening of Nav channels allowing a rapid influx of Na^+ ions. This is a brief event lasting just a few milliseconds because Nav channels are rapidly inactivated and remain inactive until repolarisation is almost complete. This has the effect that the cell is incapable of generating another action potential for this period, which is known as the effective refractory period. This applies to all cardiac cells and means that the heart is in effect protected from further stimulation. Thus, neither an abnormal pacemaker in the heart nor an electrical stimulus can induce a response during this time. The end of repolarisation is followed by a period called the relative refractory period whereby action potentials can be generated by an electrical stimulus but require a larger stimulus to do so. This is because repolarisation often results in transient hyperpolarisation, meaning that the membrane is initially repolarised to a more negative potential before settling to the normal resting value. During this time, although cells are excitable, they require a larger stimulus to reach the firing threshold.

The property of refractoriness plays a vital role in preserving the normal heart rhythm. It has the effect that as an impulse is being conducted through the heart, from cell to cell, the direction of propagation is always one-way. It can only go forward because all cells to the rear are now unexcitable. Even in the case of ectopic beats, although the direction of propagation in the heart is changed, it can only proceed into excitable tissue.

Orderly conduction

The pattern of impulse conduction through the heart is carefully arranged to ensure coordinated contraction of the atria and ventricles. In addition, the speed of conduction through the heart is well suited to maintain a normal rhythm. Although the conduction velocity in each region of the heart differs greatly, it is fairly uniform within each. Thus, conduction in the His bundles is about five times faster than in the ventricles and atria and about 40 times that in the AV node. However, the velocity is uniform in the atria and ventricles. This means that the wave of propagation in each tissue proceeds along an unbroken front. This also matters for maintaining a normal rhythm in that it limits any possibility of propagation in any direction other than forward.

The problem of re-entry

Thus, refractoriness and orderly conduction combine to ensure forward momentum of impulse propagation through the heart, and most serious arrhythmias result from disturbances in either or both of them. Scarring or myocardial infarction (damage due to a blocked coronary artery) can result in impaired conduction through a section of heart tissue. If severe, it is possible that propagation of the impulse through it may be so delayed that it finally exits when propagation through the surrounding normal cells has moved on and they are once again excitable, Figure 4.6. In this case, there is the possibility of reactivating them, a process known as re-entry. In some cases this can result in a rapid repetitive revolution of electrical impulses through or around the diseased area, which can propagate

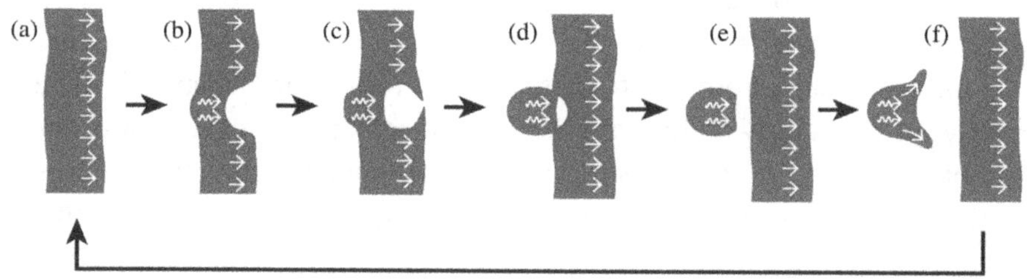

Figure 4.6: Schematic diagram of re-entry. (a) Impulse propagation proceeds in the direction of the arrows with a linear front. The grey area illustrates tissue which has been activated and is now refractory. (b) The impulse front encounters an area of tissue in which conduction is slow, which alters the leading edge of the propagation front. (c) Propagation proceeds in normal cells around the area of impaired conduction and has almost encircled it. (d) and (e) The impulse has gradually overtaken the area of slow conduction leaving the latter following behind. (f) Eventually the trailing edge of the refractory period of the normal impulse overtakes the area of slow conduction. This creates the possibility of propagation from the area of slow conduction into the now excitable normal tissue, a process known as re-entry. In some cases, this can lead to a repetitive re-entry cycle resulting in waves of excitation propagating through the heart causing an abnormal rhythm. In this way, re-entry can be the basis for many types of arrhythmia. What is remarkable is how well the heart is normally protected from such events by its systems of orderly uniform conduction and refractoriness.

through the heart causing an abnormal tachycardia. Impaired conduction is the basis for this occurring; however, refractoriness also has a bearing on the risk of it occurring in that a reduction in the refractory period of the surrounding normal tissue increases the likelihood of re-entry being possible. Thus, the combination of orderly conduction and refractoriness protect the normal heart rhythm and most serious arrhythmias are due to a disturbance in either one or both of them.

Conclusions

To describe the heart as an *intelligent* pump is not consistent with a formal definition of that word. The heart does not understand or reason except in the metaphorical sense. However, the heart has an extraordinary capacity for automatic self-regulation and the ability to communicate with and respond to other systems in the body. The molecular processes which underpin many of these functions emerged in early single-celled organisms and have been conserved over billions of years of evolution. Many of them are also ubiquitous; we find the same components or variants of them operating in the brain and other muscles. The primitive sodium pump maintains the transmembrane potential, which is key to all electrical signalling processes in biology, including the heart and appeared in the earliest cells, probably over three billion years ago. Voltage gated ion channels made possible not only the action potentials which regulate contraction of the heart, but also the signals in the brain that allow us to think.

Efforts to understand how the heart works have long intrigued humans. William Harvey's discovery of the circulation is still celebrated and I would suggest that this in part relates to the metaphorical concepts of heart discussed in Chapter 1. During the second half of the 20th century, understanding of how the heart rhythm is generated came to be seen as a major challenge. This arose when it became clear that myocardial infarction due to coronary thrombosis was a major cause of sudden early deaths due to ventricular fibrillation (a life-threatening arrhythmia). The struggle to contain this led to a wide range of research that contributed to a dramatic reduction in cardiac mortality [35], including sudden cardiac deaths [36] as better early treatments and prevention programmes were introduced. It also dramatically increased our understanding of the electrical properties of the heart and abnormalities that cause arrhythmias. This led to the discovery of antiarrhythmic drugs as well as the propensity of some other drugs to cause arrhythmias and how they do so.

The introduction of detailed recording of ECGs from inside the heart allowed clinicians to map the spread of impulse propagation and to better understanding of how re-entry occurs in humans. This made it possible to locate areas of abnormal conducting pathways and to extinguish them by a process known as ablation. Ablation involves the localised application of energy in the form of radio waves to generate heat or by freezing to cause a localised injury to an area of abnormal conduction to prevent it acting as a re-entry pathway. These are remarkable achievements that benefit many patients, but perhaps also serve to remind us how crude they appear in comparison to the efficiency and economy of evolutionary innovation.

References

[1] Krogh A. The active and passive exchanges of inorganic ions through the surfaces of living cells and through living membranes generally. *Proceedings of the Royal Society of Medicine*. 1946;133:140–200.

[2] Skou JC. The influence of some cations on an adenosine triphosphatase from peripheral nerves. *Biochimica et Biophysica Acta*. 1957;23(2): 394–401.

[3] Saez AG, Lozano E, Zaldivar-Riveron A. Evolutionary history of Na, K-ATPases and their osmoregulatory role. *Genetica*. 2009;136(3):479–490.

[4] Carafoli E, Santella L, Branca D, Brini M. Generation, control, and processing of cellular calcium signals. *Critical Reviews in Biochemistry and Molecular Biology*. 2001;36(2):107–260.

[5] Chen W. Electroconformational denaturation of membrane proteins. *Annals of the New York Academy of Sciences*. 2005;1066:92–105.

[6] Steinmetz PR, Kraus JE, Larroux C, Hammel JU, Amon-Hassenzahl A, Houliston E, *et al.* Independent evolution of striated muscles in cnidarians and bilaterians. *Nature*. 2012;487(7406):231–234.

[7] Abreu-Blanco MT, Verboon JM, Parkhurst SM. Single cell wound repair: Dealing with life's little traumas. *Bioarchitecture*. 2011;1(3):114–121.

[8] Anderson PA, Greenberg RM. Phylogeny of ion channels: Clues to structure and function. *Comparative Biochemistry and Physiology Part B, Biochemistry & Molecular Biology*. 2001;129(1):17–28.

[9] Strong M, Chandy KG, Gutman GA. Molecular evolution of voltage-sensitive ion channel genes: On the origins of electrical excitability. *Molecular Biology and Evolution*. 1993;10(1):221–242.

[10] Galione A, Evans AM, Ma J, Parrington J, Arredouani A, Cheng X, *et al.* The acid test: The discovery of two-pore channels (TPCs) as NAADP-gated endolysosomal Ca(2+) release channels. *Pflugers Archiv: European Journal of Physiology*. 2009;458(5):869–876.

[11] Zakon HH. Adaptive evolution of voltage-gated sodium channels: The first 800 million years. *Proceedings of the National Academy of Sciences of the United States of America*. 2012;109(Suppl 1):10619–10625.

[12] Chanda B, Bezanilla F. Tracking voltage-dependent conformational changes in skeletal muscle sodium channel during activation. *The Journal of General Physiology*. 2002;120(5):62–645.

[13] Brunet T, Arendt D. From damage response to action potentials: Early evolution of neural and contractile modules in stem eukaryotes. *Philosophical Transactions of the Royal Society of London Series B, Biological Sciences*. 2016;371(1685):20150043.

[14] Harz H, Nonnengasser C, Hegemann P. The photoreceptor current of the green alga *Chlamydomonas*. *Philosophical Transactions of the Royal Society of London: Series B, Biological Sciences*. 1992;338(1283):39–52.

[15] Postma N. On the effect of the second and third ligatures of stannius on the frog heart. *Experientia*. 1951;7(2):67.

[16] Keith A, Flack M. The form and nature of the muscular connections between the primary divisions of the vertebrate heart. *Journal of Anatomy and Physiology*. 1907;41(Pt 3):172–189.

[17] von Knorre GH. The 125th anniversary of the His bundle discovery. *Herzschrittmachertherapie & Elektrophysiologie*. 2018;29(1):116–121.

[18] Mazurak M, Kusa J. Jan Evangelista Purkinje: A passion for discovery. *Texas Heart Institute Journal*. 2018;45(1):23–26.

[19] Giovannone S, Remo BF, Fishman GI. Channeling diversity: Gap junction expression in the heart. *Heart Rhythm*. 2012;9(7):1159–1162.

[20] Schwiening CJ. A brief historical perspective: Hodgkin and Huxley. *The Journal of Physiology*. 2012;590(11):2571–2575.

[21] Sah R, Ramirez RJ, Oudit GY, Gidrewicz D, Trivieri MG, Zobel C, *et al.* Regulation of cardiac excitation-contraction coupling by action potential repolarization: Role of the transient outward potassium current (I(to)). *The Journal of Physiology*. 2003;546(Pt 1):5–18.

[22] Noble D, Tsien RW. Outward membrane currents activated in the plateau range of potentials in cardiac Purkinje fibres. *The Journal of Physiology*. 1969;200(1):205–231.

[23] Rudy Y. Molecular basis of cardiac action potential repolarization. *Annals of the New York Academy of Sciences*. 2008;1123:113–118.

[24] Cheng H, Cannell MB, Hancox JC. Differential responses of rabbit ventricular and atrial transient outward current (Ito) to the Ito modulator NS5806. *Physiological Reports*. 2017;5(5):e13172.

[25] Hodgkin AL, Huxley AF. A quantitative description of membrane current and its application to conduction and excitation in nerve. *The Journal of Physiology*. 1952;117(4):500–544.

[26] Anderson RH, Yanni J, Boyett MR, Chandler NJ, Dobrzynski H. The anatomy of the cardiac conduction system. *Clinical Anatomy (New York, NY)*. 2009;22(1):99–113.

[27] Kokubun S, Nishimura M, Noma A, Irisawa H. Membrane currents in the rabbit atrioventricular node cell. *Pflugers Archiv: European Journal of Physiology*. 1982;393(1):15–22.

[28] Noble D, Tsien RW. The kinetics and rectifier properties of the slow potassium current in cardiac Purkinje fibres. *The Journal of Physiology*. 1968;195(1):185–214.

[29] Noma A, Irisawa H. Membrane currents in the rabbit sinoatrial node cell as studied by the double microelectrode method. *Pflugers Archiv: European Journal of Physiology*. 1976;364(1):45–52.

[30] DiFrancesco D. The role of the funny current in pacemaker activity. *Circulation Research*. 2010;106(3):434–446.

[31] Lakatta EG, Vinogradova TM, Maltsev VA. The missing link in the mystery of normal automaticity of cardiac pacemaker cells. *Annals of the New York Academy of Sciences*. 2008;1123:41–57.

[32] Nikmaram MR, Liu J, Abdelrahman M, Dobrzynski H, Boyett MR, Lei M. Characterization of the effects of ryanodine, TTX, E-4031 and 4-AP on the sinoatrial and atrioventricular nodes. *Progress in Biophysics and Molecular Biology*. 2008;96(1–3):452–464.

[33] Ashton JL, Trew ML, LeGrice IJ, Paterson DJ, Paton JF, Gillis AM, *et al.* Shift of leading pacemaker site during reflex vagal stimulation and altered electrical source-to-sink balance. *The Journal of Physiology*. 2019; 597(13):3297–3313.

[34] Yeh SJ, Yamamoto T, Lin FC, Wang CC, Wu D. Repetitive sinoatrial exit block as the major mechanism of drug-provoked long sinus or atrial pause. *Journal of the American College of Cardiology*. 1991;18(2):587–595.

[35] Moran AE, Forouzanfar MH, Roth GA, Mensah GA, Ezzati M, Murray CJ, *et al.* Temporal trends in ischemic heart disease mortality in 21 world regions, 1980 to 2010: The Global Burden of Disease 2010 study. *Circulation*. 2014;129(14):1483–1492.

[36] Fox CS, Evans JC, Larson MG, Kannel WB, Levy D. Temporal trends in coronary heart disease mortality and sudden cardiac death from 1950 to 1999: The Framingham Heart Study. *Circulation*. 2004;110(5):522–527.

5

A Pump for Life: In Good Times and in Bad

Our last car was 17 years old when we replaced it. I was fond of its until then steady reliability. Its commodious boot had served well as family haulage space, and it was comfortable on long drives. By then I had come to the view that cars could be boring but should be reliable. I had once owned a sports car but then discovered that riding a bicycle provided more excitement, and the only occasion when I was ever stopped by a police patrol car was while riding a bike. After 17 years, the radiator hose joint snapped, signalled by a large eruption of steam. A temporary fix involving shoe laces allowed removal of the stricken vehicle to a nearby garage and although replacing the radiator proved straightforward, the episode undermined our confidence in its long-term durability. It might have gone on for many more miles, but when would another part fail? I felt disloyal to let it go after such long service and although the car salesman seemed to genuinely admire its old-fashioned durability, that was not reflected in the part exchange offer on its replacement.

The average "life" of a car in the UK is 11 years, an increase from 8.5 years in 2008. A goal of reaching 200,000 miles is seen as achievable, which would correspond to the average life of a car, around 15 years. This means that all our vehicles spend most of their lives "resting". In fact if an engine propelled a car continuously at 30 miles per hour, it would cover 200,000 miles in less than 40 weeks; or Formula 1 racing engines typically last for one or two races, a matter of hours.

Hearts work continuously throughout our lives; they cannot rest or take time out for servicing. The ability of a human heart to pump continuously for over a hundred years is surely remarkable. Although its durability is, in reality, no greater than that of other organs, its function is more perceptible to us. So, we are more aware of its workings than the electrical intricacies of our brains or the metabolism of our liver. How does it continue to function for so long? Part of the answer relates to the fact that it has evolved to work within the limitations of its abilities and is controlled by mechanisms that help it to stay within those limits. In contrast, Formula 1 racing engines run at speeds that far exceed the physical tolerances of their component parts. The rotational speeds they achieve of up to 20,000 revolutions per minute rapidly wear contact surfaces and degrade performance. At the other extreme, an antique engine can be preserved over decades by careful maintenance and by very little use. How does the heart differ from these?

Working Within Its Limitations

There are natural limitations to the performance of the heart and exceeding them can be harmful. In fact, the heart is constantly monitored to ensure that it always works within safe limits. This is achieved by a system of sensors which measure its performance and by the fact that it always works in coordination of the circulation as a whole so that its output optimally matches the needs of the body. This integration of the working heart and circulation as a whole is crucial for its durability and long working life. This chapter then discusses the heart as a soloist in a large orchestra of interconnected players.

The heart can of course vary its output to match demand and therefore its normal function can be considered as a range, described in terms of performance at rest and at maximum ability. Unlike a man-made engine which can be evaluated in isolation against controlled loads, cardiac function in man can only be assessed physiologically. For this reason, maximum performance of the heart is usually assessed in the context of peak exercise capacity, based on the assumption that it is responding to maximal stimulation at that point.

The simplest measures of performance are its rate and the cardiac output (the volume of blood ejected per minute). Allowing that there is of course considerable variation between individuals, average heart rate in normal adults increases from about 65 beats per minute (BPM) to 160 BPM at peak exercise and cardiac index (cardiac output expressed per metre squared of body surface area) from 2.7 L/min/m^2 to 8 L/min/m^2 [1]. Achieving these depends on normal function of circulation as a whole. For example, blood pressure at rest and during exercise represents a load on the heart which impacts its performance and that in turn is dependent on the function of the blood vessels into which it pumps. Blood flow during exercise increases to the working muscles and this occurs because their arterioles (small arteries) dilate in response to nervous and local stimuli (discussed in detail in Chapter 8). This reduces resistance to flow and contributes to rebalancing blood pressure and the load against which the heart works. In normal adults, mean blood pressure

$$\text{Mean blood pressure} = \frac{\text{Diastolic pressure} \times 2 + \text{Systolic pressure}}{3}$$

increases from 85 mmHg to 109 mmHg at peak exercise. The working load on the heart is described in terms of the resistance imposed by the circulation. Known as the systemic vascular resistance (SVR), this is measured as mean arterial pressure divided by the cardiac output.

Normal performance of the heart also depends on the regulatory mechanisms which control it also behaving normally. In heart transplant patients, it is not technically possible to connect the donor heart to the nervous system. So the autonomic nervous system which normally regulates heart rate during exercise is absent. As a result, these hearts cannot achieve a normal increase in heart rate during exercise. Typically, they have a higher heart rate at rest, due to lack of the inhibiting effect of parasympathetic control and a smaller increase with exercise due to lack of sympathetic nerve stimulation [2]. The autonomic nervous system then can be seen as mechanism to regulate cardiac function to meet the demands of the body but it also keeps it within safe working limits. Even at peak exercise when the heart is maximally stimulated, it avoids excessive stimulation of the heart.

Overstimulation of the heart can occur in some abnormal circumstances and is harmful. Exercise is an intermittent activity interspersed by periods of rest. Our natural sense of tiredness and exhaustion help to ensure this. Adrenaline is a hormone produced by the adrenal gland which stimulates the heart as part of the fight or flight mechanism. Continuous unregulated high levels of circulating adrenaline occur in patients with a tumour called phaeochromocytoma and can damage the heart. Similarly, cocaine blocks the body's normal mechanism for limiting sympathetic activity and thereby greatly increases its effect on the heart. Hearts that are weakened and unable to meet the demands of the body are said to be failing. In patients with heart failure, this often results in increased and prolonged sympathetic stimulation in an attempt to increase cardiac function. This too seems to be harmful over time. Treatment with β-blocking drugs, which inhibit the effect of sympathetic stimulation, improves survival in such patients. Thus the heart's longevity relates not only to its inherent properties, but also to its careful and continuous regulation so that it always functions within safe limits. All of this is automatic and beyond our conscious control. In addition, the heart never functions in isolation, but as part of an integrated system in which the nervous system, blood vessels and many sensors around the body work together to actively regulate the circulation as a whole for optimum performance and efficiency. This is the basis for its remarkable adaptability and in turn to its long life.

Physiological adaptability

Demands on the heart vary not only in terms of their degree but also in the kind of activity which provokes them. Mental activity, digesting food and exercise all require energy which ultimately is delivered by an increase in cardiac output. However, the response of the circulation is specific in each case so that the increase in flow is distributed in a pattern to meet the required need. In other words, the heart does not simply increase its output as an isolated response, but as part of a wider integrated system involving the circulation as a whole, which directs the increase in flow appropriately. This confers a major efficiency advantage and also minimises the work required by the heart.

Circulatory responses to exercises

Physical exercise requires energy and that is provided by increased blood flow to the muscles involved. An increase in heart rate and blood pressure is a constant feature of exercise. Activation of the autonomic nervous system occurs rapidly and may precede the onset of physical movement, for example, prior to a race in competing athletes. The heart rate increases rapidly due to withdrawal of parasympathetic and increased sympathetic stimulation. The increase in heart rate raises cardiac output and this is accompanied by greater force of ventricular contraction due to sympathetic stimulation. The mechanisms underlying these changes are discussed in detail in Chapters 3 and 4. These provide an increased blood flow and therefore oxygen and nutrients to the working muscles; however, the volume of blood that reaches them depends on (i) how the increased cardiac output is distributed around the body and (ii) how the local blood vessels which supply working muscle respond to the increased demand.

Distribution of blood flow during exercise

Increasing blood flow to working muscle is only partly achieved by increasing cardiac output. This is because high levels of exercise require a much greater blood flow than an increase in cardiac output alone can provide. The remainder is made available by actively redistributing the cardiac output within the circulation. The background to this is that while blood volume in an adult is around 5 L, the total capacity of the circulation is many times greater. At any given time therefore, only parts of the circulation are filled with blood. This is achieved by active regulation of how blood flow is distributed. The importance of this is well illustrated by the sudden collapse that can occur in anaphylactic shock. This is the result of an abrupt fall in blood pressure such that effective organ perfusion becomes impossible. In severe cases, impaired perfusion of the brain occurs with loss of consciousness. The cause of this is a rapid immune reaction resulting in widespread dilatation of blood vessels around the body and leaking of fluid through tiny capillary vessels. This results in fluid loss and an expansion of the vascular space to an extent that exceeds the circulating blood volume. Blood pools in a kind of reservoir instead of returning to the heart and cardiac output and blood pressure rapidly decrease. Treatment of such cases requires urgent administration of fluids to "refill" the vascular space and drugs to reverse the vasodilatation and support the heart.

The urgency of treatment in such cases rules out any possibility to assess the extent of vascular expansion. In one case report, however, an estimate was possible by chance [3]. The patient had anaphylaxis while undergoing a cardiopulmonary bypass operation. This meant that an external pump, oxygenator and reservoir were used to replace the heart and lungs. As a result, it was possible to maintain cardiac output, despite a decline in venous return by rapidly infusing fluids to "refill" the circulation. The fluid volume needed to do so amounted to 75% of the preoperative blood volume. This gives some indication of the severity of vascular expansion that occurred. It also emphasises that the potential capacity of the vascular space far exceeds the total blood volume and that restricting it to match the blood volume requires active regulation of blood vessels. By the same token, the potential to increase the vascular space in working muscle is what allows blood flow to increase when required. However, it will also be obvious that when this occurs it requires that blood flow must be reduced to other areas. In reality, activities such as exercise involve a complex redistribution of blood flow to all parts of the circulation.

Regional blood flow during exercise

Measurements of blood flow distribution during exercise in humans have been possible using magnetic resonance imaging (MRI). Presently available technology is not capable of measuring flow to all organs simultaneously but has provided useful information about flow to the lungs and upper and lower parts of the body. During bicycle exercise which resulted in a doubling of cardiac output, the distribution of flow clearly favours the lower body [4]. This is consistent with increased flow to the arm and leg muscles engaged in cycling. Blood flow to organs not directly involved in exercise such as the liver, stomach, intestines and spleen declines. In contrast, flow to the kidneys remains stable (Figure 5.1). Thus, the circulation is constantly regulated to ensure that flow is directed to areas where demand is highest. During exercise, two areas are priorities, the

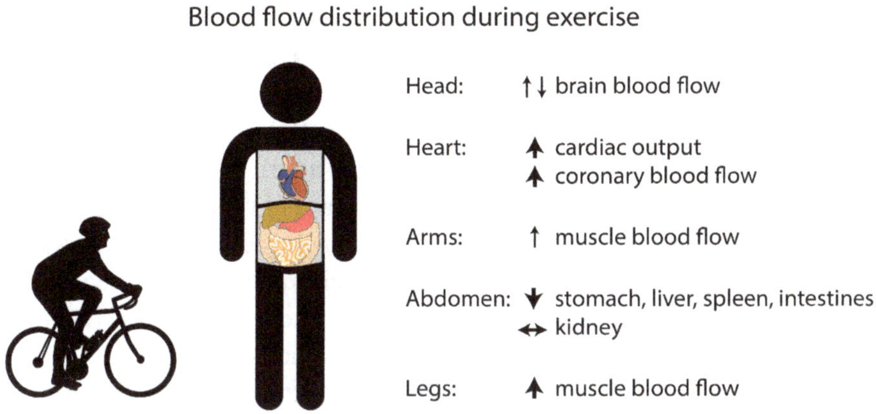

Figure 5.1: Blood flow distribution during cycling exercise. The heart responds to exercise by increasing cardiac output. This is part of a coordinated response of the circulation as a whole, which results in the redistribution of blood flow in favour of the working muscles in the legs and to a lesser extent in the arms. This is achieved by constricting blood vessels in the abdomen so that flow to the liver, stomach, intestines and spleen is reduced. Perfusion of the kidneys is maintained. Blood flow to the large front area of the brain increases at moderate levels of exercise but declines with intense exercise, while perfusion of the back of the brain where movement is coordinated is maintained. Blood flow through the heart muscle also increases to provide the energy to sustain its greater workload.

exercising muscles and the heart. This is achieved in part by mechanisms that regulate flow locally in each vascular bed.

Local Regulation of Muscle Blood Flow

At the local level, blood flow to muscle depends on the effects of contraction and relaxation of the muscles themselves and on local vascular conductance (the flow capacity of the blood vessels).

Direct muscle pump effects

In humans undertaking forearm or knee flexion/extension exercise, muscle blood flow increases almost immediately, reaching a steady level by about 6 s, followed by a later, more gradual, increase from around 15–20 s [5]. A direct pump-like action of muscle contraction on blood vessels contributes to this. Blood pressure normally increases during exercise related to the work being done and this is a major contributor to the increase in muscle blood flow. Venous outflow pressure is normally substantially less than arterial pressure, however pressure in the small veins which drain muscle also increases somewhat during exercise due to the increase in flow. Because flow is related to the arterio-venous pressure difference across any vascular bed, such an increase in venous pressure could act to impede flow. However, there is convincing evidence that muscle contraction has a direct action on veins which mitigates any such effect. This occurs because muscle contraction exerts a squeezing action on blood vessels which can both impede and increase venous flow. So while muscles are relaxed, veins within them fill and are then emptied by a subsequent contraction. Normally, arterial inflow means that upstream resistance is greater than outflow so that the overall effect is to assist forward flow via venous drainage. The result is that rhythmic muscle action as in walking or repetitive hand grip can act like a muscle pump [5], contributing to increased venous flow and thereby to muscle perfusion. The concept of external muscles propelling blood flow is not uncommon in biology. For example, hagfish have two accessory hearts near the head that rely on contraction of an external muscle to propel the flow of haemolymph.

Vascular conductance

Vascular conductance is a measure of the flow capacity of blood vessels and is related to their cross-sectional area. Far from being simple conduits, the structure of arteries contains two components that play a major role in regulating the blood flow they carry. (a) The walls of arteries of the size which perfuse muscles have a substantial layer of smooth muscle (a form of muscle found in blood vessels, intestines, bladder and the womb that is not under voluntary control). By contracting and relaxing, this can greatly modulate the diameter of arteries and therefore the blood flow they carry. (b) The interior of all blood vessels is lined by a single layer of thin flat cells, called endothelial cells. Research during the past four decades has revealed that these cells play an important role in regulating the function of the vessels by interacting with the bloodstream and the haemodynamic forces which propels it.

Haemodynamic forces acting on blood vessels

Blood vessels are affected by the forces generated by the heart in two ways. First, blood pressure exerts a pulsatile radial pressure which stretches and induces tension on the vessel wall. Laplace's law tells us that the tension generated in a vessel wall is directly related to the pressure across its wall and to its diameter. So, larger vessels have greater tension for a given pressure. Second, blood vessels are also exposed to the frictional effect of passing blood flow, known as shear stress. Shear is related to the velocity of blood flow and to the viscosity of blood. Thus, blood vessels are exposed to two forces, a radial force across the vessel wall which causes stretch in all directions and a frictional force which causes stretch in the same direction as blood flow. Importantly, endothelial cells are exposed to shear stress and act as transducers of it, meaning they sense it and respond to it by activating secondary processes in the vessel wall which lead to dilatation.

Flow-mediated dilatation

In 1933, Schretzenmayr first observed that muscle activity causes dilatation of arteries proximal to them [6]. This was a remarkable observation because it suggested that working muscles can signal their increased energy needs to arteries at some distance proximal to them. Dilatation commenced within seconds of the start of activity and plateaued at around 35 s. It also persisted for about 60 s after muscle activity stopped. At that time, the mechanism for this effect was unknown but since it occurred in arteries proximal to the exercising muscle it was thought to be produced by a substance originating upstream, possibly in the heart. The concept that endothelial cells, which line the interior of all blood vessels, could provide a mechanism for regulating vessel diameter through an interaction with the bloodstream was proposed in 1975 [7]. This hypothesis suggested that an increase in the shear stress can be detected by endothelial cells, which respond by producing substances that act on the arterial smooth muscle causing it to relax, resulting in dilatation and reduction in flow velocity and shear stress (Figure 5.2(a)).

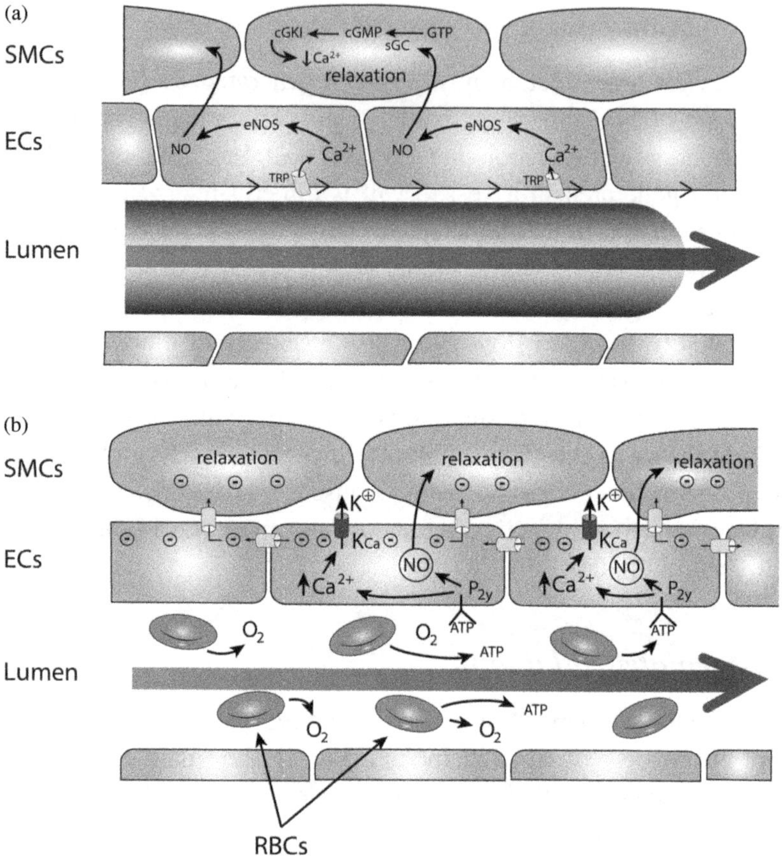

Figure 5.2: Local regulation of blood flow. (a) Blood flow exerts a drag effect, known as shear stress on the luminal surface of endothelial cells (ECs) lining blood vessels. This also slows blood flow close to the luminal surface of the vessel (indicated here by the darker shading of the blood column). Shear stress is detected by mechanosensitive receptors known as transient receptor potential (TRP) ion channels. When activated TRP channels open to allow an influx of Ca^{2+} which activates nitric oxide synthase, an enzyme that produces nitric oxide form L-arginine and a number of co-factors. Nitric oxide is a small molecule which readily diffuses to adjacent smooth muscle cells (SMCs) where it activates soluble guanylate cyclase (sGC) to produce cyclic GMP which in turn causes relaxation and dilatation of the arteriole. Flow-mediated dilatation provides a means to mitigate shear stress on the vessel wall and facilitate increased blood when energy needs increase. (b) The oxygen level in venous blood acts as a feedback control signal to regulate blood flow in arterioles. Oxygen carried bound to haemoglobin in red blood cells (RBCs) is released as blood flows through small vessels. When the oxygen content falls below a threshold level, red cells release ATP. This binds to P_{2y} purinergic receptors on endothelial cells, activating intracellular signalling pathways which cause production of nitric oxide and an increase in Ca^{2+}.

Then in 1980 Robert Furchgott published his paper showing that the ability of acetylcholine to dilate arteries depends on the presence of intact endothelium [8]. He proposed that endothelium mediated this action by producing a substance initially labelled EDRF (endothelium-derived relaxing factor) and later shown to be nitric oxide (NO). It was a discovery which has turned out to be immensely important for regulating flow in blood vessels and for which Furchgott received the Nobel Prize in 1989.

EDRF and nitric oxide

Furchgott tells the story of his discovery of EDRF and NO in his Nobel lecture [9]. He had been studying the behaviour of blood vessels since the mid-1950s and had consistently observed that isolated arteries contracted in response to acetylcholine (ACh) and yet ACh was well known to cause arterial dilatation *in vivo*. Then in 1978, the reverse unexpectedly occurred when ACh dilated an isolated artery. It was a notable observation and on looking into the contradictory findings further, it transpired that the different responses related to the way isolated arterial strips were prepared. Gentle rubbing of the inner lining of the artery prior to exposure to ACh resulted in a contraction response, but if this was carefully avoided, then the arteries relaxed in response to ACh. It was then shown that even gentle rubbing of the lining removed endothelial cells and that this converted the ACh response to contraction.

Since relaxation is mediated by smooth muscle in artery walls, it led to the conclusion that the effect was due to the release of a substance by endothelium which diffused through the arterial wall and acted on smooth muscle. It was initially named endothelium-derived relaxing factor (EDRF) and during the early 1980s, intense efforts focused on discovering what it was. At that time, elevation in smooth muscle cyclic guanosine

Figure 5.2: (*see figure on facing page*) The latter activates calcium-sensitive potassium channels resulting in an efflux of K^+ ions. The resulting loss of positive charge renders the cells negative (hyperpolarised) and this state can be conducted to adjacent endothelial cells through gap junctions as well as to smooth muscle cells through myo-endothelial junctions where it causes relaxation and dilatation. The conduction of the hyperpolarised state in effect spreads the relaxation in both directions along the vessel.

monophosphate (cGMP) was known to be associated with relaxation and subsequent experiments showed that it rapidly increased in smooth muscle exposed to EDRF in association with relaxation. At this time also, a number of vasodilator drugs were known to work by increasing cGMP levels in smooth muscle. For example, glycerol trinitrate (GTN), which had long been used as a treatment for angina (pain associated with reduced myocardial blood flow), was known to relax smooth muscle and to increase cGMP levels. It had also been proposed that a number of vasodilator drugs, including GTN, may work by producing nitric oxide (NO) which was also known to be a potent stimulus of cGMP. In 1986, Furchgott speculated that EDRF might be NO and began a series of experiments to compare the activity of EDRF with NO using various known inhibitors and found them to be identical [9]. Later during the 1980s, the ability of endothelium to cause arterial vasodilatation was shown to be present in several vascular beds, including coronary arteries [10]. Flow-mediated arterial dilatation was also confirmed in man [11].

Nitric oxide is a small molecule of just one oxygen and one nitrogen atom, in effect a gas, which is readily diffusible through tissues and so can quickly reach arterial smooth muscle when released from endothelium. It is highly unstable and therefore has a short duration of action, lasting a matter of seconds. Its importance is because it provides a mechanism whereby arteries have the ability to self-regulate their response to increased blood flow. The shear stress of the blood stream on blood vessels is related to flow velocity. The flow velocity of fluid in a tube is related to the flow rate (volume/time) and to the tube diameter (flow velocity is faster in a smaller tube). Therefore dilatation in the response to increased flow not only reduces the increased shear stress that it causes, but also facilitates organ perfusion.

Transduction of shear stress

The ability of endothelium to respond to shear stress means that it has to have a way of sensing it. During the past two decades this has been the focus of intense research and many processes appear to be involved in it. In summary, shear sensing involves mechano-sensitive calcium ion channels (called piezo ion channels), several membrane-bound receptors and a specialised layer of sugar-containing proteins (glycoproteins) which line the luminal surface of endothelial cells, called glycocalyx. The glycocalyx is rather akin to the slippery substance which lines the skin of many fish. Through these cell surface sensors, the effect of shear stress is transmitted to the cell interior by a series of signalling processes, which collectively activate nitric oxide synthase, the enzyme that generates NO from the amino acid L-arginine, oxygen and a number of additional cofactors.

In addition to NO, endothelium produces other vasodilator substances including prostacyclin, which act in consort with NO in response to increased flow. Activated endothelial cells also have the capacity to communicate with adjacent cells through gap junctions (Figure 5.2) in a manner that has some similarities with myocardial conduction as discussed in Chapter 4. This results from shear stress-induced activation of intermediate signalling proteins and opening of mechano-sensitive ion channels. This in turn causes endothelial cell membrane hyperpolarisation (EDH), which can spread to adjacent cells via gap junctions. This means that a local response to increased shear stress can spread proximally and distally along the vessel wall and that vascular beds as a whole can be activated when flow increases. The process of mechano-transduction and its associated signalling cascades is discussed in greater detail in Chapter 8.

This ability to transduce and respond to shear stress is a likely contributor to the observations made by Schretzenmayr back in 1933 [6]. It is also important for maintaining a healthy circulation and long-term cardiovascular health. In humans, it can be measured as flow-mediated dilatation (FMD) and is improved by regular exercise. In contrast, FMD is impaired in diabetes and other known cardiovascular risk factors.

Perfusion of the Heart During Exercise

Perfusion of the heart presents a number of unique features and these are discussed here and illustrate additional mechanisms by which blood flow is regulated at the tissue level. The heart is of course a pulsatile pump and this is reflected in the way blood is propelled through the circulation. It produces the pressure waves we can feel in the pulse and corresponding increase in blood flow to tissues and organs with each pulse. Perfusion of the heart is an exception to this; the contraction of heart muscle, which generates the cardiac output and pulsatile flow, squeezes the coronary vessels within it and impedes flow through them. Unlike other tissues therefore, myocardial perfusion is restricted during systole and takes place mostly during diastole. As discussed in Chapter 4, the timing of contraction and relaxation of heart muscle are important to allow not only adequate emptying, but also sufficient time for ventricular filling and this also applies to myocardial perfusion. Heart rate and the force of contraction increase during exercise due to autonomic nerve stimulation and this increases energy usage. To meet this need, myocardial perfusion must also increase during exercise and at a time when the duration of diastole is reduced, once again emphasising the importance of preserving an adequate diastolic interval.

Oxygen consumption by the heart (measured by subtracting oxygen content in coronary venous from coronary artery blood) is a measure of its energy usage and increases about fourfold during exercise. The mechanisms for this are also unique in the heart. Unlike skeletal muscle, which has the ability to function temporarily without oxygen (anaerobic metabolism) using an alternative metabolic pathway (glycolysis), heart muscle has very little capacity to do this. Consequently, it is highly dependent on rapidly increasing coronary blood flow to meet increased energy needs. This is possible because coronary vasodilatation also occurs during exercise and causes a marked increase in blood flow.

Feed-forward control

One mechanism for this is sympathetic stimulation [12]. During exercise, sympathetic stimulation increases the heart rate and force of contraction. Coronary arteries are also innervated by sympathetic nerves which regulate their response to exercise. Sympathetic stimulation works through release of noradrenaline which binds to adrenergic receptors. The small resistance vessels which control coronary flow contain β-adrenergic receptors that cause vasodilatation when activated. Consequently, the increased sympathetic stimulation that drives the heart to work faster and harder also modulates the coronary arteries to increase its blood supply. This is known as a feed-forward mechanism, meaning that the response which occurs has no influence on the stimulating mechanism.

Feedback control

Oxygen consumption is derived from the difference between oxygen content of coronary arterial and coronary venous blood and these are measured as the partial pressure often referred to as oxygen tension and expressed as mmHg. The heart has a much greater ability to extract oxygen from blood than other tissues, typically extracting around 75% of oxygen contained in coronary blood compared with 25–30% in resting skeletal muscle. For this reason, heart muscle has relatively little further reserve capacity to extract oxygen when demand increases and so has to rely on rapidly increasing blood flow to meet greater energy needs. By the same token, a comparison of oxygen availability and consumption by the heart provides a powerful signal to control blood flow. In other words, if oxygen availability falls below consumption, the difference can provide a measure by which to adjust availability in the form of blood flow. This is an example of feedback control, as found, for example, in the thermostat of central heating systems. The heart appears to do this by regulating coronary flow in response to the oxygen content of blood [12] (Figure 5.2(b)).

As oxygen consumption by the heart increases during exercise, oxygen content of coronary venous blood initially declines and this acts as a signal to increase coronary flow. Oxygen is transported bound to haemoglobin in red blood cells (erythrocytes) and is released during transit through capillaries. Erythrocytes also contain significant amounts of adenosine triphosphate (ATP), which is released in the presence of low oxygen tension [13]. This efflux of ATP is a regulated process and is related to a change in the conformation of haemoglobin following release of oxygen. ATP and its metabolites, ADP, AMP and adenosine, are all potent coronary vasodilators. They do this by acting on purinergic receptors on endothelial cells which causes relaxation of smooth muscle in the arterial wall. Purinergic receptors are so named because they bind purine containing compounds (purines are chemicals containing a pyrimidine ring joined to an imidazole ring). Adenine is one of several purines and is a constituent of ATP and its metabolites, generally referred to as adenine nucleotides. Thus a decline in oxygen tension in coronary capillaries causes the release of adenine nucleotides which in turn activate the vasodilator properties of endothelial cells lining them.

The fact that oxygen is released from haemoglobin in capillaries means that they offer an ideal site for using the oxygen tension of blood as a signal to regulate coronary blood flow. Capillaries are the smallest vessels in the circulation; their walls are very thin, which facilitates diffusion of nutrients to the tissues and uptake of metabolites such as carbon dioxide and other waste products. However, capillary walls do not contain smooth muscle and therefore cannot dilate to increase blood flow. For this reason, an additional mechanism is required to complete oxygen tension mediated feedback control of blood flow. This is achieved by retrograde conduction.

Retrograde conducted response

Capillaries are downstream of the resistance arteries, which regulate blood flow. The signal generated by activation of endothelial cell purinergic receptors must therefore be communicated retrogradely to these vessels. This is possible due to a cascade of signals that are activated following the interaction of ATP and its derivatives with purinergic receptors on endothelial cells and is described in greater detail in Chapter 8. Briefly, ATP binding to its endothelial purinergic receptor (P2y) stimulates membrane-bound phospholipase C, an enzyme that synthesises inositol triphosphate (InsP3). InsP3 binds to a specific receptor on endoplasmic reticulum (ER) causing a release of calcium (ER acts as an internal store of calcium). The raised intracellular calcium that results activates the production of vasodilator substances, including prostacyclin, nitric oxide synthase derived NO and factors which activate calcium-sensitive potassium ion channels (K_{Ca^+}). K_{Ca^+} extrude K^+ ions, and the resulting loss of positive charge hyperpolarises (makes it more negative) the cell.

The precise nature of substances which activate K_{Ca^+} channels remains unknown but are generally referred to as endothelium hyperpolarising factors (EDHFs). Endothelial cell hyperpolarisation is important because it provides the means for signal conduction between adjacent cells. As discussed in Chapter 4, gap junctions are specialised connections between cells where the membranes of two adjacent cells become very closely attached. They contain pores that allow ion movements between cells. Endothelial cells contain gap junctions composed of connexin 40 located in cell-to-cell contact areas. Ion fluxes through them conduct the hyperpolarised state from cell to cell, allowing conduction of the vasodilator stimulus retrogradely along the length of the vessel. Once the hyperpolarisation signal reaches the proximal resistance arteries, it is conducted to the smooth muscle they contain causing relaxation resulting in dilatation. In this way a decrease in the oxygen content of capillary blood acts as a signal which is transmitted through the vessel wall to proximal arterioles. This causes them to relax which increases blood flow to restore oxygen availability.

Thus, blood flow to muscles engaged in exercise is carefully controlled by mechanisms operating locally at the level of the arteries directly involved in its supply. From an engineering perspective, the concept of tubes that can sense the pressure and flow of their contents and use this information to modify their physical properties and dimensions is remarkable. These properties are also important for the long-term health of blood vessels and failure of them reduces their ability to adapt to varying conditions and reduces their longevity.

Circulatory responses to increased energy demand, as in exercise, involve a wide range of complex and integrated mechanisms. The heart is just one part of this and its proper functioning depends on its precise interaction with complementary mechanisms in the circulation as a whole. The responses to exercise illustrate how cardiac activity is precisely regulated and adjusted to meet varying demands in consort with matching changes in blood vessels throughout the body so as to target increased perfusion to the tissues calling for increased energy. It is an astonishingly efficient system and a key factor in the remarkable adaptability and efficiency of the heart and to its long-term survival.

Physical Adaptability

We have seen how the heart is able to match the varying demands of the body by adapting its rate and cardiac output continuously. These adjustments in performance are physiological, meaning they are relatively short-term reversible responses to regulating signals. The heart also has the ability to respond physically to long-term changes in demand by altering its size and shape. The simplest and most common expression of this is the way heart size varies with body weight. Originally, measurements of heart size were obtained during post mortem examinations, but because most of these were carried out to confirm the cause of death, they were often of people who had been ill and less mobile for some time. Until the 1960s, normal heart size (measured as weight) was thought to be fixed. Linzbach reported that heart weight in an average person was about 300 g and could increase to 500 g with physical training and values above that were considered abnormal [14].

Regulation of heart size

When more modern non-invasive techniques for assessing heart weight became available, such as echocardiography and MRI, the values obtained were greater than those based on autopsy weights. The discrepancy was explained by later studies which showed that post mortem heart weights of people who died without previous illness were larger than the traditional values. It also became clear that heart weights were greater in males than females and that they varied according to body size [15]. Thus, the left ventricle, which is the main pumping chamber of the heart, was 170 g and 145 g, respectively, in males and females of 70 kg compared to 224 g and 182 g in males and females of 100 kg [15].

A constant aspect of cardiac workload is combating the effects of gravity and the effect of this is illustrated by the changes that occur following prolonged inactivity or exposure to microgravity. When previously sedentary men were exposed to long-term bedrest, LV mass decreased 8% after 6 weeks and a further 7% after 12 weeks. Similarly when the effect of gravity was removed during 10 days of spaceflight, LV mass declined by 12% [16]. These changes are also reversible when normal levels of activity are resumed. So, the size of the heart is not fixed but is dynamically responsive to the demands it meets. It can change during the course of our lives, so that heart weights differ between individuals depending on their size, lifestyle and average daily activity.

Hypertrophy

When exposed to prolonged periods of increased demand, the heart enlarges in a process called hypertrophy. Cardiac hypertrophy occurs in response to the increased stress associated with increased workload on the heart. It is described as physiological when the increased demands are part of normal activity. Pathological hypertrophy is used to describe the abnormal cardiac growth that occurs in response to chronic increased workloads secondary to illness, for example, in hypertension or an obstructed outlet valve. In both of these forms of hypertrophy, the physical size of the heart increases and so do the individual cardiac myocytes. However, in other respects they differ in terms of their morphology (structure) and function. They also have different long-term consequences for health and will be described separately. Pregnancy and physical training in athletes are two common examples of physiological hypertrophy, both occurring in response to increased demands as part of normal human activity.

Pregnancy-induced hypertrophy

Pregnancy is associated with rapid and dramatic changes in cardiovascular function. This can be a major challenge for hearts with reduced functional capacity due to illness and often requires careful monitoring, rest and treatment. Otherwise, they are well within the range of normal cardiac function and its ability to adapt to changing conditions. Cardiac output increases early in pregnancy with detectible changes present within 2–3 weeks of fertilisation. It continues to increase thereafter, reaching a maximum of around 45% above pre-pregnancy values at 16–18 weeks and this is maintained until delivery [17]. The increase is generated by an increase in heart rate and stroke volume. Heart rate rises fairly steadily throughout pregnancy by about 12 beats per minute. Stroke volume also increases reaching a maximum of about 30% above pre-pregnancy values at 16–20 weeks and is maintained until delivery. These changes commence at a time when the foetus is just a few millimetres in size and are part of wide range of adaptations in the maternal circulation in response to hormones and other stimuli released from the growing placenta. These include

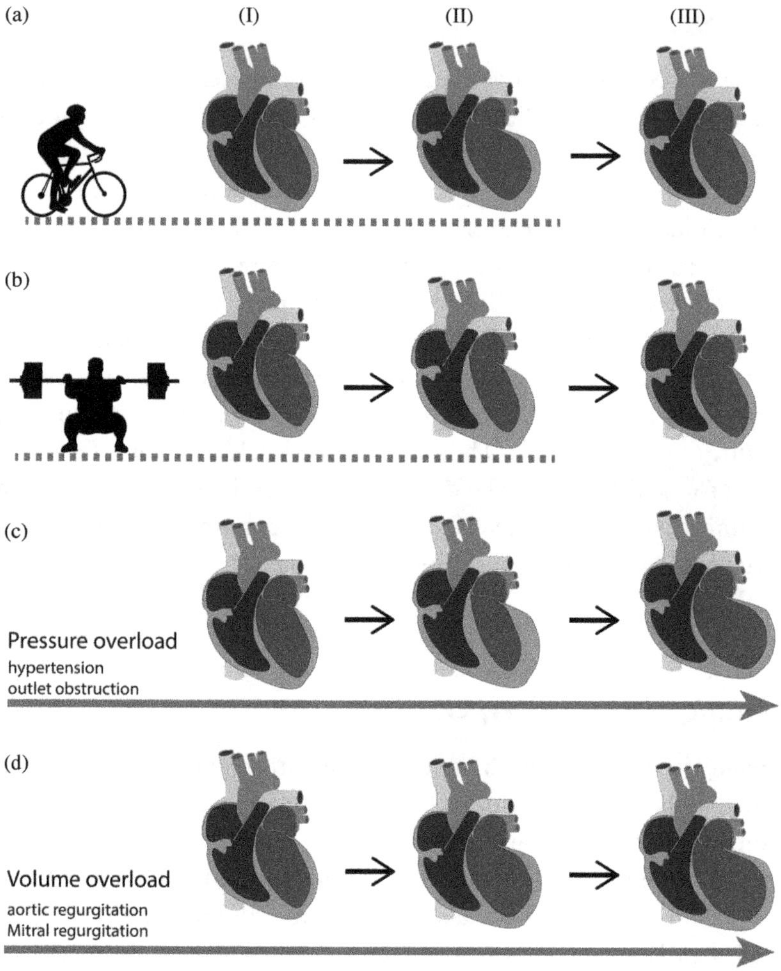

Figure 5.3: Progression of LV hypertrophy. Sustained increased workload provokes the growth of heart muscle. The form this takes depends on the type of load that stimulates it. (a) and (b) are examples of physiological hypertrophy in response to intermittent increased stresses. (a) Athletes who train by cycling or running must increase their cardiac output and with repeated challenges this results in an increase in LV chamber size with a modest increase in LV wall thickness (II). When the training program stops, the heart regresses to its previous size (III). (b) Weightlifters experience very high levels of blood pressure which produces marked increases in LV wall stress. The heart responds to this by increasing LV wall thickness (II), which also regresses to normal on cessation of training (III). (c) and (d) are examples of pathological LV hypertrophy in response to sustained increases in work-load. In (c), the heart is required to work against a high output pressure, such as high blood pressure or an obstructed aortic (outlet) valve. This is similar to the weightlifter, but in this

vasodilatation, which lowers vascular resistance and expansion of the blood volume.

Plasma volume (the volume of blood excluding all of the cellular components such as red blood cells, white cells and platelets) increases steadily during pregnancy from around 2.4 L to 3.6 L at term [18]. The initial cardiovascular response to pregnancy is a general hormone-related vasodilatation. This tends to lower blood pressure which in turn triggers release of renin, angiotensin and aldosterone, causing retention of fluid and expansion of plasma volume. The increase in plasma volume then increases venous return to the heart which results in an increase in stroke volume by activating the Frank–Starling mechanism (discussed in Chapter 3). These haemodynamic changes are also accompanied by physical changes in the heart. Left ventricular (LV) size grows substantially and continuously during pregnancy, one study reporting an increase in LV mass from 120 g to 183 g at 38 weeks [19] due to increases in both LV wall thickness and cavity dimension. All of these adaptations by the heart revert to pre-pregnancy values following delivery. The haemodynamic and blood volume changes do so within a few days of delivery, while the physical changes take several weeks to resolve completely [17]. The increase in the size of the heart during pregnancy is similar to the cardiac changes that occur in response to athletic training. Both are physiological and reversible.

Exercise-induced hypertrophy

Athletes who train regularly subject their hearts to intermittent periods of increased stress and develop a compensatory increase in heart size as a result. However, the form that takes depends on the type of stress they experience (Figure 5.3). In broad terms, exercise is described as isotonic

Figure 5.3: *(see figure on facing page)* case the increased effort is constant rather than intermittent. This results in persistently increased LV wall stress and the heart responds by increasing LV wall thickness (II). Over time if unrelieved, the heart muscle function declines and fails, a stage characterised by progressive LV dilatation (III). (d) is an example of volume overload hypertrophy where the LV is over-filled during diastole. This can occur due to a leaking aortic or mitral valve. Here the LV receives its normal filling from the left atrium but with extra filling from the leaking valves. The heart responds by increasing LV chamber size (II) and if unrelieved over time causes progressive LV dilatation and failure (III).

and isometric depending on the type of work done. Isometric exercise involves increased muscle tension with relatively little movement and weight lifting would be a typical example. Isotonic exercise involves a great deal of movement and relatively little static stress, e.g., long-distance running. In both forms of exercise, LV mass increases but this is accounted for by a greater increase in wall thickness in weight lifters, whereas runners develop a bigger increase in the LV cavity size [20]. These differences are likely to be related to the form of stress which each type of exercise imposes on the heart and its impact on the left ventricle.

Isometric and isotonic exercise

The principal cardiovascular demand of endurance running is sustained periods of increased cardiac output to perfuse the working muscles. Increased cardiac output is provided by an increase in heart rate and stroke volume. So a ventricle which can hold a larger volume in diastole and eject it in systole best suits such a requirement. In other words, the increased cavity size in runners facilitates the greater stroke volume and cardiac output they need. In weight lifters also, the form of hypertrophy which develops matches the loads on the heart imposed by isometric exercise. Weight lifting is known to cause striking increases in blood pressure and systolic and diastolic values as high as 345/245 mmHg have been reported [21]. This is likely to be due in part to the marked increases in intra-abdominal and intrathoracic pressures which occur, secondary to increased abdominal muscle tension. The need to develop such high pressures places substantial stress on the walls of the left ventricle and the increased LV wall thickness reported in weight lifters is a response to this. LV wall stress expresses the forces between the individual cells in the wall and is expressed in the Law of Laplace:

$$\text{Wall stress} = \frac{\text{Pressure} \times \text{LV radius}}{\text{Wall thickness}}$$

LVwall stress is *directly* related to (a) the pressure within the ventricle and (b) the diameter of the ventricle and is *inversely* related to (c) the thickness of the wall of the ventricle. In other words, wall stress increases with cavity pressure and diameter and decreases with wall thickness (a thicker wall allows the forces within it to be spread through its greater depth). So the increased LV wall thickness that occurs in weight lifters serves to reduce the high levels of LV wall stress associated with the pressures they develop. What is remarkable about this is that the growth response in each case matches the demands which stimulate it and that the heart can sense each type of load and respond appropriately.

Hypertrophy as a compensatory mechanism

By growing in this way, hypertrophy can be seen as a beneficial response in helping the heart to cope with increased demands. As would be expected, hypertrophy in response to training increases cardiac performance [22]. Indeed most forms of hypertrophy increase the contractile power of the heart, at least in the early stages. Consequently, hypertrophy was considered a useful compensatory response to increased loads on the heart. That view changed in the 1960s when population-based studies showed that hypertrophy could have serious negative long-term consequences [23]. Globally, hypertension is the most frequent cause of cardiac hypertrophy. This was an important discovery because it focused attention on the harmful effects of hypertrophy and on hypertension. It also stimulated research to understand the effects of hypertrophy on cardiac function, why it can be harmful and whether all forms of hypertrophy are similar in this respect. The consequences of training-induced hypertrophy were of particular interest (a) because of concerns that it might be harming healthy people and (b) differences between hypertrophy due to training and conditions like hypertension might shed light on why and how it can be harmful.

Is exercise-induced hypertrophy harmful?

Over the past few decades, many studies have looked at cardiac function in trained athletes and the overwhelming evidence is that the growth of the heart which occurs in response to training does not have harmful long-term consequences. It undoubtedly increases the performance of the heart [22] and this is sustained over many years of continuous training [24]. A question often asked about athletic training is, does it have any long-term adverse consequences? This raises two issues, (i) what are the long-term consequences of hypertrophy that develops in young athletes and (ii) is training equally beneficial or harmful at different ages and is there an optimal age for our bodies ability to respond to training?

Development and regression of hypertrophy

Assessment of hypertrophy relies on measurements of LV mass and this is obtained by multiplying the volume of the left ventricle by the density (the weight of 1 cm^3) of the heart. Ventricular volume is measured using non-invasive imaging such as echocardiography or MRI. Studies in training athletes have shown that increases in LV mass develop rapidly (within a few weeks) of regular intense training. Thus, in a group of previously non-training volunteers aged 17–19 years, who swam for 2 hours daily, 6 days per week, for 9 weeks, LV mass was already increased by week one [25]. This was accompanied by increases in LV cavity size and in stroke volume. The same study examined the effect of detraining for 3 weeks in a group of cross country runners, who had previously been running 60–70 miles per week at 8–10 miles per hour for more than 3 months. The results demonstrated rapid regression of LV mass and dimensions. Thus, at the end of three weeks without training, LV mass was reduced by almost 40% and was accompanied by reductions on LV cavity size and stroke volume [25]. Many other studies have confirmed that training-induced LV hypertrophy is readily reversible.

Exercise hypertrophy in older hearts

The impact of age on our capacity for physical activity is of interest because of increased life expectancy. It looks certain that we will be working later in life than earlier generations and exercise is increasingly recommended for older people as a way of staying healthy. In addition, physical training is increasingly popular among older people. Studies of exercise performance in people of different ages have shown a linear decline in aerobic capacity with time from the age of 20 onwards [26].

However, these changes were also associated with increased body fat composition and inversely with the amount of regular physical activity undertaken. Therefore the findings may not relate directly to age but to lifestyle confounding factors.

Marathon running is an increasingly popular endurance sport among all ages. Among men who ran in three city marathons from 2001 to 2016, the average age of winners was 28.3 for men and 30.8 for women. Finishing times were grouped by age in 5-year increments for runners between 35 and 74 years of age and showed a steady decline in group winners' race time with age [27]. It seems clear then that the best performances among participants in these races declined with age. City marathons are of course competitive races and winning depends not just on physical aptitude and potential for endurance running, but also on preparation for it. Therefore they may not provide the best way to distinguish between the impact of biological ageing and other associated changes with age, such as an increasingly sedentary lifestyle. In an attempt to exclude effects due to lifestyle differences, the finishing times among over 600,000 marathon and half marathon runners aged 20–79 years of age were studied [28]. This group was mostly of individuals who trained regularly and were non-sedentary non-smokers and of normal weight, in other words a healthy physically active group. Finishing times did not differ significantly among runners aged 20–54 years and only a moderate decline occurred among participants older than that. Importantly too, older subjects appeared capable of the training required to complete these endurance challenges. It seems likely therefore that reported reductions in exercise capacity over time are heavily influenced by lifestyles and are not simply due to biological ageing. People of all ages who train regularly can retain their capacity for endurance running. Interestingly also the peak age for finishing times in ultramarathons (races of 50 km or longer) is around 40, considerably older than for marathons, 25–35 [29].

Long-term effects of physiological hypertrophy

Although cardiac hypertrophy associated with training is now well known to be reversible, there remains the question could it have lasting adverse effects on the quality of life or longevity. When Pheidippides died suddenly in 490 BCE after running from Marathon to Athens to announce victory over the Persians, did he die from acute cardiac injury as Robert Browning suggests in the lines:

> Ran like fire once more: and the space 'twixt the fennel-field
> And Athens was stubble again, a field which a fire runs through,
> Till in he broke: "Rejoice, we conquer!" Like wine through clay,
> Joy in his blood bursting his heart, — the bliss!

In a study of almost 10,000 athletes who performed in the Olympics between 1896 and 1936, the potential risks associated with different forms and intensities of exercise were compared. This found no added risks associated with the level of cardiovascular activity or between athletes engaged in static and dynamic exercise [30]. Thus, longevity is similar among athletes irrespective of the form and severity of exercise they undertake. To assess whether extreme training might affect life expectancy compared to the general population, longevity of the first 20 athletes to run a mile in less than 4 min was compared with the life expectancy of their age-related compatriots [31]. Eighteen of these lived to be 80–88 years and exceeded life expectancy by an average of 12 years, suggesting that extreme training does not reduce longevity. Thus, physiological hypertrophy appears to be a normal response of the heart to the increased workloads encountered during exercise. It can be seen as a healthy compensatory mechanism to maintain the heart's ability to work within safe limits, as, for example, by the way increased wall thickness mitigates the increased stress associated with weight lifting or the greater cavity size that of long-distance runners.

Pathological Hypertrophy

There have been several well-publicised tragic cases of sportsmen becoming suddenly unwell or dying during competitive sports, and severe hypertrophy has been a common finding among these. However, this is not simple physiological hypertrophy of the heart, but rather an abnormal growth response to some previously undetected abnormality of the heart, which in many cases is due to an inherited genetic abnormality. This is an example of pathological hypertrophy, a condition in which the heart grows in an effort to compensate for abnormalities in the heart or circulation which impose increased loads on it. These can be broadly characterised in terms of the kind of load they impose.

Pressure overload

Afterload refers to work imposed on the heart once the main outlet valves open during systole. Hypertension is the most frequent cause of increased afterload and left ventricular hypertrophy (LVH) (Figure 5.3). Blood pressure varies constantly in normal people and in those with high blood pressure and is generally higher during the day than at night. In hypertensive people, the average levels during the day are higher, so that the heart is constantly pumping against a higher resistance. This is a distinguishing feature of pathological hypertrophy; it results from increased loads that are fixed in contrast to the intermittent nature of exercise. Increased afterload can be caused by any abnormality that obstructs ventricular outflow. Examples include hypertension, stenosis (narrowing) of the aortic valve, and coarctation (narrowing) of the aorta. Afterload-induced LVH initially results in increased wall thickness (referred to as concentric hypertrophy) and later, if unrelieved by LV dilatation.

Volume overload

Increased volume load describes conditions where a ventricle is subjected to increased filling during diastole. Examples affecting the left ventricle include aortic incompetence when the aortic leaflets fail to oppose correctly on closure and blood leaks back from the aorta into the ventricle during diastole. This, together with normal inflow from the left atrium results in overfilling of the ventricle. Mitral incompetence also causes increased LV filling during diastole. In this case, the mitral leaflets fail to close correctly during systole resulting in some blood leaking back into the left atrium during LV ejection. Consequently, the left atrium receives an excess volume of blood and in turn overfills the left ventricle during the subsequent diastole. LV volume overload results in LVH, which is expressed primarily as an increased LV cavity size (also called eccentric hypertrophy) (Figure 5.3).

Hypertrophic cardiomyopathy

Hypertrophic cardiomyopathy (HCM) describes a condition characterised by hypertrophy of heart muscle caused by faulty genes in the heart rather than any external factor. To date, around 1,400 such gene abnormalities have been discovered in 11 genes which carry the code for making proteins in the sarcomeres of cardiac muscle [32]. Most of the defects result in abnormal synthesis of myosin heavy chains (35%), myosin binding protein (35%) and the troponin-tropomyosin group (12–20%) (Figure 5.4). For more detailed information and the function of these myofibrillar proteins, see Chapter 3. Many of these faulty genes are inherited and it is estimated that about 1 in 500 people have one of them. The effect of each varies depending on the function of the protein it encodes and the extent to which the abnormality interferes with it. In some cases, the effect can be mild and unknown to the carrier. In severe cases, the muscle of the

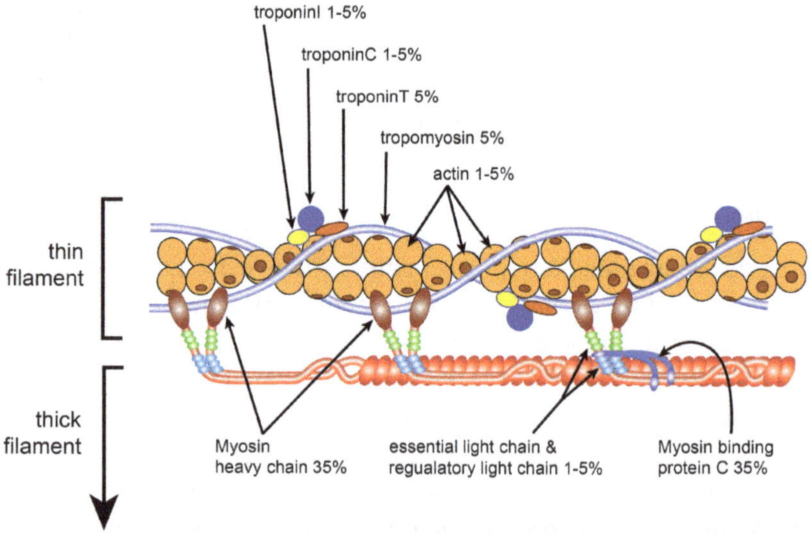

Figure 5.4: Diagrammatic illustration of myofibrils showing the location of gene abnormalities which cause HCM and the frequency of each among affected patients. There are an estimated 1,400 gene abnormalities that cause HCM. Almost all of these carry the code for sarcomere proteins. The genes which encode myosin heavy chains and myosin binding protein-C are most frequently involved but HCM can be caused by abnormalities in several other myofibrillar genes.

heart grows dramatically to produce very thickened walls. In a subgroup of patients, the muscle around the LV outflow grows disproportionately and can impede the ejection of blood during systole, a condition called hypertrophic obstructive cardiomyopathy (HOCM). Like other forms of LVH, HCM can also be seen as a compensatory response, but in this case secondary due to an intrinsic weakness in myocardial performance due to a defect in the contractile machinery. And like all forms of pathological hypertrophy, it can result in disturbances in the structure and function of the heart.

Long-term effects of overload hypertrophy

The forms of hypertrophy which occur due to increased afterload and volume load resemble the physiological hypertrophy seen with isotonic and isometric training, respectively, and this is not surprising since they are provoked by similar demands. They are also similar in that regression can occur when the causes of pathological hypertrophy are relieved. Thus LVH in patients with hypertension or aortic stenosis can regress with successful treatment to remove the underlying cause. Many clinical trials have demonstrated that regression of LVH can be achieved in hypertensive patients following successful treatment to lower blood pressure [33]. Similarly, aortic valve replacement in patients with aortic stenosis is very effective in reversing LVH [34]. Despite these similarities, pathological and physiological LVH differ in other respects, most importantly in their long-term consequences.

Early epidemiological studies clearly demonstrated that LVH is associated with significantly higher risks of cardiovascular disease and death [23]. In these studies, LVH was almost always due to hypertension, which has many adverse effects on the circulation as a whole. Its most devastating consequences relate to its harmful effects on blood vessels throughout the body, which increases the risk of stroke and coronary artery disease. Initially, LVH can be seen as a compensatory response to the increased wall stress caused by raised blood pressure. However, if unrelieved over time, additional changes occur in the structure of the heart when it becomes resistant to regression and at risk of failure, a late phase that is usually accompanied by dilatation. In rich western countries, hypertension is now recognised and treated far more frequently than when the risks were identified in the 1960s and this is one reason for the reduction in cardiovascular deaths in those populations. Unfortunately, economic development in many regions of the world has been associated with the adoption of unhealthy lifestyles and an increasing prevalence of obesity and hypertension. As a result, hypertension has become the most important risk factor for early cardiovascular death in the world. Hypertension, the most common cause of LVH, also increases the risk of coronary heart disease and stroke. Consequently angina pectoris, myocardial infarction, arrhythmias and heart failure are all more prevalent with LVH. However, even in the presence of normal coronary arteries, LVH causes abnormalities in electrical and mechanical functions of the heart and impaired myocardial perfusion [35].

Morphology of LVH

Cardiac muscle cells (myocytes) constitute about 80% of the volume of the heart. In prenatal and early postnatal life, cardiac myocytes grow in number by cell division. Thereafter growth takes place by increasing cell size only; in some mammals, myocytes increase in volume around five-fold between birth and adulthood and the increase in myocyte size corresponds to the increase in heart size. Changes in myocyte size also account for variations in heart size with body weight. Myocyte size also increases in all forms of hypertrophy, but with differences in myocyte shape. In severe pressure overload hypertrophy, the predominant change is an increase in diameter, whereas with volume overload, myocytes become longer [36]. In severe and prolonged LVH myocyte nuclei exhibit increasing polyploidy, meaning that the nuclei of myocytes contain increasing numbers of each chromosome. Most cell nuclei are diploid, meaning they contain a single pair of each chromosome. Cardiac myocytes differ in this respect in that polyploidy increases from zero at birth to around 33% of cells in early childhood. Later this decreases to about 10% in adults. Hypertrophy is associated with a resurgence of polyploidy and may be present in 90% of cells in severe cases [36]. Polyploidy increases the number of copies of each gene which increases a cell's capacity to synthesise new proteins.

Inter-myocyte connections

As discussed in Chapters 3 and 4, normal function of the heart is based on coordinated conduction of the pacemaker impulse and synchronised contraction of cardiac myocytes and their contractile proteins. These in turn depend on how myocytes are organised and on the specialised connections between them. Gap junctions provide the electrical coupling between myocytes which enables the rapid spread of depolarisation through the heart. Intercalated discs form the strong mechanical coupling between myocytes and ensure that they function as a continuous series of contracting units. These structures appear to be well maintained in physiological hypertrophy but fail to develop normally in severe hypertrophy [37] resulting in disturbances in the electrical connections between myocytes and a risk of rhythm disturbances as well as impaired pump function.

Coronary circulation

Normal cardiac function depends on an adequate blood supply by the coronary circulation, which is carefully regulated to match demand. As the heart grows in size, so too does its energy needs and this in turn requires that the coronary circulation should grow in a manner to match the increased demand. Angina pectoris describes chest pain which occurs when heart muscle is deprived of an adequate blood supply and most often occurs during exercise in patients with one or more narrowed coronary arteries. However, angina can occur in the absence of coronary artery disease in patients with severe LVH due to hypertension, HCM or aortic stenosis. Assessment of coronary flow dynamics in humans is possible by measuring coronary flow reserve. The latter refers to the difference between flow at rest and the maximum possible flow and represents the capacity to increase flow to meet demands. Several studies have demonstrated reduced coronary flow reserve in patients with hypertension and aortic stenosis despite having apparently normal coronary arteries [38, 39]. This suggests that such patients experience angina because of an inability to increase coronary flow to meet elevated demand. In contrast, coronary flow reserve remains normal in physiological hypertrophy as, for example, in elite athletes [40]. Reduced coronary flow reserve in pathological LVH is due to (a) changes in the structure of coronary vessels and (b) an increased compressive effect of contraction on them.

Coronary morphology

The coronary circulation is a branching circuit and like others it divides into smaller and smaller vessels. From a functional perspective, it may be divided into three sections. (a) The large coronary arteries which run along the surface of the heart are subject to the atheromatous deposits that cause coronary artery disease. (b) Coronary resistance vessels (called arterioles) are branches of the former that permeate the muscle of the heart. Their walls contain smooth muscle that allows them to constrict and dilate in response to various stimuli to regulate myocardial perfusion. (c) The smallest vessels (capillaries) which permeate throughout the heart. These have very thin walls that allow nutrients and oxygen to diffuse through the surrounding tissue. Morphological studies suggest that with severe hypertrophy, the coronary circulation fails to develop at the same rate as the myocardium it supports. This results in a reduction in the density of arterioles and capillaries in LVH [41, 42]. A reduction in capillary density means that nutrients and oxygen must traverse a greater diffusion distance to reach the tissues and fewer resistance arterioles reduce the maximal capacity to deliver coronary flow to capillaries. Thus, structural changes in the coronary microcirculation in LVH limit the capacity for myocardial perfusion. While these may not be sufficient to cause ischemia at rest, they increase the risk of this occurring when demand increases.

Haemodynamic effects on coronary perfusion

The coronary circulation is unique in that it is compressed during systole by contraction of surrounding myocardium with the result that flow is impeded during systole and primarily takes place during diastole. LV hypertrophy causes a prolongation of the ejection phase of systole, meaning the heart takes longer to empty. This lengthens the period during which the coronary circulation is compressed and reduces the diastolic period when most coronary flow occurs. Experimental studies have shown that the increased muscle bulk and prolonged ejection time associated with LVH increases the compressive effect on the coronary circulation and impedes coronary flow [42]. In patients with LVH due to aortic stenosis, a reduction in coronary flow reserve occurs and is directly related to this reduction in diastolic perfusion time [43]. Aortic valve replacement results in regression of LVH, normalisation of diastolic perfusion time and recovery of coronary flow reserve [34].

Thus, impaired myocardial perfusion is an important consequence of pathological LVH and results from a failure of the coronary circulation to grow at the same rate as the myocardium. But it also results from a greater compressive impact of LV contraction and a reduction in the time available for perfusion during diastole. Together they account for a greater vulnerability to myocardial ischaemia and angina pectoris even in the absence of atheromatous coronary disease.

Heart Failure

The long-term consequences of LVH depend on the severity and duration of the increased workload imposed on the heart. Physiological hypertrophy is reversible and has no long-term adverse effects. Pathological hypertrophy, if protracted, results in heart failure when the heart is unable to meet the energy needs of the body. Heart failure usually presents with increasing difficulty with physical exertion due to an inability to increase cardiac output and this is caused by deterioration in LV function. In other words, if LVH and its underlying causes become chronic, ultimately the compensatory effect of LVH gives way to deteriorating function and heart failure (Figure 5.3). This transition from compensatory LVH to LV failure has several causes.

Hypertrophy transition to heart failure

The hypertrophic compensatory response is not without a cost. As discussed above, pathological hypertrophy comes with alterations in the structure and function of the coronary microcirculation that impair perfusion of heart muscle. In the short term, the heart benefits from the increased muscle power of hypertrophy, but repeated exposure to inadequate blood perfusion is harmful and over time leads to a reduced contractile force. The reductions in gap junctions and intercalated disks which coordinate electrical and mechanical functions are also likely contributors to this. Recent studies have also confirmed that chronic hypertrophy is associated with death of myocytes [44]. This can occur by a process called autophagy (discussed in more detail in the following), which is normally used in recycling and renewing cell components such as proteins released from damaged cells. However, if damage to a cell's contents is excessive, then autophagy can result in destruction of the cell. Cell death can also occur through apoptosis and necrosis. The former is a natural mechanism for removing and replacing cells in organs. It is normally triggered by several signalling mechanisms, which, for reasons yet to be clarified, is increased in hypertrophied myocardium [45]. Necrosis refers to the death of cells due to injury or trauma.

As discussed in Chapters 3 and 4, calcium ions (Ca^{2+}) play a vital role in triggering contraction in cardiac myocytes and its concentration is carefully regulated for this purpose. Several lines of evidence indicate that abnormal Ca^{2+} homeostasis contributes to the transition from hypertrophy to failure. Provisional evidence suggests that some forms of hypertrophy may be mediated by increased intracellular levels of Ca^{2+} and that inhibition of Ca^{2+} entry through sarcolemmal calcium channels and release from the endoplasmic reticulum could mitigate the development of hypertrophy [46]. With prolonged LVH, calcium stores in the endoplasmic reticulum appear to be depleted due to leakage through ryanodine receptors [47], which may result in lower Ca^{2+} transients during activation and a reduction in contractility. However, ryanodine receptor leakage may also cause higher background cytosolic levels of Ca^{2+} and the risk of Ca^{2+} overload, which in turn can activate a process of necrosis resulting in cell death [48]. This line of research is still very active, and a final picture has yet to emerge, but as so often in biology, compensatory mechanisms which deviate from normal function may be advantageous in the short term but always come at a cost. Similarly hypertrophy of remaining cells if prolonged results in deteriorating function and heart failure.

Heart failure is a common end point for many cardiovascular illnesses. It is almost always accompanied by dilatation of the left ventricle. This confers some short-term advantages in coping with loss of ejecting power, but it too comes at a cost. An increase in the diameter of the ventricle means lengthening the circumference. This requires a rearrangement of cells in the LV wall resulting in an increased number of cells connected end to end around the circumference and fewer cells across the wall thickness. This means that contraction of a ventricle with a larger circumference can be greater and achieve a greater ejection volume in systole. While this confers some immediate advantage, ultimately it contributes to further deterioration in function. This is because LV wall stress (according to the Laplace law) is directly related to the radius of the LV cavity and therefore increases with dilatation. Unfortunately, wall stress is also directly related to oxygen consumption. This means that LV dilatation increases myocardial oxygen requirement and since this is already compromised in LVH, it is ultimately harmful.

Maintenance and Regeneration of the Heart

Servicing the heart

We are familiar with the need to service our cars regularly. The heart is also serviced but of course must continue working while this is done. The efficiency of the heart is maintained by a process of protein turnover whereby older proteins are constantly being degraded and replaced by newly synthesised copies. The rates of synthesis and breakdown determine the weight of the heart and are normally in balance when this is stable. Protein turnover in the heart is very carefully controlled by a highly evolved and complex system of interacting components. These operate on both the synthetic and degradation pathways. The activity of these systems in the heart is quite astonishing. Troponins T, I and C have half-lives of 3–5 days, actin and tropomyosin of around 7–10 days and myosin 5–8 days [49]. In other words, about half of all the molecules in the heart's contractile machinery are degraded and replaced every week. To achieve this, newly synthesised proteins are carefully inserted into the working sarcomere without disruption of contractile function using special "minder" proteins (called chaperones) which protects their correct folded structure [50]. In addition, the mechanism of insertion appears to be specific for each protein. This is suggested by the finding that whereas troponin I is inserted evenly along the actin filament, tropomyosin is preferentially inserted at the end [50]. Details of how actin and myosin filaments are replaced have yet to be worked out in detail, however it is known that both involve several chaperone proteins which bind to them and maintain their correct folded structure until they can be safely inserted into a working sarcomere.

Proteins inevitably become damaged by partial degradation or misfolding and if they were to accumulate would cause impaired organ function. It is important therefore that they can be removed and replaced efficiently. There are three separate systems for targeting, removing and degrading defective proteins.

The ubiquitin-proteasome system

The ubiquitin-proteasome system (UPS) operates in two stages (Figure 5.5, lower left panel). First, defective proteins are identified and tagged for degradation. This involves three enzymes; ubiquitin-activating enzymes (also known as E1) activate a small protein called ubiquitin and deliver it to a second ubiquitin carrier enzyme (E2). The ubiquitin-E2 complex then binds to a ubiquitin ligase enzyme (E3). The E3 ligases act by targeting the ubiquitin complex to a specific protein (actin in the example illustrated in Figure 5.5). The human genome encodes over 800 of them, including a number known to be specific for cardiac sarcomere proteins. In other words, the ubiquitin-E3 complex provides a mechanism for marking specific proteins for degradation and the large number of substrate specific E3 enzymes means that degradation of individual proteins can be very precisely controlled. Binding of E3 to its protein results in the transfer of ubiquitin to that protein and once a single ubiquitin molecule is attached it quickly attracts others resulting in polyubiquitination. This prepares the protein for degradation in the next step which takes place in proteasomes.

Proteasomes are large multi-molecular structures which degrade proteins. They are composed of four rings forming a stack resembling a cylinder with a central core. The inner part of the core contains protease enzymes which break down proteins into small 6–8 amino acid fragments.

At each end of the cylinder, a cap, also composed of proteins, regulates the introduction of ubiquitin tagged proteins to the core. Proteasomes are found in all eukaryote cells and in some bacteria also. There are a number of proteasomes of which the most widely studied is known as S26, which has a molecular mass of around 2,000 kilodaltons. Proteasomes function in collaboration with the ubiquitination process described above, forming a major system for regulating cellular functions by removing unwanted or damaged proteins. The ubiquitin-proteasome mechanism has great importance in cell biology, for example, they regulate mitosis by rapidly destroying key enzymes involved once their task is complete to ensure that cell division proceeds in an orderly manner. This was recognised in the 2004 Nobel prize to its discoverers, Aaron Ciechanover, Avram Hershko and Irwin Rose.

The UPS accounts for most protein degradation in cells and therefore the proteolytic side of protein turnover. It remains an area of intense research, particularly in identifying components which control the degradation of specific proteins. A recent example was the identification of an E3 ligase located within the Z disk of sarcomeres [51]. This included a well-known component of E3 ligases, Fbx122, which tags specific sarcomere proteins. In other words, sarcomeres themselves contain the machinery necessary for identifying and tagging damage.

Autophagy

Autophagy provides a second system for degrading proteins but in a more gradual and less specific manner (Figure 5.5, right panel). In this case, larger complexes of protein material are enveloped in a membrane and transported to a lysosome with which it docks so that the contents are exposed to enzymes contained within the lysosome. These are proteases that can attack a wide range of bonds within proteins, meaning that they can degrade almost any intracellular protein efficiently. There is some evidence that ubiquitination has a role in autophagy however it is generally considered as a "catch-all" backup mechanism to the highly specific UPS.

Calpain system

Calpains are proteolytic enzymes that are activated by calcium (Figure 5.5, left upper panel). They are composed of two elements, calmodulin and a cysteine protease. The name calpain derives from the calmodulin component. Calmodulin (CAM) is a protein that acts as a Ca^{2+} sensor and regulator in many cells. Upon activation by calcium, calpains increase the

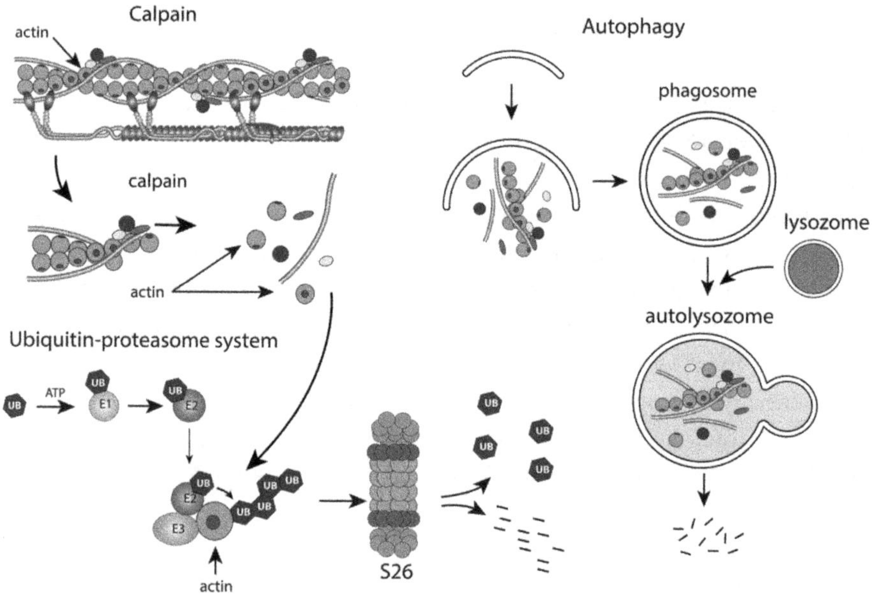

Figure 5.5: Protein degradation pathways. About half of cardiac myofibrillar proteins are removed and replaced each week. Degradation involves three distinct systems. Calpain (upper left) is an enzyme which removes sections of myofibrils from the sarcomere and breaks them into their separate constituent proteins. These proteins are then accessible to further degradation by the Ubiquitin–Proteasome System (UPS) and autophagy. The UPS (lower left panel) targets and degrades specific proteins. Ubiquitin (UB) is a tagging protein which designates damaged or misfolded proteins for degradation. Initially it is bound to the enzyme E1 which acts by transferring it to a second enzyme E2. The E2-UB complex then binds to a third enzyme E3. There are over 800 known E3 ligases which target specific protein substrates (in this example actin). The E2-E3 complex then transfers UB from E2 to the target protein, which makes it accessible to the proteasome (S26). Proteasomes are large multi-protein barrel-shaped machines which specialise in degrading ubiquitinated proteins into 6–8 amino acid peptides, which are then broken into individual amino acids and recycled for new protein synthesis. The UPS is highly specific in targeting particular proteins once they become damaged or misfolded and in this way acts as a quality control

degradation of a wide range of myofibrillar proteins. They do this, not by degrading individual proteins, but by disassembling protein complexes from sarcomeres which makes them available to the UPS for tagging and destruction [49].

Protein turnover and muscle growth

Protein turnover is a reflection of their rates of synthesis and breakdown. Heart muscle grows when the rate of synthesis exceeds breakdown due to either enhanced synthesis or suppressed breakdown. Early studies of overload-induced hypertrophy indicated that LVH results from suppression of protein degradation rather than enhanced synthesis [52]. More recently, pressure overloading the heart was shown to down-regulate Fbx122, the substrate-specific component of the E3 ligase. In other words, mechanical stress induces hypertrophy due to suppression of protein degradation and this is linked to inhibition of the UPS. By removing misfolded or damaged proteins, the UPS and autophagy provide a vital quality control mechanism for preserving normal cardiac function and several experimental studies have shown that when they are inhibited in heart muscle defective proteins accumulate and impair contraction [51]. In addition, a decline in autophagy has been implicated in ageing of the heart [53], meaning that loss of quality control results in impaired function due to accumulation of defective proteins. In contrast, exercise has been shown to induce autophagy capacity and is known to preserve cardiac function in older individuals.

Can the heart regenerate?

Many treatments have emerged during the past half century that have significantly improved outcomes in patients with heart failure, mostly by

Figure 5.5: (*see figure on facing page*) mechanism. Autophagy (upper right) acts independently of the UPS although ubiquitinated proteins may also be degraded by it. Autophagy degrades single proteins and protein aggregates by enclosing them within a membrane called a phagosome which then merges with a lysosome. Lysosomes contain several protein-splitting enzymes that are capable of degrading a wide variety of proteins. Autophagy and the UPS are complementary in that if the UPS is inhibited or faced with a large amount of substrate, autophagy activity increases to degrade the excess material.

reducing the workload of the heart and by modifying cardiovascular regulatory mechanisms. In addition, better treatment and prevention of the causes of heart failure have reduced its incidence and mortality [54, 55]. However, disability and mortality remain high in heart failure and although the incidence is in decline in Western countries, it is on the rise globally. This has prompted interest in whether damaged hearts can be regenerated by encouraging some remaining cells to replicate and replace lost tissue or by inserting pluripotent cells (also called stem cells) which might have the potential to form mature cardiac myocytes.

The concept that many cell types are derived from a common ancestor was initially proposed early in the 20th century and in 1963 a stem cell (called a haemopoietic stem cell) in bone marrow was found to give rise to all cells contained in blood. This led to the use of bone marrow transplantation as a treatment for certain types of leukaemia [56]. The concept here was to eradicate the entire cell content of a patient's bone marrow including the leukaemic cells and then replace it with healthy bone marrow stem cells from a donor in the hope that it would lead to the generation of all blood cell types. The treatment is complex and requires careful management of the immune system, but the basic concept worked. This in turn led other scientists to wonder if stem cells could be used to replace damaged cells in other organs including the heart. The concept attracted a great deal of interest and research. This has generated valuable information about how stem cells function and new sources of them, however clinical trials have not demonstrated significant benefits in heart patients [57]. There are many possible reasons for this, not least the fact that, in contrast to circulating blood cells, cardiac myocytes are fixed and have complex electrical and mechanical connections to adjacent cells. They are continuously active, both electrically and mechanically. The challenge of introducing undifferentiated stem cells into the heart which would then have to mature and be inserted safely into the complex network of existing working cells is great. Various other growth-promoting factors have also been investigated, but with little success as yet.

Regeneration by proliferation of existing myocytes

Many animals have the ability to regenerate their tissues. It has been known for more than a century that newts can regenerate their limbs [58].

Zebrafish, which have been widely studied, can regenerate liver, retina, spinal cord and fins [59]. A report in 2002 that they could regenerate heart tissue within 2 months after 20% had been amputated [60] led to intense interest in whether this could be applied to heart disease in humans. In zebrafish, this process involves the rapid formation of a blood clot that seals a hole made in the heart. Cells lining the inner and outer surfaces of the heart (called endocardium and epicardium, respectively) are then activated and begin to express genes normally involved in the embryonic growth of the heart. They revert to a non-specialised cell type and divide, forming two new cells by mitosis and then grow over the inner and outer surfaces of the clot. New blood vessels then grow into the newly formed heart muscle, and by four weeks its regeneration is almost complete. The distribution of this ability among animals lends a clue as to why it is present in some and not in others. Warm-blooded animals, including humans, rats, mice, rabbits, sheep and chickens, all have this ability, but only at their embryo stage and for a few days after birth. In contrast, many cold-blooded species retain the ability into adult life [61]. So human cardiac myocytes withdraw from the cell cycle around the time of birth and can no longer divide by mitosis thereafter. As a result, damage caused by heart attacks cannot be repaired by regrowth. Instead, scar tissue forms in the area and although strong enough to maintain the integrity of the heart, it cannot participate in contraction.

Endothermic (warm-blooded) land-based animals have much greater energy needs than fish or amphibians due to the need for heat generation and countering the effects of gravity. Humans have to maintain blood pressure 5–6 times greater than fish and had to evolve four-chambered hearts to separate the systemic and pulmonary circulations to allow this. Consequently, their hearts must be capable of generating much greater force and withstanding greater internal pressures. Regeneration under these circumstances is a much more challenging prospect. A simple blood clot could not seal a hole in the heart working at such pressures. A further difficulty arises from the fact that dividing cells must first undergo a process of extensive degradation of their working contents to facilitate mitosis. This would be a major operation in human adult cardiac myocytes which are packed with contracting sarcomeres. In addition, it would mean the loss of its contribution to the contacting performance of the heart and if undertaken on any scale could compromise its ability to function normally. It seems likely therefore that the ability to regenerate the heart became a less practicable solution to injury as the heart evolved to become a more powerful pump in association with the transition to endothermy and terrestrial life. Recent research efforts to reverse this withdrawal from regeneration by cell division in cardiac myocytes in the context of repair is discussed in Chapter 6.

Polyploidy

Loss of regenerative capacity of mammalian heart cells is because several genes which regulate mitosis become epigenetically repressed during early postnatal life [62]. However, variants of the cell cycle do survive, which resemble a form of incomplete mitosis. This involves duplication of chromosomes but without division of the cell and takes two forms. Acytokinetic mitosis describes the duplication of chromosomes and division of the nucleus but without division of the cell, resulting in a binuclear cell. Endoreduplication involves duplication of chromosomes without division of the nucleus or the cell, resulting in a polyploid cell, meaning one which has a nucleus containing more than one pair of each chromosome. Polyploidy is a common feature in postnatal human heart cells and has been shown to increase in myocytes adjacent to injured tissue. Polyploidy increases the capacity for gene expression and protein synthesis and thereby facilitates the growth of individual myocytes and hypertrophy. This provides a way to increase the contractile power of the heart to match increasing demands. In contrast, cell division would first require cells to degenerate their contractile machinery and would result in at least a temporary loss of contractile power. It seems likely therefore that polyploidy may have evolved as a more effective way to compensate for injury in animals with high-pressure circulations. A link between the evolutionary transition from mitotic regeneration to polyploid-induced hypertrophy is suggested by recent research demonstrating that induction of polyploidy in zebrafish heart cells blocked their ability to divide by mitosis [63].

In humans living now, cardiac injury most often results from conditions such as coronary heart disease and hypertension and can result in serious debility. However, these illnesses tend to be present after reproductive age and therefore do not directly affect it and since evolutionary success is based on passing on successful genes to the next generation, they may not impact evolutionary progression either. Of course, some would argue against this, pointing out that individuals beyond reproduction age continue to make important contributions to following generations through child-rearing. In addition, humans appear to be unique among primates in that the reproductive life of women is far less than life expectancy [64] and these illnesses are less prevalent among females. So perhaps evolution has begun to grapple with these survival challenges; if so, it appears to have taken a preventive route rather than a response to injury approach.

Conclusion

The heart was the first organ to develop at the embryo stage of our lives and for most of us, its silence will be the sign that marks its end. During its long life, it sustains us at rest and during our most intense physical and emotional experiences. We are only occasionally aware of its working, but on reflection its long life is remarkable. There are several reasons for its durability.

Its ability to continue working for so long is a reflection of the success that comes from millions of years of evolutionary testing and conserving what works best so that the heart is extremely well suited to the tasks demanded of it. It is able to do this due to the intrinsic and external regulatory mechanisms which ensure it always works within safe limits.

When required to work harder to meet the increased needs of the body, its energy supply is also increased automatically. And it does not work alone; when increased blood flow is required by a particular organ, the circulation as a whole participates in the response by adjusting the distribution of flow to the area in need and thereby mitigates the work of the heart. When we experience a sustained period of increased physical activity in our lives, the heart can modify its size and structure to match the increased demand in a manner that keeps it within safe working limits.

This extraordinary adaptability is further supported by a continuous system for servicing it. Research in recent years has revealed complex and precise quality control mechanisms that replace its working proteins regularly so that almost half the myofilament proteins which power the heart are replaced each week. This is done while the heart continues working and it is achieved in a way that recycles the amino acids of degraded proteins to synthesise new copies. And so, the heart has evolved into a highly adaptable, integrated and self-renewing pump through countless millennia of evolutionary innovation to become for us a pump for life!

References

[1] Kao AC, Van Trigt P, 3rd, Shaeffer-McCall GS, Shaw JP, Kuzil BB, Page RD, *et al.* Central and peripheral limitations to upright exercise in untrained cardiac transplant recipients. *Circulation.* 1994;89(6):2605–2615.

[2] Tucker WJ, Beaudry RI, Samuel TJ, Nelson MD, Halle M, Baggish AL, *et al.* Performance limitations in heart transplant recipients. *Exercise and Sport Sciences Reviews.* 2018;46(3):144–151.

[3] Clarke R, Sadleir P, Van Niekerk AW, Platt P. Quantification of volume loss and haemodynamic changes of Gelofusine-induced anaphylaxis during cardiopulmonary bypass. *Anaesthesia and Intensive Care.* 2011;39(3): 492–495.

[4] Sanders TM, Werner RA, Bloor CM. Visceral blood flow distribution during exercise to exhaustion in conscious dogs. *Journal of Applied Physiology.* 1976;40(6):927–931.

[5] Tschakovsky ME, Sheriff DD. Immediate exercise hyperemia: Contributions of the muscle pump vs. rapid vasodilation. *Journal of Applied Physiology (Bethesda, MD: 1985).* 2004;97(2):739–747.

[6] Schretzenmayr AJ. Über kreislaufregulatorische Vorgänge an den großen Arterien bei der Muskelarbeit. *Pflüger's Archiv für die gesamte Physiologie des Menschen und der Tiere.* 1933;232(1):743–748.

[7] Rodbard S. Vascular caliber. *Cardiology.* 1975;60(1):4–49.

[8] Furchgott RF, Zawadzki JV. The obligatory role of endothelial cells in the relaxation of arterial smooth muscle by acetylcholine. *Nature.* 1980; 288(5789):373–376.

[9] Furchgott RF. Endothelium-derived relaxing factor: Discovery, early studies, and identification as nitric oxide. *Bioscience Reports.* 1999;19(4): 235–251.

[10] Holtz J, Forstermann U, Pohl U, Giesler M, Bassenge E. Flow-dependent, endothelium-mediated dilation of epicardial coronary arteries in conscious

dogs: Effects of cyclooxygenase inhibition. *Journal of Cardiovascular Pharmacology*. 1984;6(6):1161–1169.
[11] Sinoway LI, Hendrickson C, Davidson WR, Jr., Prophet S, Zelis R. Characteristics of flow-mediated brachial artery vasodilation in human subjects. *Circulation Research*. 1989;64(1):32–42.
[12] Gorman MW, Feigl EO. Control of coronary blood flow during exercise. *Exercise and Sport Sciences Reviews*. 2012;40(1):37–42.
[13] Ellsworth ML, Ellis CG, Goldman D, Stephenson AH, Dietrich HH, Sprague RS. Erythrocytes: Oxygen sensors and modulators of vascular tone. *Physiology (Bethesda, MD)*. 2009;24:107–116.
[14] Linzbach AJ. Heart failure from the point of view of quantitative anatomy. *The American Journal of Cardiology*. 1960;5:370–382.
[15] Hangartner JR, Marley NJ, Whitehead A, Thomas AC, Davies MJ. The assessment of cardiac hypertrophy at autopsy. *Histopathology*. 1985;9(12):1295–1306.
[16] Perhonen MA, Franco F, Lane LD, Buckey JC, Blomqvist CG, Zerwekh JE, *et al*. Cardiac atrophy after bed rest and spaceflight. *Journal of Applied Physiology (Bethesda, MD: 1985)*. 2001;91(2):645–653.
[17] Hunter S, Robson SC. Adaptation of the maternal heart in pregnancy. *British Heart Journal*. 1992;68(6):540–543.
[18] de Haas S, Ghossein-Doha C, van Kuijk SM, van Drongelen J, Spaanderman ME. Physiological adaptation of maternal plasma volume during pregnancy: A systematic review and meta-analysis. *Ultrasound in Obstetrics & Gynecology: The Official Journal of the International Society of Ultrasound in Obstetrics and Gynecology*. 2017;49(2):177–187.
[19] Robson SC, Hunter S, Boys RJ, Dunlop W. Serial study of factors influencing changes in cardiac output during human pregnancy. *The American Journal of Physiology*. 1989;256(4 Pt 2):H1060–H1065.
[20] Morganroth J, Maron BJ, Henry WL, Epstein SE. Comparative left ventricular dimensions in trained athletes. *Annals of Internal Medicine*. 1975;82(4):521–524.
[21] Palatini P, Mos L, Munari L, Valle F, Del Torre M, Rossi A, *et al*. Blood pressure changes during heavy-resistance exercise. *Journal of Hypertension Supplement: Official Journal of the International Society of Hypertension*. 1989;7(6):S72–S73.

[22] Di Bello V, Talarico L, Di Muro C, Santoro G, Bertini A, Giorgi D, *et al.* Evaluation of maximal left ventricular performance in elite bicyclists. *International Journal of Sports Medicine.* 1995;16(8):498–506.

[23] Kannel WB, Gordon T, Offutt D. Left ventricular hypertrophy by electrocardiogram. Prevalence, incidence, and mortality in the Framingham study. *Annals of Internal Medicine.* 1969;71(1):89–105.

[24] Suzic Lazic J, Dekleva M, Soldatovic I, Leischik R, Suzic S, Radovanovic D, *et al.* Heart rate recovery in elite athletes: The impact of age and exercise capacity. *Clinical Physiology and Functional Imaging.* 2017;37(2):117–123.

[25] Ehsani AA, Hagberg JM, Hickson RC. Rapid changes in left ventricular dimensions and mass in response to physical conditioning and deconditioning. *The American Journal of Cardiology.* 1978;42(1):52–56.

[26] Jackson AS, Beard EF, Wier LT, Ross RM, Stuteville JE, Blair SN. Changes in aerobic power of men, ages 25–70 yr. *Medicine and Science in Sports and Exercise.* 1995;27(1):113–120.

[27] Zavorsky GS, Tomko KA, Smoliga JM. Declines in marathon performance: Sex differences in elite and recreational athletes. *PLoS One.* 2017; 12(2):e0172121.

[28] Leyk D, Ruther T, Wunderlich M, Sievert A, Essfeld D, Witzki A, *et al.* Physical performance in middle age and old age: Good news for our sedentary and aging society. *Deutsches Arzteblatt International.* 2010;107(46):809–816.

[29] Nikolaidis PT, Knechtle B. Age of peak performance in 50-km ultramarathoners — Is it older than in marathoners? *Open Access Journal of Sports Medicine.* 2018;9:37–45.

[30] Zwiers R, Zantvoord FWA, Engelaer FM, van Bodegom D, van der Ouderaa FJG, Westendorp RGJ. Mortality in former Olympic athletes: Retrospective cohort analysis. *BMJ: British Medical Journal.* 2012; 345:e7456.

[31] Maron BJ, Thompson PD. Longevity in elite athletes: The first 4-min milers. *Lancet (London, England).* 2018;392(10151):913.

[32] Maron BJ, Maron MS. Hypertrophic cardiomyopathy. *Lancet (London, England).* 2013;381(9862):242–255.

[33] Soliman EZ, Ambrosius WT, Cushman WC, Zhang ZM, Bates JT, Neyra JA, *et al.* Effect of intensive blood pressure lowering on left ventricular hypertrophy in patients with hypertension: SPRINT (Systolic Blood Pressure Intervention Trial). *Circulation.* 2017;136(5):440–450.

[34] Rajappan K, Rimoldi OE, Camici PG, Bellenger NG, Pennell DJ, Sheridan DJ. Functional changes in coronary microcirculation after valve replacement in patients with aortic stenosis. *Circulation*. 2003;107(25):3170–3175.

[35] Sheridan DJ, Kingsbury MP. Mechanisms of reduced coronary reserve in cardiac hyperttrophy. In: Sheridan DJ, editor. *Left Ventricular Hypertrophy*. London: Churchill Communications; 1998. pp. 135–143.

[36] Rakusan A. Left ventricular hypertrophy: Alterations in myocyte number, size and shape. In: Sheridan DJ, editor. *Left Ventricular Hypertrophy*. London: Churchill Communications; 1998. pp. 23–28.

[37] dos Santos DO, Blefari V, Prado FP, Silva CA, Fazan R, Jr., Salgado HC, *et al*. Reduced expression of adherens and gap junction proteins can have a fundamental role in the development of heart failure following cardiac hypertrophy in rats. *Experimental and Molecular Pathology*. 2016;100(1):167–176.

[38] Opherk D, Mall G, Zebe H, Schwarz F, Weihe E, Manthey J, *et al*. Reduction of coronary reserve: A mechanism for angina pectoris in patients with arterial hypertension and normal coronary arteries. *Circulation*. 1984;69(1):1–7.

[39] Marcus ML, Doty DB, Hiratzka LF, Wright CB, Eastham CL. Decreased coronary reserve: A mechanism for angina pectoris in patients with aortic stenosis and normal coronary arteries. *The New England Journal of Medicine*. 1982;307(22):1362–1366.

[40] Radvan J, Choudhury L, Sheridan DJ, Camici PG. Comparison of coronary vasodilator reserve in elite rowing athletes versus hypertrophic cardiomyopathy. *The American Journal of Cardiology*. 1997;80(12):1621–1623.

[41] Tomanek RJ, Palmer PJ, Peiffer GL, Schreiber KL, Eastham CL, Marcus ML. Morphometry of canine coronary arteries, arterioles, and capillaries during hypertension and left ventricular hypertrophy. *Circulation Research*. 1986;58(1):38–46.

[42] O'Gorman DJ, Thomas P, Turner MA, Sheridan DJ. Investigation of impaired coronary vasodilator reserve in the guinea pig heart with pressure induced hypertrophy. *European Heart Journal*. 1992;13(5):697–703.

[43] Rajappan K, Rimoldi OE, Dutka DP, Ariff B, Pennell DJ, Sheridan DJ, *et al*. Mechanisms of coronary microcirculatory dysfunction in patients with aortic stenosis and angiographically normal coronary arteries. *Circulation*. 2002;105(4):470–476.

[44] Nadal-Ginard B, Kajstura J, Leri A, Anversa P. Myocyte death, growth, and regeneration in cardiac hypertrophy and failure. *Circulation Research.* 2003;92(2):139–150.

[45] Foo RS, Mani K, Kitsis RN. Death begets failure in the heart. *The Journal of Clinical Investigation.* 2005;115(3):565–571.

[46] Wang GJ, Guo LY, Wang HX, Yao YS. IP3R and RyR calcium channels are involved in neonatal rat cardiac myocyte hypertrophy induced by tumor necrosis factor-alpha. *American Journal of Translational Research.* 2017;9(2):343–354.

[47] Bers DM. Altered cardiac myocyte Ca regulation in heart failure. *Physiology (Bethesda, MD).* 2006;21:380–387.

[48] Nakayama H, Chen X, Baines CP, Klevitsky R, Zhang X, Zhang H, *et al.* Ca2+- and mitochondrial-dependent cardiomyocyte necrosis as a primary mediator of heart failure. *The Journal of Clinical Investigation.* 2007; 117(9):2431–2444.

[49] Willis MS, Schisler JC, Portbury AL, Patterson C. Build it up-Tear it down: Protein quality control in the cardiac sarcomere. *Cardiovascular Research.* 2009;81(3):439–448.

[50] Michele DE, Albayya FP, Metzger JM. Thin filament protein dynamics in fully differentiated adult cardiac myocytes: Toward a model of sarcomere maintenance. *The Journal of Cell Biology.* 1999;145(7):1483–1495.

[51] Spaich S, Will RD, Just S, Spaich S, Kuhn C, Frank D, *et al.* F-box and leucine-rich repeat protein 22 is a cardiac-enriched F-box protein that regulates sarcomeric protein turnover and is essential for maintenance of contractile function in vivo. *Circulation Research.* 2012;111(12): 1504–1516.

[52] Magid NM, Borer JS, Young MS, Wallerson DC, DeMonteiro C. Suppression of protein degradation in progressive cardiac hypertrophy of chronic aortic regurgitation. *Circulation.* 1993;87(4):1249–1257.

[53] Miyamoto S. Autophagy and cardiac aging. *Cell Death and Differentiation.* 2019;26(4):653–664.

[54] Levy D, Kenchaiah S, Larson MG, Benjamin EJ, Kupka MJ, Ho KK, *et al.* Long-term trends in the incidence of and survival with heart failure. *The New England Journal of Medicine.* 2002;347(18):1397–1402.

[55] Paren P, Schaufelberger M, Bjorck L, Lappas G, Fu M, Rosengren A. Trends in prevalence from 1990 to 2007 of patients hospitalized with heart

failure in Sweden. *European Journal of Heart Failure*. 2014;16(7): 737–742.

[56] Thomas ED, Buckner CD, Clift RA, Fefer A, Johnson FL, Neiman PE, *et al*. Marrow transplantation for acute nonlymphoblastic leukemia in first remission. *The New England Journal of Medicine*. 1979;301(11):597–599.

[57] Hansson EM, Lindsay ME, Chien KR. Regeneration next: Toward heart stem cell therapeutics. *Cell Stem Cell*. 2009;5(4):364–377.

[58] Byrnes EF. Regeneration of the anterior limbs in the tadpoles of frogs. *Archiv für Entwicklungsmechanik der Organismen*. 1904;18(2):171–177.

[59] Lien CL, Harrison MR, Tuan TL, Starnes VA. Heart repair and regeneration: Recent insights from zebrafish studies. *Wound Repair and Regeneration: Official Publication of the Wound Healing Society [and] the European Tissue Repair Society*. 2012;20(5):638–646.

[60] Poss KD, Wilson LG, Keating MT. Heart regeneration in zebrafish. *Science (New York, NY)*. 2002;298(5601):2188–2190.

[61] Vivien CJ, Hudson JE, Porrello ER. Evolution, comparative biology and ontogeny of vertebrate heart regeneration. *NPJ Regenerative Medicine*. 2016;1:16012.

[62] Quaife-Ryan GA, Sim CB, Ziemann M, Kaspi A, Rafehi H, Ramialison M, *et al*. Multicellular transcriptional analysis of mammalian heart regeneration. *Circulation*. 2017;136(12):1123–1139.

[63] Gonzalez-Rosa JM, Sharpe M, Field D, Soonpaa MH, Field LJ, Burns CE, *et al*. Myocardial polyploidization creates a barrier to heart regeneration in zebrafish. *Developmental Cell*. 2018;44(4):433–446.e7.

[64] Alberts SC, Altmann J, Brockman DK, Cords M, Fedigan LM, Pusey A, *et al*. Reproductive aging patterns in primates reveal that humans are distinct. *Proceedings of the National Academy of Sciences of the United States of America*. 2013;110(33):13440–13445.

6

Broken Hearts: Repair and Recovery

The heart like any other tissue in the body is subject to disease and injury but its role in maintaining constant perfusion of other organs means that serious malfunction of the heart can rapidly put the entire body in jeopardy. However, in many instances, it is also capable of recovering, either partially or fully. To discuss this, I will use two examples. The first, "broken heart syndrome", is an unusual illness in which the pump function of the heart is greatly diminished following an episode of severe emotional or physical stress, but nearly always recovers fully. The second, myocardial infarction (heart attack), which is caused by a blocked coronary artery, results in permanent damage to part of the heart. In most of these cases, it is capable of healing sufficiently well to sustain a normal life.

Broken Heart Syndrome: A Heart Metaphor with Cardiac Consequences

In Chapter 1, the use and origin of heart metaphors to describe emotional experiences was discussed and while these often involve cardiac physiological responses to emotion, they originate in the brain rather than the heart. However, it would be quite wrong to suggest that the experience of a broken heart is itself "just a metaphor". For most of us, the term broken heart brings to mind the sense of emotional stress most often related to the loss of a loved one. In literature and art, it is encountered as the pain associated with a failed romance or the loss of a partner following a long relationship as a result of death or separation. These frequently evoke stressful and painful emotional experiences that are described using heart metaphors. Although these experiences occur in the brain rather than the heart, there are several ways in which powerful emotional stress can directly and indirectly be harmful to the heart.

Takotsubo cardiomyopathy also known as "broken heart syndrome" is characterised by an acute onset of transient left ventricular weakness (also referred to as stunning) that is triggered by severe emotional or physical stress. In Chapter 1, I pointed out that in such patients, other forms of heart disease are also frequently present and elucidating the specific contribution of emotional stress can be difficult. It is also the case that degrees of emotional stress are common experiences as part of normal life and the overwhelming majority of these are physically harmless. Nevertheless, severe emotional stress is known to trigger sudden left ventricular dysfunction in a small percentage of people. In reports of such patients, emotional stresses cited include the death of someone close, car accidents, acute onset of fear (facing a court appearance or a medical procedure) and armed robbery [1]. To avoid confusion with other forms of heart disease, particularly coronary artery disease (CAD), a set of diagnostic criteria have been agreed which includes (i) the presence of a transient abnormality in left ventricular wall motion and (ii) the absence of obstructive CAD that could cause such an abnormality [2]. The left ventricle becomes dilated and thin walled at the apex but remains relatively normal at the base (Figure 6.1). The apical region contracts poorly and may even balloon outwards during systole. The name *takotsubo* is derived from the Japanese name for octopus traps because the altered shape of the left ventricle was thought to resemble them. In one reported series of 19 patients seeking medical help due to cardiac symptoms following acute emotional stress, clinical features included chest pain, hypotension and heart failure. The average age was 63, most were female and pre-existing cardiovascular risk factors were present in all except three. Evidence of severe left ventricular dysfunction was present in all these patients on admission but had returned to normal by four weeks [1], indicative of severe transient cardiac injury. The fact that the heart is the most severely affected organ does not mean that the primary disturbance causing it occurs there; indeed, most evidence points to the brain for this.

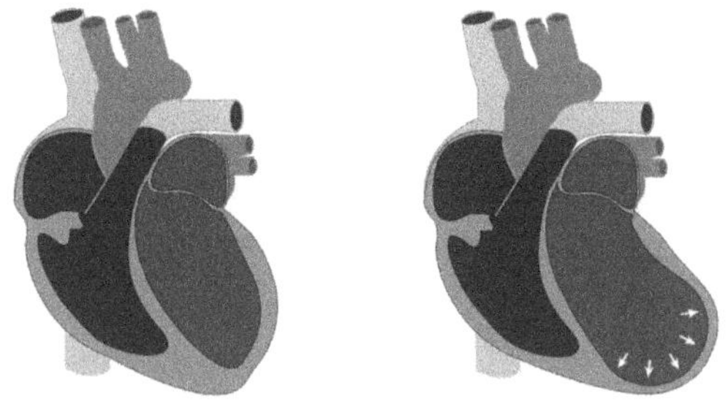

Takotsubo cardiomyopathy

Figure 6.1: Diagram to illustrate the altered appearance of the heart in takotsubo cardiomyopathy, otherwise known as broken heart syndrome. The apical region (white arrows) of the left ventricle is weakened and contracts poorly causing it to dilate. In contrast, the basal region (close to the outflow valves) is less affected or normal. In severe cases, the apex may be so weakened that it bulges during contraction giving it a balloon-like appearance. The name takotsubo syndrome derives from the Japanese word for octopus trap because the shape of the ventricle was thought to resemble those used in Japan where the syndrome was first described.

Interactions between the brain and the heart are important in regulating cardiac function and as discussed in Chapters 3 and 4, the autonomic nervous system plays a major role in this. Stress causes sympathetic stimulation of the heart, increasing the rate and force of contractions, mediated by the release of noradrenaline at nerve endings. This is part of the normal reaction to stresses of various kinds, known as the fight or flight response. However excessive sympathetic stimulation of the heart is also known to be harmful, as discussed in Chapter 3. This can occur with tumours such as phaeochromocytoma, which release large amounts of catecholamines (of which noradrenaline is one) into the circulation and with cocaine abuse, which amplifies the effect of normal sympathetic stimulation. In both conditions, damage to the heart can occur, which in some cases resembles takotsubo cardiomyopathy [3]. Patients with takotsubo cardiomyopathy have elevated plasma levels of catecholamines, higher not only compared to normal people, but also higher than in patients with heart failure due to other causes [1]. This is consistent with a marked increase in sympathetic stimulation sometimes referred to as a sympathetic storm. Interestingly, increased sympathetic activity is also associated with feelings of joy and happiness and recently, transient cardiac dysfunction (fulfilling the same diagnostic criteria used for broken heart syndrome) has also been reported in association with such experiences [4]. Possible mechanisms for takotsubo cardiomyopathy include (i) that the brains of affected individuals react differently to emotional stress or (ii) that their hearts are more sensitive to stress responses. It is also possible that (iii) the presence of other risk factors for heart disease renders the heart more vulnerable to the effects of stress. Preliminary studies of the brain using functional magnetic resonance suggest that regions of the brain involved in regulating autonomic function may be altered in such patients [4, 5] and that they react differently to stress [6]. This together with findings of higher levels of circulating catecholamines in response to stress [1] suggest that takotsubo cardiomyopathy originates in the brain and is caused by an abnormal increase in sympathetic stimulation.

Left ventricular function recovers in most cases of takotsubo cardiomyopathy. However, complications are reported in around 20%. These are more likely to occur in cases triggered by physical stresses, acute neurological conditions such as stroke or acute psychiatric illness and in patients with very poor left ventricular function at the time of diagnosis [7]. Recovery is spontaneous and gradual over weeks. Treatment is aimed at alleviating the underlying stress and supporting the heart and circulation during recovery. There is no specific treatment, rather it is as if the heart has been over-driven by a period of excessive sympathetic stimulation beyond its normal range of function to a point where it fails to contract normally. Recovery occurs gradually as the adverse stimulation declines.

Takotsubo cardiomyopathy is a very uncommon condition and in order to learn more about it, an international registry of cases was established in 2011. This had collected clinical details of 1,750 patients in 15 countries by 2016. Of these only 27% were considered to be triggered by an emotional event, so it is important to emphasise that such reactions to stress are rare and that the vast majority of people cope with them normally. These usually involve physiological increases in heart rate and force of contraction that we may feel without any adverse effects. Takotsubo cardiomyopathy serves to remind us of the close links between the heart and the brain and why the heart features so frequently in metaphors for emotional experiences. Such metaphors are used to describe normal emotional experiences. However, takotsubo cardiomyopathy (broken heart syndrome) is not just a metaphor, rather it is a rare and transiently harmful effect of emotions on the heart.

Lonely Hearts

In contrast to the acute and transient nature of broken heart syndrome, long term stresses related to chronic emotional loss can also adversely affect health. Studies in different ethnic groups in many parts of the world show that being married is associated with reduced all-cause mortality [8], including cardiovascular deaths. For example, in patients with established cardiovascular disease, divorce, separation or death of a spouse are known to reduce survival [9]. In addition, a protective effect of marital status on the risk of developing cardiovascular disease is also apparent. In a meta-analysis of 34 studies that included over two million people, being divorced, widowed or never married compared with those who were married was associated with increased risk of developing cardiovascular disease, including coronary heart disease and stroke. Furthermore, the protective effect of being married is greater in men than women [10]. Thus, being in a close stable relationship has important effects of health including cardiac health.

Several possible mechanisms have been considered to explain the effect of marriage on health. A lower prevalence of other cardiac risk factors such as smoking, high blood pressure and physical inactivity among married people could explain their reduced cardiovascular mortality. Some [8] but not all [11] studies have found that such risk factors are indeed lower among married people. However behaviours also differ between some groups of single subjects; for example, those never married appear to have healthier habits (drinking and smoking less) compared to those married and even more so compared with divorced subjects [12]. In addition to having a preventive effect on the development of cardiovascular disease, people with established cardiovascular disease who are living in a stable relationships also fare better [8]. This may be due to better social support and health care when a partner is present. For example, patients who are married are likely to seek medical help earlier when suffering heart attacks [13] and they are more likely to take medications correctly [14], which could contribute to better survival following heart attacks [15]. Although the evidence is somewhat conflicted, most researchers accept that lifestyle habits related to marital status do contribute to health outcomes, but that they only partly explain them. Loneliness, depression and mental ill-health are also known to carry increased risks of developing heart disease and while some of this can be linked to a higher prevalence of adverse risk factors such as smoking, substance abuse and physical inactivity, the direct burden of chronic stress also adversely affects cardiovascular health.

Heart Attack

Diseases of the heart and circulation are the most common cause of death in high income countries and this pattern is also emerging on a global scale. Developing nations have increasingly adopted western lifestyles instead of their traditional ones as their economies have grown and this has brought with it dramatic changes in the causes of illness and death. Globally, between 1990 and 2010 the number of deaths from cardiac and circulatory diseases increased from 11.9 million to 15.6 million (31%) and this includes a 35% increase in deaths from coronary artery disease (deaths due to blocked or narrowed coronary arteries) [16]. During the same period, deaths due to infections declined globally by 32%. In this context, HIV was an important exception, increasing by 50% during this period but falling later by 35% between 2010 and 2017 as treatments were developed and distributed [17]. Cardiovascular diseases are not only common, they are the most frequent cause of death globally, almost twice as common as cancer deaths (15.6 million compared to 7.9 million in 2010) and even coronary heart disease alone accounted for almost as many deaths (7.0 million) as all cancers.

Heart attacks are the most frequent form of cardiac damage in patients with coronary artery disease and are caused by the abrupt loss of blood supply to a region of heart muscle. Known by the medical term myocardial infarction (MI), meaning death of heart muscle, it describes the onset of a process which will lead to the destruction of a section of heart muscle followed in most but not all cases by its repair and healing. Dead tissue is also often referred to as necrotic or infarcted. Unlike broken heart syndrome which is almost always a transient impairment of cardiac function, heart attacks result in a permanent loss of heart muscle followed by a complex repair process.

Interestingly, the term heart attack itself also carries a certain emotional connotation, implying being a victim of some aggressor, which is perhaps why it features as the title of numerous songs, films and TV series. But a heart attack is a serious illness, a medical emergency that is often life threatening and requires urgent treatment. The heart receives its blood supply through three coronary arteries and obstruction of any one of them leads to reduction in blood flow to the myocardium. When this occurs, heart muscle is said to be ischaemic and if this is prolonged for more than 15–20 min cardiac myocytes begin to die. An estimated 15.9 million cases of MI occurred globally in 2015 [18]. Most of us will know a friend or relative who has suffered one and the overall burden to society is great. The costs, including health care costs and indirect costs due to lost productivity, of coronary heart disease in 2006 in the USA was $142 billion [19]. Comparative costs in the UK in 2004 were £8.47 billion and in South Korea in 2012 were $1.17 billion [20].

A 20th Century Epidemic

Coronary heart disease (CHD) could be said to be an illness of the 20th century. During the first half of the century, deaths from all cardiovascular diseases increased dramatically and MI was the single most frequent cause. The link between blockage of a coronary artery due to a blood clot (coronary thrombosis) and MI was known from the early 1900s and heart attacks were recognised to be the leading cause of death in the industrialised world by the 1950s. Understanding the risks of developing CHD began with the Framingham Heart Study (FHS) in the USA. This was set up in 1948 to investigate the health of residents in the town after which the study was named and to follow them over time to identify factors associated with the risk of developing heart disease [21]. At that time, the causes of coronary heart disease were very poorly understood. President Franklin Roosevelt had died from a cerebral haemorrhage in 1945 following a long period of illness due to hypertension and heart failure. In 1948, his vice president and successor president Harry Truman enacted the National Heart Act, which provided funding for the FHS. The identification of risk factors for the development of CHD followed, initially with hypertension, hyperlipidaemia and diabetes and later adding smoking, cholesterol and body weight. These were crucial steps in the development of prevention strategies that in time became a major contributor to the decline in CHD mortality during the last quarter of the 20th and early 21st centuries [22]. This commenced in the mid-1960s in the USA and Australia, but not until the mid-1970s in the UK, where the development of acute services for treating MI was initially opposed and prevention was slower to take off [22]. Since that time, CHD mortality has continued to decline resulting in 50% reduction in most western countries. This can be attributed to better prevention of CHD and substantial improvements in the medical treatment of acute heart attacks.

Heart Attack Symptoms

For the individual who suffers a heart attack, the experience is both dramatic and intense. Many heart attack patients experience a profound sense of fear. Chest pain, often described as a crushing or band-like sensation, is the most frequent symptom of an impending heart attack, but this can vary considerably. In many patients, it spreads to the left shoulder and arm, to the neck, jaw and back, while in others, it may occur in any of these without any chest pain. In addition, marked perspiration is a common feature. Traditionally, heart attacks were thought to be an illness affecting mainly men. However, as the incidence rose during the 20th century it also became clear that women were increasingly affected and importantly, that their symptoms may differ from those in men. Compared to men, women are less likely to experience chest pain and more frequently have the following symptoms: pain in the neck, right arm and jaw, fatigue, syncope, and nausea [23]. This variation in the location of symptoms of heart attacks relates to the indirect way in which heart pain is produced and sensed. Unlike the sense of touch and pain we feel in response to skin stimuli, which are mediated by direct stimulation of local sensors, myocardial ischaemia produces pain indirectly by the effect of chemicals released from injured cardiac cells on nearby sympathetic and parasympathetic nerve endings. Signals from these excite cells in the spinothalamic tract located in the spinal cord at the level of the neck and upper chest. Variations in the regional distribution of chemicals released from damaged cells (depending on the area of ischaemia) results in differences in the relative excitation of sympathetic and parasympathetic nerves. This causes different parts of the spinothalamic tract to be excited and this in turn results in differences in the areas of the chest and neck where pain is perceived [24]. Understanding how symptoms can vary with heart attacks was an important lesson in avoiding misdiagnosis and delayed treatment, particularly in women [25] and older patients.

Diagnosis

Myocardial infarction is a medical emergency for two main reasons. It can cause life threatening complications soon after its onset and the resulting myocardial damage is rapidly progressive. Early intervention increases the possibility of mitigating the injury and preventing or treating harmful complications. Rapid diagnosis is therefore important. In many cases, the clinical features are clear cut, however since treatments are often invasive and not without risk, it is always important to confirm the diagnosis.

The Electrocardiogram

Electrocardiogram (ECG) recording emerged as an invaluable tool for diagnosing heart attacks from the 1930s and became a mainstay from the 1940s. As discussed in Chapter 4, all cardiac muscle cells maintain an electrical gradient across the cell membrane and activation of heart muscle contraction is achieved by the spread of a wave of depolarisation through the heart. This electrical activity results in small voltage changes which can be detected on the surface of the body using electrodes attached to the skin. These are then amplified to produce the ECG recordings widely used in medicine today. The interruption of coronary blood flow during a heart attack deprives cardiac cells of the oxygen and nutrients needed to maintain their cell membrane potential and they begin to depolarise. This changes the distribution of voltage potentials in the heart and in turn the voltage signals detected on the body surface. In MI, the pattern of ECG changes that result are usually characteristic allowing rapid confirmation of the diagnosis.

Blood Tests

Injury of heart cells due to interruption of coronary blood flow also results in cardiac cell damage which causes them to release chemicals into the blood. Creatine kinase is an enzyme that catalyses the breakdown of high energy compounds to supply the energy needed to maintain cell functions and is found in many tissues, particularly skeletal muscle and to a lesser extent in heart muscle. It was known to be released into the blood stream in certain diseases of skeletal muscle and in 1962 was also found to be elevated in acute MI [26]. It served as a useful test for several years. However, it is not specific to heart muscle and is less useful in cases where coexisting skeletal muscle injuries are present. And so, more specific tests were sought. In the 1970s, an isoenzyme of creatine kinase found only in heart tissue was identified and gave more specific results. Troponin T is part of the troponin complex which interacts with tropomyosin in cardiac muscle to activate its contractile machinery (see Figure 4.4 in Chapter 4 and related text for more details) and is only found in the heart. It too is released from injured heart cells and is currently the most sensitive blood test for heart attacks. It has proved a valuable tool in rapidly triaging patients with chest pain caused by heart attacks from other causes. Once a diagnosis is made, appropriate management is aimed at limiting damage to the heart. This relates to the underlying cause of heart attacks and the injury that results. Understanding of the causes and development of heart attacks increased dramatically during the second half of the 20th century. A detailed discussion of this can be found in many textbooks of cardiovascular pathology. Here the immediate causes of coronary heart disease are briefly outlined.

Atherosclerosis

Atherosclerosis is a disease of the inner lining of blood vessels (Figure 6.2). It begins with abnormal endothelial function which allows increased infiltration of lipids (particularly cholesterol) and

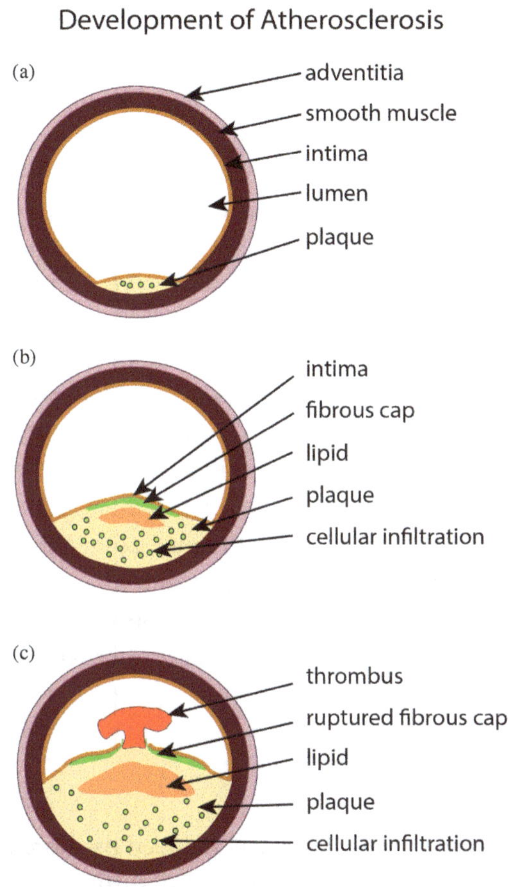

Figure 6.2: Cross section of an artery illustrating the development and progression of atherosclerosis. Atherosclerosis is a disease of the intima, the thin layer which lines the lumen of all arteries. (a) The process begins with areas of thickening of the intimal layer (called plaques) due to the accumulation of lipid containing cells derived from the blood stream. (b) Over time plaques can increase in length and thickness with greater accumulation of lipid deposits. Mature plaques have a firm fibrous cap and a covering layer of intima which is continuous with the rest of the vessel lumen and which tends to stabilise them. (c) Plaques that are rapidly progressing in size may be unstable and rupture. When this happens, the covering fibrous cap and intimal layer break, exposing the passing blood cells to the inner contents of the vessel wall which is highly thrombogenic. This can lead to the onset of clot formation and blockage of an artery. In the coronary circulation, this results in unstable coronary syndromes; if the blockage is transient, it results in unstable angina, meaning angina at rest. When the occlusion and deprivation of oxygen and nutrients is prolonged, it results in a heart attack also known as myocardial infarction meaning the death of heart muscle supplied by the artery.

macrophages (derived from white cells that normally circulate in blood) into the vessel wall. Macrophages become filled with lipids and cause inflammation within the vessel wall. As they accumulate, thickening of the inner layer (called intima) results. Over time (usually several years) this can progress resulting in progressive narrowing of the vessel lumen. Such segments of intimal thickening are known as plaques. Generally, this causes no symptoms until the vessel lumen is reduced by more than 50%. Atherosclerosis can affect any artery and narrowing can cause many illnesses, including ischaemic heart disease, stroke and intermittent claudication (a cramp like pain in leg muscle brought on by walking), depending on which vessels are affected. Understanding the nature of atherosclerosis and finding ways to prevent and slow its progression have contributed to the marked reduction in deaths from cardiovascular disease during the past half century. In particular, reducing blood levels of cholesterol through diet and statin therapy, controlling blood pressure, avoiding tobacco, remaining physically active, avoiding obesity and diabetes all help to maintain healthy endothelium and limit intimal thickening. In many cases however, atherosclerosis tends to progress with important effects on the heart.

Myocardial Ischaemia

When the lumen of a coronary artery is reduced by more than 50%, blood flow to heart muscle begins to be impaired sufficiently to cause symptoms. This usually begins with chest discomfort with exercise, known as angina pectoris or simply angina; while blood flow is still adequate at rest, it cannot increase to meet the demand required with exertion. The myocardium affected is deprived of adequate oxygen and nutrients, a condition known as myocardial ischaemia. Patients with angina come to understand its relationship with exercise and that it is relieved by resting so that the discomfort rarely lasts longer than a few minutes.

Coronary Thrombosis

Blood has evolved in a manner that it rapidly coagulates, meaning it forms a thrombus (also known as a clot), if it leaks outside the circulation. Clot formation is part of a natural mechanism for sealing such leaks to maintain the integrity of the circulation. The process of thrombosis is a complex one involving several factors in blood and is described in greater detail in Chapter 9. In order to allow blood to flow normally through blood vessels, it must always be maintained in a liquid state within the circulation and almost all heart attacks are caused by inappropriate thrombus formation in one of the coronary arteries. The thin layer of endothelial cells which line all blood vessels plays a vital role in maintaining blood in its liquid state, by inhibiting mechanisms that trigger the onset of thrombosis. Blood naturally clots if it comes in contact with almost any other surface. The presence of a normal endothelial lining is therefore important for maintaining normal blood flow, including to the heart through the coronary arteries.

Intimal thickening associated with atherosclerosis involves the accumulation of macrophage cells beneath the intimal layer, meaning that as the vessel lumen is reduced it usually remains lined by endothelial cells. As the condition progresses however, there is a risk that the endothelial lining may rupture, exposing blood to the inner contents of the vessel wall and this can lead to a cascade of interactions between the vessel wall and blood factors resulting in thrombus formation (Figure 6.2(c)). In particular, a protein that is normally restricted to the inner lining of blood vessels (known as tissue factor) can come into contact with and bind to clotting factor VII (a protein in circulating blood, one of several factors involved in the initiation of thrombus formation). In addition, platelets, small circulating blood cells (2–3 μm in size or around 25% that of red blood cells), play an important role in thrombus formation. They are derived from bone marrow cells called megakaryocytes and are formed as sections of these cells detach and are released into the circulation. Platelets circulate in an inactive form and the presence of an intact endothelium is crucial for maintaining that state. If the endothelial cell layer is ruptured or missing, platelets are exposed to subendothelial tissue to which they can adhere and become activated. This adhesion occurs because proteins exposed on the surface of platelets (called adhesion molecules) react with and bind to specific proteins associated with the connective tissue of blood vessels. In this way, platelets become firmly attached to the vessel surface and are then rapidly (within seconds) activated. Activated platelets are highly thrombogenic. They change shape, projecting extensions of the cytoplasm which further increase their adhesiveness to the vessel wall as well as to other platelets and to proteins such as fibrinogen which consolidate clot formation. This process can progress to occlude the vessel and, if it occurs in a coronary artery, it results in MI. Thus, atherosclerosis is a disease of arteries that causes gradual narrowing of the lumen over many years. As it progresses, it can damage and rupture the inner vessel lining, causing rapid thrombus formation and partial or complete occlusion of the artery affected. Atherosclerotic plaques can be stable for years without causing any symptoms and it is not always the largest that causes intimal rupture. Plaques more likely to do so (called vulnerable plaques) tend to have a thinner intimal covering (called a cap) [27] and more lipid filled macrophages. There has long been interest in detecting and treating vulnerable plaques to protect patients with established CAD and a recent study suggests that they can be identified using near infrared spectroscopy in patients undergoing cardiac catheterisation [28].

Acute Coronary Syndromes

The natural history of atherosclerosis is that of a stable chronic disease interspersed by episodes of acute life-threatening events. Angina pectoris, a transient chest pain provoked by exertion and relieved by rest, represents the stable phase of the disease. This is caused by progressive vessel narrowing to a point when blood flow no longer meets the increased demand associated with exertion and may be present for months and years. It is the occurrence of thrombosis which rapidly changes the course of the disease from a stable condition to a medical emergency. Following the development of coronary angiography (an X-ray technique for visualising the lumen of coronary arteries by injecting radiopaque dye), it became clear that while coronary thrombosis was an important cause of vessel occlusion, such thrombi could sometimes spontaneously resolve. This is possible because blood also contains anti-thrombotic systems to prevent minor clots from progressing and sometimes may be sufficient to dissolve small clots. When this occurs, the extent of myocardial injury that results depends on how quickly it does so. Therefore, in some cases, thrombi may occur for relatively short periods such that no permanent heart muscle damage results. In other words, it is the duration of myocardial ischaemia which determines whether permanent myocardial injury results.

Chest pain is the most frequent symptom of myocardial ischaemia; however, its prognostic significance varies depending on how it is caused. In order to distinguish between these different clinical manifestations of coronary disease, the concept of unstable coronary syndromes was introduced to indicate those needing emergency care. These include thrombotic coronary occlusions resulting in heart attacks at one extreme and transient self-resolving occlusions causing transient chest pain at the other. The latter were recognised to be important because they indicate an increased risk that more prolonged episodes may follow. Distinguishing between stable angina and acute coronary syndromes is therefore important for the correct management of patients. A detailed discussion of the diagnosis and treatment of ischaemic heart disease is beyond the scope of this chapter and can be found in numerous medical texts. Briefly, diagnosis rests on carefully recording the patient's history of chest pain (particularly whether it occurs on exertion or at rest, its duration and severity), ECG recordings and blood tests such as troponin T levels. Treatments include medication to relieve chest pain, to dissolve thrombus causing coronary occlusion and interventions to either bypass narrowed artery segments or insert scaffolds (called stents) to maintain vessel patency. In patients with acute coronary syndromes, the primary goal is to limit the extent of myocardial loss due to the death of cells.

Myocardial Infarction

Myocardial infarction occurs over a period of several hours following the onset of coronary occlusion. It is irreversible meaning that such cells cannot recover. The extent of myocardium lost varies depending on the size of the coronary vessel involved and the site of the occlusion. Humans have three coronary arteries. Together they supply the entire heart, but the size of each varies and therefore the volume of heart muscle each supplies also differs. Consequently, occlusion of a larger coronary artery results in greater loss of myocardium. In addition, coronary arteries give off side branches to supply the heart so the site of an occlusion also matters;

greater damage results when occlusions are close to the origin of the coronary artery involved. In practice, the loss of myocardium can vary from as little as 5% to 30% of left ventricular volume. Most patients recover well from heart attacks and management of such patients has improved greatly in recent decades. Prevention of progression of the underlying coronary disease and revascularisation of existing disease with stents or bypass surgery have increased survival.

Complications of MI

In a minority of patients, MI has serious adverse consequences (Figure 6.3) and not surprisingly, infarct size is a major determinant of prognosis including the risk of death and heart failure [29].

Sudden Cardiac Death

Sudden cardiac death (SCD) is usually defined as an unexpected death from cardiac causes within 1 h of the onset of symptoms. It is caused by coronary artery disease (CAD) in around 75% of cases and SCD accounts for about 50% of all deaths due to CAD [30]. Sudden death during MI occurs most frequently during the early stages. About 10% of heart attack patients are victims of SCD and many cases occur out of hospital with no opportunity to determine the immediate cause. However, in cases which have been observed, cardiac arrest meaning the absence of heart pump function was almost always present. This can be the result of asystole (when the heart simply stops beating) or ventricular fibrillation (VF), a profound disturbance in the heart rhythm resulting in rapid disorganised and uncoordinated contractions that are incapable of any meaningful pump function (Figure 6.3(a)). Asystole is most often the result of severe pump failure and cardiogenic shock due to extensive myocardial damage and is almost always fatal. The risk of ventricular fibrillation is not related to infarct size and if observed early can often be corrected by DC cardioversion (discussed below) with restoration of sinus (normal) rhythm. The occurrence of VF adversely affects long term prognosis; nevertheless, many such patients do survive and recover, depending on how quickly treatment was administered and the extent of infarction. Survival at

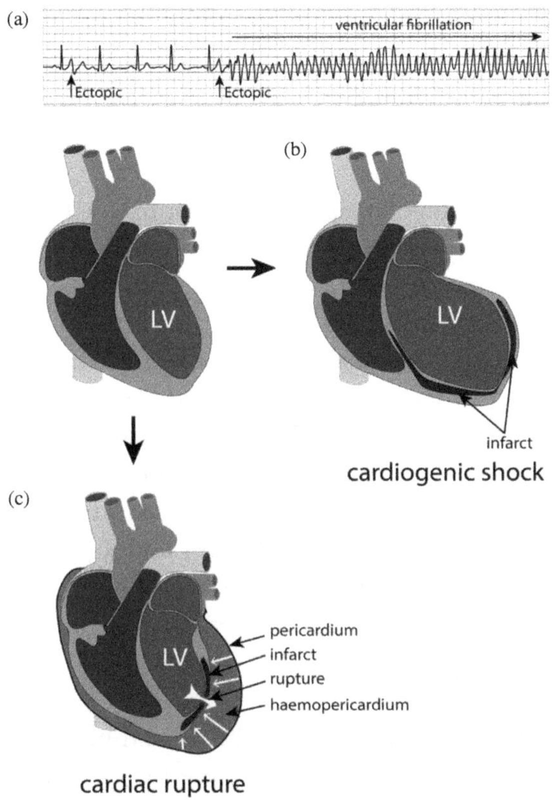

Figure 6.3: Complications of myocardial infarction. (a) An ECG rhythm strip showing sinus rhythm on the left leading to the onset of ventricular fibrillation, which is a form of cardiac arrest. The first sinus beat is followed by a single ventricular ectopic beat. The fifth beat is also followed by a similar ectopic beat which then triggers the onset of ventricular fibrillation. This is a rapid and highly disorganised rhythm in which the heart quivers but with no effective pump action. (b) Cardiogenic shock describes a syndrome in which cardiac output and blood pressure are severely reduced. Typically, patients are sweating with a rapid pulse as well as other compensatory circulatory responses. In cases due to myocardial infarction, the primary cause is marked reduction in cardiac performance due to severe left ventricular damage. In such cases, the left ventricle is extensively infarcted and contracts very poorly. The left ventricular chamber also becomes markedly dilated. (c) Left ventricular rupture and cardiac tamponade. An extensive full thickness myocardial infarction can result in rupture of the left ventricular free wall. The heart is contained in a tough membranous sac called the pericardium and with rupture blood escapes into this sac, called a haemopericardium. Human pericardium is non-elastic so accumulating blood within the pericardial sac restricts and compresses the heart (white arrows). This prevents adequate filling of the ventricles causing a reduction in cardiac output, a condition known as cardiac tamponade.

30 days in heart attack patients is around 75% in those who had VF compared with 90% in those who did not [31].

Cardiogenic Shock

An estimated one billion or a quarter of all cardiac cells can be lost during a large MI, resulting in a permanent loss of contractile function. Such a large loss of muscle can weaken the heart such that it cannot maintain normal blood pressure and organ perfusion, a condition called cardiogenic shock (Figure 6.3(b)). The incidence of cardiogenic shock in heart attack patients has fallen in recent years. One series reported a decrease from 10% to 6% between 2010 and 2017, due to improved rescue interventions such as thrombolysis (clot dissolving) and immediate revascularisation [32]. However cardiogenic shock in such patients still carries a poor prognosis with a 40% mortality within 30 days and almost all who survive beyond that need long-term treatment for heart failure.

Cardiac Rupture

Rupture of the left ventricular wall is the second most frequent cause of early death in heart attack patients. Rupture may occur through the free wall or through the interventricular septum; Figure 6.3(c) shows an illustration of the former. Both are caused by extensive infarction extending through the full thickness of the ventricular wall, weakening it to such a degree that it cannot withstand left ventricular pressure. Free wall rupture is usually and rapidly fatal because blood escaping from the ventricle enters the pericardial sac, a tough membranous lining which contains the heart, and consequently compresses the heart. This restricts filling of the left ventricle (a condition known as cardiac tamponade) causing a dramatic reduction in cardiac output, hypotension and collapse. The incidence of rupture has also decreased during recent decades. In one large study of MI patients observed between 1977 and 2006, cardiac rupture occurred in 425 of 6,678 patients (6.4%). However, the incidence was lower (4.8%) in patients treated between 2001 and 2006. Again, this is attributed to improved early rescue treatments such as thrombolysis and revascularisation with coronary angioplasty (dilatation of a narrowed artery by inflating a balloon within it) [33]. Cardiac rupture carries a high mortality, but this too has declined over several decades from 95% in the 1970s to 75% in the early 2000s [33]. This is due to the introduction of coronary revascularisation and drugs such as β-blockers [34] which reduce sympathetic stimulation and energy needs of the heart [33].

Evolving Infarction

The onset of myocardial ischaemia occurs rapidly following coronary occlusion due to the high energy needs of contracting muscle. Initially, pathophysiological disturbances (functional abnormalities) occur and, if the occlusion persists, cells begin to die. A third phase then follows in which the body reacts and initiates a healing response.

Pathophysiology

Interruption of coronary blood flow to heart muscle leads to myocardial ischaemia and rapid disturbances in cardiac function.

(i) ***Pump dysfunction*:** Pump function declines to a degree depending on the volume of muscle affected. In most patients, the heart continues to maintain an adequate cardiac output. However, in a minority of cases, it can be severe enough to cause an abrupt fall in blood pressure, resulting in syncope (loss of consciousness). In the most severe cases, the loss can lead to heart failure or cardiogenic shock. Management of pump function plays an important part of the care of such patients, by avoiding unnecessary cardiac work (by resting in the early stages and preventing hypertension) and supporting the circulation in those with severely weakened hearts.

(ii) ***Electrophysiological dysfunction*:** Myocardial ischaemia also has important effects on the electrical properties of heart muscle. As discussed in Chapter 4, maintaining normal cardiac electrophysiology is crucial for regulating the heart rhythm. Ventricular fibrillation (VF) is an important cause of sudden death in the early stages of MI.

If detected, VF can be corrected by defibrillation. This involves passing an electrical current through the heart (known as DC cardioversion) which depolarises and resets all the previously unco-ordinated activity allowing the normal pacemaker to resume control. The risk of VF is highest in the early stages of heart attacks, most occurring within one hour of symptoms and as a result many cases occur in the community. Reported prodromal symptoms (within the preceding hour) can be minor and about 20% may have none [35]. VF is of particular clinical interest because if detected and treated promptly, some patients can recover and survive many years.

(iii) ***Ventricular fibrillation*:** Disturbances in cardiac electrophysiology that can lead to VF have been the subject of intense interest during the past century and this became possible with the development of methods to study the electrical properties of individual cells. Early work on cellular electrophysiology was done using squid axons (single nerve fibres) because their large size facilitated inserting electrodes and led to major new discoveries for which Alan Hodgkin, Andrew Huxley and John Eccles were awarded the Nobel Prize in 1963. The development of the glass microelectrode (made by pulling a heated pipette to a fine point) in 1946 [36] allowed much smaller cells, including cardiac and skeletal muscle cells to be studied. It was well known from the mid-1940s that ECG recordings were markedly altered by MI, suggesting abnormalities in the behaviour of the affected cardiac myocytes. This was confirmed in experimental models of myocardial ischaemia in which action potentials were recorded directly from affected cells. Within minutes of the onset of myocardial ischaemia, cells became depolarised and action potentials were reduced in amplitude and duration. In addition, the speed of cell-to-cell conduction was reduced [37]. All these changes preceded the onset of VF.

Normal cardiac electrophysiological function is dependent on a wide range of cellular processes which require a constant supply of oxygen and nutrients. In particular, maintaining the correct intracellular and extracellular concentrations of ions is crucial. These in turn are regulated by several ionic pumps and voltage sensitive ion channels that are energy dependent (discussed in Chapter 4). Interruption of the energy supply to cardiac muscle cells causes rapid failure of these processes. Early evidence of this came from experiments which demonstrated a rise in potassium ion (K^+) concentration in blood draining from ischaemic myocardium, indicating that ischaemic cells were leaking K^+ [38]. In addition, ischaemic cells can no longer maintain resting membrane potential resulting in depolarisation and inability to produce normal action potentials. These abnormalities occur within minutes of the onset of myocardial ischaemia [37] and are progressive, often followed by rhythm disturbances including VF.

Loss of cell membrane integrity also leads to leakage of chemicals such as the diagnostic enzymes discussed above and others such as noradrenaline which may be harmful. Sympathetic nerve endings which normally regulate cardiac function contain a store of noradrenaline, some of which is released in response to signals from the brain to stimulate the heart (discussed in Chapters 3 and 4). However, myocardial ischaemia also causes inappropriate release of noradrenaline from nerve endings in the heart due to disruption of its storage mechanism, resulting in harmful myocardial stimulation [39]. Together, these changes result in abnormalities in conduction and refractoriness that are arrhythmogenic.

Thus, VF is major complication of MI and the most important cause of sudden death in heart attack patients, often occurring early after the onset. It is clear from observation of VF in hospitalised patients that it can be self-terminating in some cases. Short episodes lasting more than a few seconds are likely to produce syncope due to a rapid fall in blood pressure. VF that lasts longer than about 3 min starts to cause serious brain injury. Hence VF is a medical emergency requiring immediate DC cardioversion. In recognition of the fact that many cases of VF occur outside hospital wards, defibrillators are now frequently placed in locations where large numbers of people gather such as sports stadia, entertainment venues and transport hubs. In addition, patients with established coronary disease who are thought to be at high risk of developing VF receive implantable devices similar to a pacemaker but which can also detect and defibrillate VF if necessary.

Myocardial Response to Injury

The ultimate result of coronary occlusion is necrosis (death) of myocardial tissue. Most patients who have heart attacks survive and many go on to lead long healthy lives. This implies that heart muscle must be capable of repairing the injury at least to some degree if not completely. How does this occur? Regeneration of lost cardiac myocytes would be an ideal solution. As discussed in Chapter 5, many cold-blooded species can do this [40]. Several warm-blooded animals such as humans, rats, mice, rabbits, sheep and chickens can also do so, but only at the embryonic stage of their lives and for a few days after birth. This may have occurred as part of an evolutionary compromise in which cardiac myocytes lost their ability to proliferate in exchange for increased specialisation to meet the greater cardiac workload required by terrestrial warm-blooded animals. As a result, humans cannot repair myocardial necrosis by regrowth. Instead, a process of inflammation is initiated leading to scar tissue formation.

Inflammation

Few of us will not recall having experienced a minor injury, infection or irritation of the skin resulting in a warm, painful red swelling. Acne spots (infected hair follicles) are common examples of localised inflammatory responses, familiar to many teenagers. The first clinical description of inflammation is attributed to Aulus Cornelius Celsus (25 BCE–50 CE). Celsus was a prolific author whose works included De Medicina, an eight-volume medical treatise which gives a detailed account of fevers and records the four classic signs (often referred to as cardinal signs) of inflammation, calor (warmth), dolor (pain), tumour (swelling) and rubor (redness). A fifth sign, loss of function, was later added which by custom has been attributed to Galen (129–200 CE), but without reliable evidence [41]. These signs are caused by tissue reactions that are characteristic of inflammation; warmth and redness are caused by increased blood flow (blood temperature is normally warmer than skin) to the area. Swelling is due to the accumulation of fluid and blood cells in the affected area. Pain results from the release of substances such as bradykinin and histamine from damaged tissues which stimulate local pain-sensitive nerve endings (called nociceptors) sending pain signals to the brain and also by modifying the sensitivity of nociceptors [42]. Increased blood flow and increased extravasation of fluid and blood cells are key features of an inflammatory response, and these are triggered by a highly sophisticated and organised series of events initiated by the release of chemicals from injured cells. Inflammation has been known for centuries to occur in response to infection and sometimes inflammation and infection are used as synonymous terms. However, inflammation is now known to occur with any type of tissue injury as a defence and repair mechanism and this includes MI. In most cases, inflammation is short lived; however, when the injury or infection persist, the inflammatory response can become chronic, manifesting as illness. In many chronic human diseases, persistent activation of such inflammatory processes are important contributors, for example, to osteoarthritis, cancers, Parkinson's and Alzheimer's disease.

Inflammation in Myocardial Infarction

Myocardial infarction is also followed by an inflammatory response that is a vital part of the healing process. This takes place in two phases: an initial intense acute inflammatory response digests and removes severely damaged or dead cells, followed by a second phase which mitigates the initial inflammation and initiates repair and scar formation (Figure 6.4). The fundamental processes which initiate and regulate inflammatory responses are similar in all forms of inflammation, differing only in their time course and tissue distribution. This also applies to MI; all the characteristic features of inflammation are present with the exception that the response is initiated at the edge of the ischaemic area where blood flow is still intact. The overall response to MI is initiated and regulated by several molecules and cell types in a highly orchestrated manner. Both phases involve inflammatory and immune responses that are closely interdependent. The myocardium contains many cell types and most if not all of these are affected by or involved in the inflammatory response. In the following sections, those for which a role is well established will be described.

Prolonged loss of blood flow to myocardium results in the death of cardiac myocytes and other cells due to necrosis and apoptosis. The former

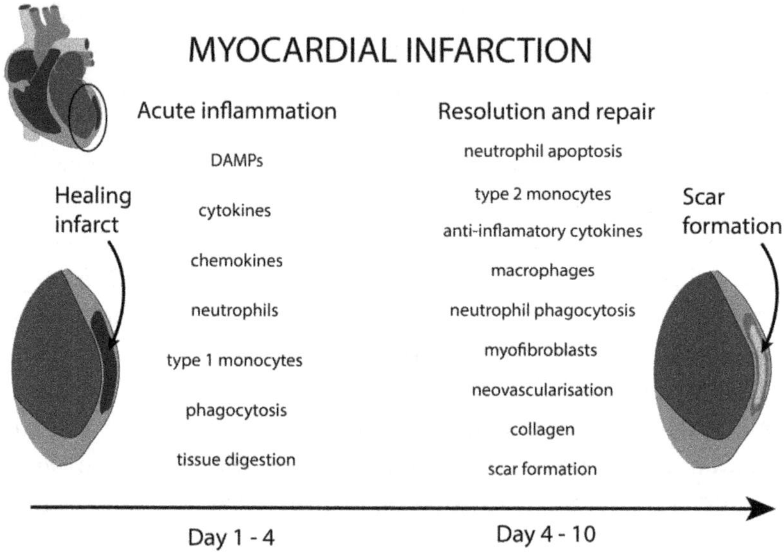

Figure 6.4: Healing response to myocardial infarction (MI). Day 1–4: An acute inflammatory reaction occurs rapidly following MI, initiated by the release of danger associated molecules (DAMPs) from injured cells. This triggers the production of chemokines and cytokines, which attract circulating neutrophils and pro-inflammatory type 1 monocytes to infiltrate the infarct zone. Phagocytosis and tissue digestion follows, clearing away damaged cells and material. Day 4–10: The initial inflammation is actively suppressed from about Day 4. Neutrophils undergo apoptosis and are phagocytosed by type 2 macrophages. This also results in the production of anti-inflammatory cytokines. Fibroblast cells from surviving border areas migrate into the infarct zone, proliferate and differentiate into myofibroblasts leading to the production of matrix proteins and collagen, which later matures to form a scar.

describes death from injury or lack of oxygen and nutrients; apoptosis on the other hand is a highly regulated process by which internal mechanisms of threatened cells initiate the killing process. It differs from necrosis in that the cell contents of apoptotic cells are engulfed by scavenging phagocytic cells and thereby most of their contents, which could otherwise be harmful to neighbouring cells, are safely removed and made available for recycling. Apoptosis is a normal process for removal of cells throughout our lives. It is widely used during embryonic development in shaping tissue and organs, as, for example, in forming the ostium primum during atrial septation (discussed in Chapter 2). One estimate suggests that as many as 0.5% of an adult's cells are lost each day in this way. During myocardial ischaemia, damaged and threatened cells breakdown and release a host of molecules, many of which play a key role in triggering the inflammatory response to MI. Collectively these are known as a danger associated molecular pattern (DAMP).

PAMPs and DAMPs

DAMPs form part of our immune system called innate immunity. Innate immunity is a ready-to-go mechanism for responding to diverse adverse challenges such as infection or tissue injury. Immunity is characterised as innate when it is available prior to attack and differs in this respect from adaptive immunity which works by developing specific protective responses, such as antibodies, to an invading organism following initial exposure to it. Innate immunity was originally thought of as a system activated in response to invading pathogens such as bacteria and viruses by recognising and responding to a wide range of molecules associated with them, known as pathogen-associated molecule patterns (PAMPs). PAMPs bind to cognate receptors of the immune system resulting in the initiation of an inflammatory response. From around the mid-1990s, the view of immunity as a defence against invading organisms was broadened to one which considers it a protective regime against danger more generally and this led to the concept of DAMPs which act in a similar way to PAMPs. Many molecules released from injured cells including cardiac myocytes are now known to be such DAMPs and play a key role in initiating the inflammatory response.

DAMPs and their corresponding cognate receptors.

DAMP	Receptor
HMGB1	TLR2, TLR4, TLR9, RAGE
S100A1	Endolysosomal TLR4
S100A8/A9	TLR4, NLRP3, RAGE
Fibronectin-EDA	TLR2, TLR4
Interleukin-1α	IL-1 receptor
HSP-60, HSP-70	TLR2, TLR4, TLR6

(*Continued*)

(*Continued*)

DAMP	Receptor
LMW hyaluronic acid	TLR2, TLR4, NLRP3
ATP	P2X7/NLP3
Uric acid	TLR2, TLR4, NLRP3
Mitochondrial DNA	TLR9, NLRP3
Double-stranded RNA	TLR3
Single-stranded RNA	TLR7, TLR8

Just like PAMPs, DAMPs also bind to cognate receptors known as pattern recognition receptors (PRRs) (the term cognate is used to indicate that each DAMP has specific receptor targets). The table shows a list of DAMPs and their PRRs involved in MI and further details about individual DAMPs and PRRs are provided in Appendices I and II. PRRs are located on immune cells, surviving myocardial cells and blood cells which have infiltrated the infarct zone. DAMP-PRR binding results in the production of powerful inflammatory mediators such as cytokines (small signalling proteins that modulate inflammation), chemokines (a group of cytokines that attract certain cells) and cell adhesion molecules (molecules that are produced on cell membranes to which other cells adhere) resulting in an intense inflammatory response. Myocardial infarction results in elaboration of numerous DAMPs, the main PRR targets for which are TLRs, RAGE, NLRP3, P2X7 and IL-1R. Together, these receptors activate a wide range of intracellular signalling pathways [43–46] resulting in synthesis of several pro-inflammatory mediators.

Pro-inflammatory Phase

Dying cardiac cells are the main source of DAMPs which stimulate the inflammatory response during MI. Surviving cells in the area bordering the MI may also contribute to it as a result of stimulation of TLRs by DAMPs released centrally. The immune system which produces the inflammatory response is composed of many different cell types. Some are local to the heart while others are present in circulating blood and must be attracted to and infiltrate the infarct tissue. For this to occur, circulating cells are drawn

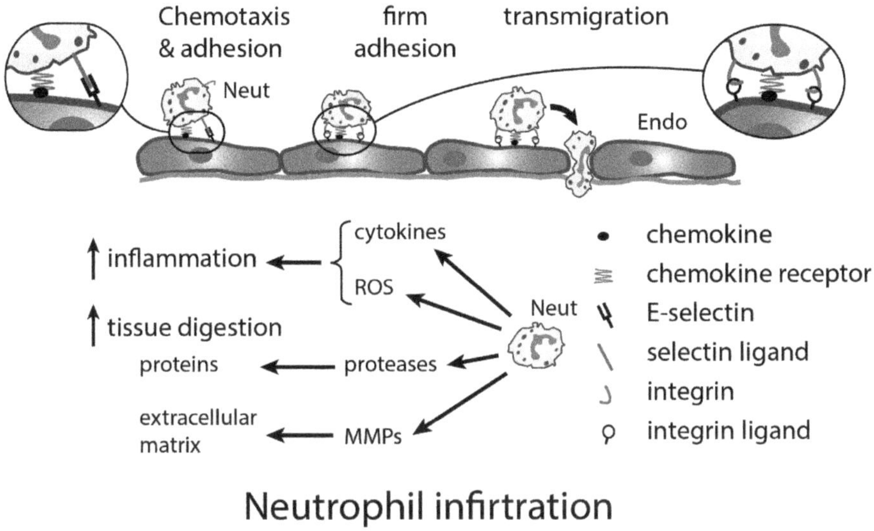

Figure 6.5: Neutrophil infiltration is key to the early inflammatory response following MI and peaks around Day 1. Danger associated molecules released by injured cells stimulate endothelial cells to produce chemokines and express E-selectin on the cell surface, attracting neutrophils, which bind to E-selectin. Chemokines also activate integrins expressed on neutrophils, which then bind more firmly to endothelial cells and transmigrate through adjacent cells to infiltrate the infarct. Neutrophils enhance the inflammatory response by producing more cytokines and ROS (reactive oxygen species). They also secrete proteases and MMPs (Matrix metalloproteinases), which degrade cellular and extracellular proteins. Neut = Neutrophil; Endo = Endothelial cell.

to the injured area and stimulated to penetrate into it by a set of signalling and adhesion molecules. These processes are illustrated in Figure 6.5.

Cytokines and Chemokines

Cytokines are molecules that regulate inflammation. A great many are secreted early during MI, including interleukins (IL1, IL6 and IL18) and tumour necrosis factor (TNF), and play a key role in regulating the inflammatory response that occurs by (i) initially amplifying the response and later by suppressing it, (ii) spreading the inflammatory activity to adjacent cells and (iii) stimulating production of chemokines and expression of adhesion molecules on endothelial cells, which attract leukocytes (circulating white blood cells) and facilitate their penetration into the infarcted myocardium. Chemokines are chemotactic cytokines, meaning they are able to attract certain cells towards them. Chemotaxis is widely present in nature, for example, in assisting bacteria to move towards food. Chemokines elaborated in response to MI adhere to the outer layer of endothelial cells, which attracts leukocytes to where their concentration is highest and thereby assist their recruitment to the infarct zone. At least four types of chemokines are known, two of which are referred to by the pre-fix CC and CXC. CXC chemokines have two cysteine amino acids separated by a single other amino acid and these attract neutrophils. CC chemokines have two adjacent cysteine amino acids near one end which preferentially attract monocytes. Both have important roles in MI by promoting and resolving the inflammatory process.

Endothelial Cells

Endothelial cells are the second most numerous cell types in the heart. They form the lining of all blood vessels including the tiny capillaries which ramify the heart. Their normal function is to maintain a stable blood flow and facilitate movement of nutrients and waste products to and from tissues. When activated however they play a crucial role in inflammation. DAMPs released from injured cardiac myocytes induce chemokine production by endothelial cells early in MI, which bind to aminoglycans contained in the outer coat of these cells. DAMPs also cause rapid expression of the adhesion molecule, E-selectin, on the luminal surface of endothelial cells. Chemokines and adhesion molecules act collectively, the former attract leukocytes to the surface of blood vessels in the injured tissue while the latter bind them. Initially, leukocytes are loosely attached to endothelial cells and roll along the surface. Chemokines also activate the production of integrins on the surface of leukocytes which then bind more firmly to endothelial cell adhesion molecules like vascular cell adhesion molecule-1 (VCAM-1), intercellular adhesion molecules (ICAM-1 and ICAM-2) as well as junctional adhesion molecules (JAMs). JAMs normally provide a tight junction between adjacent endothelial cells to maintain the lumen integrity of vessels; however, leukocytes are also able to bind them and by doing so navigate through the endothelial barrier to infiltrate the injured tissue.

Neutrophils

Neutrophils are the most abundant leukocytes (white blood cells) in blood and are the first line of defence against attack by invading organisms and injury. They release cytokines, reactive oxygen species (ROS), which are potent inflammatory molecules and enzymes that degrade damaged cells and material. Neutrophils are also the first immune cells to infiltrate infarcted myocardium and do so within 24 hours (Figure 6.5). This is enabled by CXC-chemokines and cytokines produced by endothelial cells in response to the actions of released DAMPs, predominantly in the border of infarct zones. Neutrophils can extend the inflammatory response to adjacent areas by the release of cytokines. Their proteolytic enzymes help degrade and digest damaged cells and extracellular proteins as a first step in its clearance. Neutrophils are short lived initial inflammatory responder cells. Their function is to digest and clear away damaged tissue by releasing potent chemicals which would also be harmful to healthy tissue. It is important therefore that their activity is targeted at and confined to injured cells. Therefore, prolonged neutrophil infiltration post MI may contribute to extending the region of injury, for example, by release of ROS, which are toxic to native cells as well as to invading organisms. So, the timing and duration of neutrophil infiltration is important for optimal infarct repair. Indeed, neutrophil accumulation in infarct tissue is actively suppressed following the initial period of intense infiltration.

Monocytes

Monocytes are the largest leukocytes in blood and have important roles in innate and adaptive immunity. They have three main functions. (i) They can extend their cell membrane to engulf external particles, to internalise them as captured vesicles, which they then degrade and digest, a process called phagocytosis. (ii) They can also produce cytokines in response to microbial products and (iii) they have a role in antibody production. These functions are undertaken by subpopulations (also called phenotypes) of monocytes which can be identified by the presence of specific proteins present on their cell surface. This is important in relation to MI because each plays a particular role at different stages in evolving MI. Under normal conditions, monocytes have a monitoring function that involves patrolling the circulation and responding to local endothelial cell activation. Like other leukocytes, they are produced in bone marrow but are also stored in the spleen which acts as a reservoir. Splenic monocytes are mobilised soon (during Day 1) after MI resulting in increased numbers in circulating blood for around seven days, during which they accumulate in the infarct zone. They are attracted by chemokines attached to glycosaminoglycans on endothelial cell membranes. They then infiltrate the infarct zone in a manner similar to neutrophil chemotaxis.

Although the total number of circulating monocytes and those infiltrating the infarct remain fairly stable during this period, recent experiments have shown that monocyte recruitment to infarcted myocardium occurs in two sequential waves and involves two subsets of these cells which undertake different functions. The monocyte subsets involved in each phase are identified by the expression of specific proteins contained on the cell surface, but for this discussion are named simply type 1 and type 2. Recruitment of type 1 monocytes peaks about 3 days and declines thereafter, whereas type 2 cells are recruited later and only peak at Day 7. This biphasic recruitment is highly orchestrated:

(i) MI induced mobilisation of monocytes from the spleen occurs during phase 1 but not during phase 2 and those released are predominantly type 1 monocytes. As a result, circulating type 1 monocytes are more abundant than type 2 cells during phase 1.
(ii) The profiles of chemokines synthesised by infarcted myocardium and their respective target receptors on monocytes also change over time to sequentially attract type 1 cells initially and later type 2 cells [47]. There is also evidence that some type 1 monocytes may be transformed into type 2 cells within the infarct environment [48].

Monocyte Functions

Monocytes contribute to healing MI in two stages that broadly correspond to their bi-phasic recruitment. (i) Initially injured cells and material are cleared away. Type 1 monocytes contribute to this by secreting cytokines which maintain the inflammatory response. They also produce enzymes such as proteinases which degrade damaged material and by phagocytosing injured cells and digesting them internally. (ii) Regrowth and repair follow the formation of scar tissue and new vessel growth. Type 2 monocytes contribute to this. Their ability to secrete proteinase enzymes is greatly reduced compared to type 1 cells. Instead, they produce vascular endothelial growth factor (VEGF) which stimulates the formation of new blood vessels. They are also important for producing healthy scar tissue. Thus, type 1 monocytes are pro-inflammatory, degrade and remove damaged tissue, whereas type 2 monocytes promote healing by new vessel growth and collagen deposition in scar tissue.

Repair and Proliferative Phase

Suppression of the early pro-inflammatory response to MI is an active process and necessary step for the repair process to begin (Figure 6.6). Several cells and molecules participate in this process. Much of the evidence for this is derived from experimental models of MI and although timing may vary somewhat, the principles are likely to be similar in man.

(i) Neutrophils which accumulate in large numbers during the pro-inflammatory phase are short lived and undergo apoptosis from around Day 3. This has two anti-inflammatory effects. First, unlike cell necrosis, apoptosis is a form of cell death in which the cell contents are retained within the cell membrane avoiding release of pro-inflammatory material. Second, apoptotic neutrophils release anti-inflammatory molecules such as transforming growth factor beta1 (TGFβ1), lactoferrin and annexin A1.
(ii) During the pro-inflammatory phase of MI, infiltrating monocytes begin to mature into macrophages and by Day 7 are the most abundant leukocytes in infarcted tissue. This differentiation is stimulated by the cytokine, macrophage-colony stimulating factor (MCSF). Macrophages are prodigious phagocytic cells meaning they engulf and digest other cells and material, including apoptotic neutrophils. The process is assisted by the release of lactoferrin and annexin A1 by apoptotic neutrophils, which act as chemokines attracting macrophages towards them. Phagocytosing macrophages also contribute

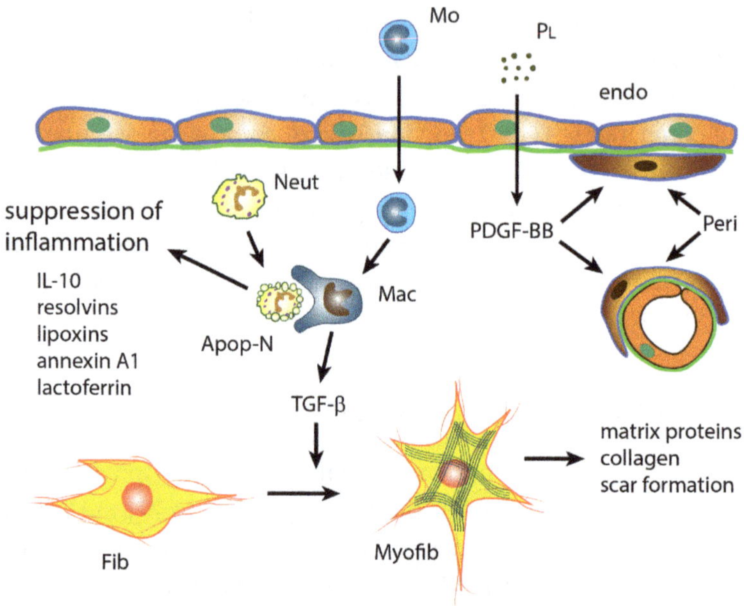

Figure 6.6: Resolution and repair following MI. After about four days of intense activity, the inflammatory response is actively suppressed. Neutrophils are short-lived cells and begin undergoing apoptosis after around three days. Type 2 monocytes infiltrate the infarct tissue and transform into macrophages, which phagocytose apoptotic neutrophils. In doing so, they release anti-inflammatory molecules such as IL-10, resolvins, lipoxins, annexin A1 and lactoferrin which suppress further neutrophil infiltration. Macrophages also release TGF-β which stimulates fibroblasts to differentiate into myofibroblasts. The latter synthesise actin and become contractile cells which contribute to wound closure. They also produce matrix proteins and collagen to replace the extracellular structure destroyed by the MI. Vascular endothelial growth factor (VEGF) is synthesised within a few hours by cells in the border regions of MI and stimulates the growth of new vessels. These tiny capillaries initially consist of endothelial cells and are highly permeable. PDGF-BB released from circulating platelets stimulates the growth of pericyte cells which provide and outer covering of capillaries and normalise their permeability. Mo = Monocyte; PL = Platelets; Endo = Endothelial cells; Neut = Neutrophil; Apop-N = Apoptotic neutrophil; Mac = Macrophage; IL-10 = Interleukin 10; TGF = Tumour growth factor; Fib = Fibroblast; Myofib = Myofibroblast; PDGF = Platelet derived growth factor; Peri = Pericyte.

to resolution of the inflammatory phase by releasing anti-inflammatory mediators such as interleukin10 (IL10) and TGF-β.

(iii) Activation of inflammation is initiated by DAMPs released from injured myocardium and removal of these assists in its resolution. This is partly achieved by binding of DAMPs to adhesion molecules expressed on neutrophils, for example, hyaluronans are removed by binding to CD44 on expressed on neutrophils.

(iv) TLRs are major components of the PRRs which bind DAMPs following injury and infection. These responses are relatively non-specific and imprecise and consequently carry a risk of over stimulation resulting in its extension and prolongation. Recent research suggests that several mechanisms have evolved which limit TLR expression, including molecules produced during inflammation.

(v) Chemokines play a key role in leukocyte infiltration of injured tissues. Consequently, suppression of chemokine synthesis contributes to resolving the inflammatory process. Cytokines, such as TGF-β and IL10, are synthesised following the initial inflammatory response and inhibit further chemokine production. In addition, some evidence suggests that non-active receptors may act as decoys by binding chemokines and neutralising their chemotactic function.

New Growth

Healing of MI requires extensive formation of new blood vessels which grow into the infarct to provide oxygen and nourishment. This takes place in response to stimulation by VEGF produced by type 2 monocytes and other cells in the border region of MI. New vessels are formed by proliferation of endothelial cells, creating tiny channels through the tissue. Initially these are highly permeable but over time they acquire a coating of additional cells, called pericytes, to form mature capillaries. This normalises their permeability and reduces extravasation and infiltration of inflammatory cells. Larger vessels acquire an additional outer coating of smooth muscle cells to take on the role of new arterioles.

Fibroblasts

The adult heart contains around 3–4 billion cardiac myocytes and an equal number of endothelial cells lining blood vessels. An additional group called mesenchymal cells, consisting of around 5–10 billion cells [49], is responsible for maintaining the structural architecture of the heart; many of these are fibroblasts. These cells synthesise the extracellular matrix (ECM), which has important roles in cell adhesion, intercellular communication and differentiation. Fibroblasts also produce collagen, a structural protein in the ECM and the most abundant protein in the body. It is formed in long fibres with variable tensile and elastic properties to suit the needs of different tissues such as tendons and expansile vessels, respectively. Rebuilding the ECM is crucial in all wound healing and fibroblasts are key to this. Recent research has shown that like myocytes, fibroblasts within an infarct zone also die. However, fibroblasts in the border area undergo rapid proliferation and migrate into the infarct area, where they continue to proliferate for 3–4 days, reaching a three to fourfold higher density than in uninjured tissue [50]. These cells then differentiate into myofibroblasts and as a result acquire two functions essential for wound healing. (i) They synthesise a form of contractile protein found in smooth muscle cells and become contractile and mobile and (ii) they produce matricellular proteins (proteins present in the extracellular space which play important roles in regulating many interactions between cells and the extracellular environment) which are needed to rebuild the extracellular space. This includes collagen synthesis which forms the basis of scar tissue. After about 10 days, they lose their motility and revert to a cell type closer to those associated with tendons. This is in keeping with the need to maintain a strong scar to replace the lost myocardial cells.

Thus, the heart's ability to repair myocardial infarction is essentially an inflammatory process involving the innate immune system. This takes place in carefully orchestrated and synchronised stages involving a rapid and intense infiltration of pro-inflammatory cells over 3–4 days which initiates digesting and clearing away damaged tissue. Infiltration by monocytes follows in two waves, the first of which continues the process of removing dead cells and actively suppressing the initial inflammatory phase, while the second initiates tissue repair and new vessel growth. Fibroblasts normally present in the border area undergo rapid proliferation and migrate into the infarct zone where they synthesise proteins and other molecules to rebuild the extracellular matrix and collagen which forms the scar needed to replace the lost myocardium and maintain integrity of the heart chamber. The intensity and timing of each of these phases are important for optimal healing. A diminished or prolonged early proinflammatory phase are both known to be harmful to subsequent repair as is failure of any of the components of later stage repair.

Human cardiac myocytes lack the ability to replicate and therefore MI repair is always less than ideal in so far as full pre-infarct function cannot be restored. Nevertheless, the ability to initiate the wound healing process in this context is remarkable. The heart continues to work throughout the healing process while it creates new tissue to replace the lost muscle and maintain the integrity of the heart structure. In addition, the remaining healthy muscle is capable of growing to replace at least some lost function.

Therapeutic Intervention in the Inflammatory Response in MI

Anti-inflammatory treatments have a long history in human disease. Aspirin was originally taken as willow bark, until salicin was extracted from it in 1829. Aspirin was later synthesised and used empirically until its mode of action was discovered in the 1970s. Corticosteroids and other non-steroidal anti-inflammatory drugs are widely used to treat conditions such as osteoarthritis, asthma, sarcoidosis, Crohn's disease and skin ailments. These tend to be chronic conditions where the underlying cause is not well understood or is not reversible and persistent inflammation becomes injurious. The inflammatory response to MI is brief and mostly beneficial. Due to its complexity, research to explore possible beneficial targeting of the inflammatory response to MI has to date proved unsuccessful.

Stem Cell Interventions in MI

Human cardiac myocytes lose the ability to proliferate a few days after birth so there is no possibility naturally to replace those lost during MI. The discovery of stem cells raised the hope that cells could be introduced with the potential to proliferate and differentiate into mature functioning replacements. To date this has not been achieved. However, attempts to do so have led to new discoveries and work in this area continues with some hopes of eventual success.

Stem cells are those which have the ability to proliferate and to differentiate into other cell types. Three types of stem cells have been explored for replenishing cardiac myocytes.

(i) Adult stem cells (also called stromal cells) are primitive cells present throughout the body that have the potential to replace damaged cells to repair injured tissues. To do this, they differentiate into the required cell type and multiply. In practice, however, stromal stem cells have a poor ability to transform into cardiac myocytes and are unable to replace lost cardiac muscle [51].
(ii) Pluripotential stem cells have the ability to transform into all three embryonic germ layers, endoderm, mesoderm and ectoderm, and therefore can differentiate into all cell types found in adults. Embryonic stem cells are pluripotent. Pluripotential cells were first isolated from mouse embryos in 1981 and from human blastocysts (formed around 5–9 days after fertilisation) in 1998. However, the use of human embryonic cells is fraught with ethical concerns and has not been widely pursued.
(iii) Inducible pluripotential cells are pluripotent cells formed by reprogramming mature cells to revert to an embryonic phenotype. Embryonic pluripotential cells transform into all cell types found in the body during development and normally this is a one-way process, i.e., skin cells cannot transform into brain cells. Research since the 1960s laid the ground work for techniques to reprogram cells to a pluripotential state, reviewed in Ref. [52]. Then in 2006, gene manipulation of mouse fibroblasts showed that they could be reprogrammed into pluripotent cells and soon after the same technique was used to form human pluripotent cells. John Gurdon and Shinya Yamanaka were awarded the Nobel Prize for these discoveries in 2012. Trans differentiation of pluripotent cells into cardiac myocytes has been achieved and non-human experiments have shown that they can be delivered to infarct tissues. Clinical trials in man have had limited success however due to poor graft survival and ongoing concerns about arrhythmogenic effects [53]. Newly introduced cardiac myocytes are vulnerable to the intense inflammatory environment of MI and electrical integration of new myocytes carries the risk of disturbing the heart rhythm. Thus, efforts to replenish cardiac

myocytes lost during MI using somatic and pluripotent cells have not been successful to date. Better understanding of the biology involved may reverse this but the goal remains a challenging one.

(iv) More recently, a novel approach has used microRNA (miRNA) to reverse the end stage non-proliferative phenotype of surviving cardiac myocytes. Cardiac myocytes of warm-blooded animals lose their ability to divide and proliferate within a few days after birth, entering a non-proliferative end stage phenotype. In 2012, Eulalio *et al.* showed that miRNAs control the proliferative ability of cardiac myocytes and that selected miRNAs could induce adult cardiac myocytes to re-enter the cell cycle phase and proliferate. miRNAs are short (around 22 nucleotides) strands of RNA which do not code proteins, but function as regulators of gene transcription. They do this by binding to complementary segments of messenger RNAs, inhibiting their transcription and thus blocking protein synthesis encoded on the parent gene. Initial experiments in mice suggested that transfection of two miRNAs into infarcted myocardium could induce proliferation of surviving myocytes to replenish lost cells and reduce the infarct size [54]. This result was later confirmed in pigs using a single miRNA (has-miR-199a-3p) transfected into the infarct tissue [55]. These results also showed reduced infarct size and increased functional recovery. However, despite these improvements, most of the animals died suddenly at 7–10 days due to severe arrhythmias. This most likely reflects electrical instability as newly grown cardiac myocytes integrated with the electrophysiological properties of the surviving myocardium. Transfection is a technique which uses a virus to place a required gene into the DNA of a cell. The gene to be transfected is first inserted into a viral genome (in this study, an adenovirus). The target cell is then infected with the modified virus which then inserts the gene into the target cell's DNA. The cell transcribes the new gene, producing, in this case, miRNA. So it is possible therefore, that prolonged exposure to this miRNA may have caused persistent cell division, which could disrupt the normal architecture of the myocardium resulting in electrical instability. It remains to be seen whether a shorter period of cardiac myocyte proliferation might be sufficient to replenish cells lost during MI without inducing arrhythmias.

Despite the failure of this research to find a way to replace infarcted myocardium, it has nevertheless been extraordinarily successful in extending understanding of how proliferation of myocardial cells is regulated, and I have no doubt that this knowledge will in time lead to new medical treatments. In original science, positive and negative results can have equal value.

Conclusions

For most of our lives, the heart functions normally and in many of us it will only finally fail due to illness elsewhere in the body. Like other organs, it is vulnerable to many illnesses whether from infections, trauma or the effects of harmful lifestyles. Provided it can maintain a sufficient cardiac output, given time, it has the ability to recover from many of these. Exceptions include severe trauma causing catastrophic haemorrhage and some toxins and infections which irreversibly weaken heart muscle. Takotsubo cardiomyopathy is an example of left ventricular dysfunction triggered by physical and emotional stress which in most patients is transient. The heart is functionally disabled and weakened, but the myocardium is still alive and functioning and so can recover.

In contrast, myocardial infarction involves the death of myocardial cells due to interruption of coronary blood flow. Here there is irreversible loss of contracting myocardium. Nevertheless around 80% of heart attack patients survive and live for many years. Healing of myocardial infarction involves rapid activation of the innate immune system. Molecules released from dying cardiac myocytes create a signalling system to alert immune cells in circulating blood and guide them to the infarcted tissue which they then infiltrate, setting up an acute inflammatory reaction. An initial phase digests and clears away the damaged tissue. Then after a few days, cells which normally maintain the structure of the myocardium migrate from the surviving border region into the affected area, proliferate to replace those lost, and produce proteins to rebuild the structure of the damaged area. This includes collagen which over time matures to form a strong scar. Although contracting cardiac myocytes cannot replicate to replace those lost, nevertheless in many cases healing is sufficient to allow the heart to sustain normal life for many years. It is yet another remarkable example of coordinated activity between the heart and other body systems, in this case the immune system, a process orchestrated by mutual molecular signalling.

References

[1] Wittstein IS, Thiemann DR, Lima JA, Baughman KL, Schulman SP, Gerstenblith G, *et al.* Neurohumoral features of myocardial stunning due to sudden emotional stress. *The New England Journal of Medicine*. 2005; 352(6):539–548.

[2] Prasad A, Lerman A, Rihal CS. Apical ballooning syndrome (Tako-Tsubo or stress cardiomyopathy): A mimic of acute myocardial infarction. *American Heart Journal*. 2008;155(3):408–417.

[3] Y-Hassan S. Clinical features and outcome of pheochromocytoma-induced Takotsubo syndrome: Analysis of 80 published cases. *The American Journal of Cardiology*. 2016;117(11):1836–1844.

[4] Ghadri JR, Sarcon A, Diekmann J, Bataiosu DR, Cammann VL, Jurisic S, *et al.* Happy heart syndrome: Role of positive emotional stress in takotsubo syndrome. *European Heart Journal*. 2016;37(37):2823–2829.

[5] Silva AR, Magalhaes R, Arantes C, Moreira PS, Rodrigues M, Marques P, *et al.* Brain functional connectivity is altered in patients with Takotsubo syndrome. *Scientific Reports*. 2019;9(1):4187.

[6] Templin C, Hanggi J, Klein C, Topka MS, Hiestand T, Levinson RA, *et al.* Altered limbic and autonomic processing supports brain-heart axis in Takotsubo syndrome. *European Heart Journal*. 2019;40(15):1183–1187.

[7] Kato K, Lyon AR, Ghadri JR, Templin C. Takotsubo syndrome: Aetiology, presentation and treatment. *Heart (British Cardiac Society)*. 2017;103(18): 1461–1469.

[8] Dhindsa DS, Khambhati J, Schultz WM, Tahhan AS, Quyyumi AA. Marital status and outcomes in patients with cardiovascular disease: A review. *Trends in Cardiovascular Medicine*. 2020;30(4):215–220.

[9] Manzoli L, Villari P, G MP, Boccia A. Marital status and mortality in the elderly: A systematic review and meta-analysis. *Social Science & Medicine* (1982). 2007;64(1):77–94.

[10] Wong CW, Kwok CS, Narain A, Gulati M, Mihalidou AS, Wu P, *et al.* Marital status and risk of cardiovascular diseases: A systematic review and meta-analysis. *Heart (British Cardiac Society)*. 2018;104(23): 1937–1948.

[11] Weiss NS. Marital status and risk factors for coronary heart disease. The United States health examination survey of adults. *British Journal of Preventive & Social Medicine*. 1973;27(1):41–43.

[12] Kaplan RM, Kronick RG. Marital status and longevity in the United States population. *Journal of Epidemiology and Community Health*. 2006;60(9):760–765.

[13] Fournier S, Muller O, Ludman AJ, Lauriers N, Eeckhout E. Influence of socioeconomic factors on delays, management and outcome amongst patients with acute myocardial infarction undergoing primary percutaneous coronary intervention. *Swiss Medical Weekly*. 2013;143:w13817.

[14] Wu JR, Lennie TA, Chung ML, Frazier SK, Dekker RL, Biddle MJ, *et al.* Medication adherence mediates the relationship between marital status and cardiac event-free survival in patients with heart failure. *Heart & Lung: The Journal of Critical Care*. 2012;41(2):107–114.

[15] Chandra V, Szklo M, Goldberg R, Tonascia J. The impact of marital status on survival after an acute myocardial infarction: A population-based study. *American Journal of Epidemiology*. 1983;117(3):320–325.

[16] Global, regional, and national incidence, prevalence, and mortality of HIV, 1980–2017, and forecasts to 2030, for 195 countries and territories: A systematic analysis for the Global Burden of Diseases, Injuries, and Risk Factors Study 2017. *The Lancet HIV*. 2019. https://doi.org/10.1016/S2352-3018(19)30196-1.

[17] Lozano R, Naghavi M, Foreman K, Lim S, Shibuya K, Aboyans V, *et al.* Global and regional mortality from 235 causes of death for 20 age groups in 1990 and 2010: A systematic analysis for the Global Burden of Disease Study 2010. *Lancet (London, England)*. 2012;380(9859):2095–2128.

[18] Global, regional, and national incidence, prevalence, and years lived with disability for 310 diseases and injuries, 1990–2015: A systematic analysis for the Global Burden of Disease Study 2015. *Lancet (London, England)*. 2016;388(10053):1545–1602.

[19] Thom T, Haase N, Rosamond W, Howard VJ, Rumsfeld J, Manolio T, *et al.* Heart disease and stroke statistics — 2006 update: A report from the American Heart Association Statistics Committee and Stroke Statistics Subcommittee. *Circulation*. 2006;113(6):e85–e151.

[20] Luengo-Fernandez R, Leal J, Gray A, Petersen S, Rayner M. Cost of cardiovascular diseases in the United Kingdom. *Heart (British Cardiac Society)*. 2006;92(10):1384–1389.

[21] Mahmood SS, Levy D, Vasan RS, Wang TJ. The Framingham Heart Study and the epidemiology of cardiovascular disease: A historical perspective. *Lancet (London, England)*. 2014;383(9921):999–1008.

[22] Dwyer T, Hetzel BS. A comparison of trends of coronary heart disease mortality in Australia, USA and England and Wales with reference to three major risk factors-hypertension, cigarette smoking and diet. *International Journal of Epidemiology*. 1980;9(1):65–71.

[23] Coventry LL, Finn J, Bremner AP. Sex differences in symptom presentation in acute myocardial infarction: A systematic review and meta-analysis. *Heart & Lung: The Journal of Critical Care*. 2011;40(6):477–491.

[24] Foreman RD. Mechanisms of cardiac pain. *Annual Review of Physiology*. 1999;61:143–167.

[25] Anand SS, Xie CC, Mehta S, Franzosi MG, Joyner C, Chrolavicius S, *et al.* Differences in the management and prognosis of women and men who suffer from acute coronary syndromes. *Journal of the American College of Cardiology*. 2005;46(10):1845–1851.

[26] Vincent WR, Rapaport E. Serum creatine phosphokinase in the diagnosis of acute myocardial infarction. *The American Journal of Cardiology*. 1965;15:17–26.

[27] Yonetsu T, Kakuta T, Lee T, Takahashi K, Kawaguchi N, Yamamoto G, *et al.* In vivo critical fibrous cap thickness for rupture-prone coronary plaques assessed by optical coherence tomography. *European Heart Journal*. 2011;32(10):1251–1259.

[28] Waksman R, Di Mario C, Torguson R, Ali ZA, Singh V, Skinner WH, *et al.* Identification of patients and plaques vulnerable to future coronary events with near-infrared spectroscopy intravascular ultrasound imaging: A prospective, cohort study. *Lancet (London, England)*. 2019;394(10209): 1629–1637.

[29] Lonborg J, Vejlstrup N, Kelbaek H, Holmvang L, Jorgensen E, Helqvist S, *et al.* Final infarct size measured by cardiovascular magnetic resonance in patients with ST elevation myocardial infarction predicts long-term clinical outcome: An observational study. *European Heart Journal Cardiovascular Imaging*. 2013;14(4):387–395.

[30] Deo R, Albert CM. Epidemiology and genetics of sudden cardiac death. *Circulation*. 2012;125(4):620–637.

[31] Jabbari R, Engstrom T, Glinge C, Risgaard B, Jabbari J, Winkel BG, *et al.* Incidence and risk factors of ventricular fibrillation before primary angioplasty in patients with first ST-elevation myocardial infarction: A nationwide study in Denmark. *Journal of the American Heart Association*. 2015;4(1): e001399.

[32] Helgestad OKL, Josiassen J, Hassager C, Jensen LO, Holmvang L, Sorensen A, *et al.* Temporal trends in incidence and patient characteristics in cardiogenic shock following acute myocardial infarction from 2010 to 2017: A Danish cohort study. *European Journal of Heart Failure.* 2019;21(11):1370–1378.

[33] Figueras J, Alcalde O, Barrabes JA, Serra V, Alguersuari J, Cortadellas J, *et al.* Changes in hospital mortality rates in 425 patients with acute ST-elevation myocardial infarction and cardiac rupture over a 30-year period. *Circulation.* 2008;118(25):2783–2789.

[34] Mechanisms for the early mortality reduction produced by beta-blockade started early in acute myocardial infarction: ISIS-1. ISIS-1 (First International Study of Infarct Survival) Collaborative Group. *Lancet (London, England).* 1988;1(8591):921–923.

[35] Jabbari R, Risgaard B, Holst AG, Nielsen JB, Glinge C, Engstrom T, *et al.* Cardiac symptoms before sudden cardiac death caused by coronary artery disease: A nationwide study among young Danish people. *Heart (British Cardiac Society).* 2013;99(13):938–943.

[36] Graham J, Gerard RW. Membrane potentials and excitation of impaled single muscle fibers. *Journal of Cellular and Comparative Physiology.* 1946;28(1):99–117.

[37] Downar E, Janse MJ, Durrer D. The effect of acute coronary artery occlusion on subepicardial transmembrane potentials in the intact porcine heart. *Circulation.* 1977;56(2):217–224.

[38] Harris AS, Bisteni A, Russell RA, Brigham JC, Firestone JE. Excitatory factors in ventricular tachycardia resulting from myocardial ischemia; potassium a major excitant. *Science (New York, NY).* 1954;119(3085):200–203.

[39] Culling W, Penny WJ, Lewis MJ, Middleton K, Sheridan DJ. Effects of myocardial catecholamine depletion on cellular electrophysiology and arrhythmias during ischaemia and reperfusion. *Cardiovascular Research.* 1984;18(11):675–682.

[40] Vivien CJ, Hudson JE, Porrello ER. Evolution, comparative biology and ontogeny of vertebrate heart regeneration. *NPJ Regenerative medicine.* 2016;1:16012.

[41] Rather LJ. Disturbance of function (functio laesa): The legendary fifth cardinal sign of inflammation, added by Galen to the four cardinal signs of Celsus. *Bulletin of the New York Academy of Medicine.* 1971;47(3):303–322.

[42] Guan Z, Hellman J, Schumacher M. Contemporary views on inflammatory pain mechanisms: TRPing over innate and microglial pathways. *F1000Research.* 2016;5. DOI:10.12688/f1000research.8710.1. PMID: 27781082; PMCID: PMC5054801.

[43] Mann DL. The emerging role of innate immunity in the heart and vascular system: For whom the cell tolls. *Circulation Research.* 2011;108(9): 1133–1145.

[44] Frangogiannis NG. Regulation of the inflammatory response in cardiac repair. *Circulation Research.* 2012;110(1):159–173.

[45] Prabhu SD, Frangogiannis NG. The biological basis for cardiac repair after myocardial infarction: From inflammation to fibrosis. *Circulation Research.* 2016;119(1):91–112.

[46] Gordon JW, Shaw JA, Kirshenbaum LA. Multiple facets of NF-kappaB in the heart: To be or not to NF-kappaB. *Circulation Research.* 2011;108(9): 1122–1132.

[47] Nahrendorf M, Swirski FK, Aikawa E, Stangenberg L, Wurdinger T, Figueiredo J-L, *et al.* The healing myocardium sequentially mobilizes two monocyte subsets with divergent and complementary functions. *Journal of Experimental Medicine.* 2007;204(12):3037–3047.

[48] Hilgendorf I, Gerhardt LM, Tan TC, Winter C, Holderried TA, Chousterman BG, *et al.* Ly-6C^{high} monocytes depend on Nr4a1 to balance both inflammatory and reparative phases in the infarcted myocardium. *Circulation Research.* 2014;114(10):1611–1622.

[49] Bergmann O, Zdunek S, Felker A, Salehpour M, Alkass K, Bernard S, *et al.* Dynamics of cell generation and turnover in the human heart. *Cell.* 2015; 161(7):1566–1575.

[50] Fu X, Khalil H, Kanisicak O, Boyer JG, Vagnozzi RJ, Maliken BD, *et al.* Specialized fibroblast differentiated states underlie scar formation in the infarcted mouse heart. *The Journal of Clinical Investigation.* 2018;128(5): 2127–2143.

[51] Hu X, Xu Y, Zhong Z, Wu Y, Zhao J, Wang Y, *et al.* A large-scale investigation of hypoxia-preconditioned allogeneic mesenchymal stem cells for myocardial repair in nonhuman primates: Paracrine activity without remuscularization. *Circulation Research.* 2016;118(6):970–983.

[52] Gurdon JB, Melton DA. Nuclear reprogramming in cells. *Science.* 2008; 322(5909):1811–1815.

[53] Okano S, Shiba Y. Therapeutic potential of pluripotent stem cells for cardiac repair after myocardial infarction. *Biological & Pharmaceutical Bulletin*. 2019;42(4):524–530.

[54] Eulalio A, Mano M, Dal Ferro M, Zentilin L, Sinagra G, Zacchigna S, *et al.* Functional screening identifies miRNAs inducing cardiac regeneration. *Nature*. 2012;492(7429):376–381.

[55] Gabisonia K, Prosdocimo G, Aquaro GD, Carlucci L, Zentilin L, Secco I, *et al.* MicroRNA therapy stimulates uncontrolled cardiac repair after myocardial infarction in pigs. *Nature*. 2019;569(7756):418–422.

[56] Bresnick AR, Weber DJ, Zimmer DB. S100 proteins in cancer. *Nature Reviews Cancer*. 2015;15(2):96–109.

[57] Ritossa F. A new puffing pattern induced by temperature shock and DNP in *Drosophila*. *Experientia*. 1962;18(12):571–573.

[58] Tissieres A, Mitchell HK, Tracy UM. Protein synthesis in salivary glands of *Drosophila melanogaster*: Relation to chromosome puffs. *Journal of Molecular Biology*. 1974;84(3):389–398.

[59] Dosch M, Gerber J, Jebbawi F, Beldi G. Mechanisms of ATP release by inflammatory cells. *International Journal of Molecular Sciences*. 2018; 19(4):1222.

[60] Ghaemi-Oskouie F, Shi Y. The role of uric acid as an endogenous danger signal in immunity and inflammation. *Current Rheumatology Reports*. 2011; 13(2):160–166.

[61] Nakayama H, Otsu K. Mitochondrial DNA as an inflammatory mediator in cardiovascular diseases. *Biochemical Journal*. 2018;475(5):839–852.

[62] Patel SJ, Jindal R, King KR, Tilles AW, Yarmush ML. The inflammatory response to double stranded DNA in endothelial cells is mediated by NFkappaB and TNFalpha. *PLoS One*. 2011;6(5):e19910.

[63] Pohar J, Lainscek D, Ivicak-Kocjan K, Cajnko MM, Jerala R, Bencina M. Short single-stranded DNA degradation products augment the activation of Toll-like receptor 9. *Nature Communications*. 2017;8:15363.

[64] Javaid N, Yasmeen F, Choi S. Toll-like receptors and relevant emerging therapeutics with reference to delivery methods. *Pharmaceutics*. 2019; 11(9):441.

Appendix I

Danger associated molecular patterns

HMGB1: High mobility group box 1 protein is a nuclear protein involved in DNA transcription. Its name derives from its rapid electrophoretic mobility and the G-box refers to a domain of the protein involved in DNA binding. HMGB1 is secreted by stressed or damaged cells, and like most proteins secreted in this way, the release mechanism differs from that of hormones (referred to as unconventional secretion) in that it does not involve endoplasmic reticulum processing and transport within the cell. HMGB1 binds with PRRs such as TLR2, TLR4 and RAGE on immune cells, triggering an inflammatory response and on endothelial cells causing an increased expression of cell adhesion molecules (ICAM-1 and VCAM-1).

S100: S100 stands for a family of small calcium binding proteins found in vertebrates. The S100 abbreviation derives from the fact that in early studies they were noted to be soluble in 100% saturated solutions of ammonium sulphate at Ph7. There are 21 known members of the S100 family and although structurally similar, they are functionally distinct, participating in a wide range of processes including proliferation, inflammation and migration of cells. They are important markers of some cancers [56] and modulators of inflammation including in MI.

Fibronectin-EDA: Fibronectin (Fn) is a large ~440 KDa glycoprotein (a protein containing a carbohydrate component) found in the material which surrounds cells called the extracellular matrix. The fibronectin gene gives rise to several fibronectin isoforms resulting from alternative splicing during gene transcription, a process in which certain sections of DNA are included or omitted during transcription giving rise to different mRNA sequences and consequently protein isoforms. Extra domain A (EDA) is one of these. Fn-EDA binds to integrins, playing important roles in cell adhesion, wound healing and inflammation.

Interleukins: Interleukins (ILs) consist of a large family of glycoproteins which act as cytokines. They were originally thought to be produced by white blood cells (leukocytes) but later found to be synthesised by other cells. The name was coined in 1979 to reflect understanding of their role at that time as signalling molecules between leukocytes. ILs have an important role in modulating inflammation.

Heat Shock Protein: Heat shock proteins (HSPs) are a family of proteins produced by all cells in response to a variety of stressful stimuli. They have important roles in assisting the folding of new proteins and refolding damaged ones. Heat shock-induced changes in chromosome activity were initially reported in 1962 [57], which were later shown to involve synthesis of specific proteins [58]. There are many different HSPs which are named according to their molecular weight, of which HSP-60 and HSP-70 are important in cardiac repair.

LMW Hyaluronan Fragments: Hyaluronans (HAs) are widely distributed extracellular matrix glycosaminoglycans (long chains of sugar molecules). They are highly polar, meaning they attract water and are able to function as lubricants in the body. At sites of inflammation, HAs are broken into smaller sugar chains called low molecular weight (LMW) HA fragments. These are potent activators of inflammatory genes leading to the production of several inflammatory mediators including chemokines and cytokines.

ATP: Adenosine Triphosphate (ATP) is a ubiquitous high energy compound that fuels numerous energy demanding processes in biology including muscle contraction (discussed in Chapter 3). Although normally contained within cells, ATP rapidly leaks from damaged cells into the extracellular space, where it acts as a DAMP [59]. ATP binds to purinergic receptors which are also PRRs (see Appendix II).

Uric Acid: Uric acid is a breakdown product of purines, which are components of DNA and RNA. They are normally present in urine and blood. High plasma concentrations of uric acid can result in deposits in joints where crystalline forms may cause gout by inducing an intense inflammation. Uric acid released form damaged cells also acts as a DAMP, binding to PRRs including TLR2, TLR4 and NLRP3 (see Appendix II) to act as a pro-inflammatory signal [60].

Mitochondrial DNA: Damaged cells release several chemicals that originate in their mitochondria and which act as DAMPs [61]. These include mitochondrial DNA which binds to PRRs (TLR9 and NLPP3) resulting in the production of inflammatory cytokines [61].

Double-stranded DNA: Double-stranded DNA (dsDNA) has recently been found to act as a DAMP and pro-inflammatory signal. Viral RNA and bacterial DNA had been known to act as a PAMP and as an important initiator of an inflammatory response to infection. More recently, dsDNA released from damaged cardiac myocytes has also been found to act as a DAMP resulting in the activation of endothelial cells [62].

Single-stranded DNA: Single-stranded DNA (ssDNA) acts as a DAMP by activating PRRs, notably TLR9 [63] resulting in the production of interleukins.

Appendix II

Pattern recognition receptors

Toll-like Receptors (TLRs): TLRs are a large group of proteins involved in activating the innate immune system. Their name derives from the German word "toll", meaning stunning or fantastic, uttered in response to first seeing a marked deformity in fruit flies by scientist Christiane Nüsslein-Volhard in 1985. The deformity was later shown to be due to a gene mutation. The gene was named the toll gene and was later found to play a role in the defence against fungal infections in fruit flies. Then in 1995, a mammalian homologue of toll, a toll-like receptor (TLR4), was discovered, activation of which induced the same NF-κB pathway as in fruit flies. Three years later, TLR4 was shown to be the receptor for the endotoxin of gram-negative bacteria, and this led to the understanding that TLRs were pattern recognition receptors (PRRs), which bind PAMPs to activate an inflammatory response. TLRs are usually expressed on the membranes of macrophages and other cells that act as a first line of defence against invading organisms. TLRs are now known to recognise and bind many molecules derived from pathogens such as bacteria and initiate an inflammatory response. More recently, TLRs have been shown to bind to endogenous molecules released from damaged or injured cells (DAMPs), including during myocardial infarction. There are 13 known TLRs labelled 1–13, of which 1–10 are found in humans [64].

RAGE: RAGE is an abbreviation for the "Receptor for Advanced Glycation End products". RAGE is a cell membrane spanning receptor in endothelial and smooth muscle cells and a member of the immunoglobulin superfamily. It is one of the PRRs and binds to proteins and lipids that are heavily glycated (meaning bound to many sugar molecules). This activates RAGE which in turn initiates an inflammatory cascade in the cell. RAGE also binds to other DAMPs including HMGB1 and S100A8 and A9.

NLRP3: NLRPs are proteins composed of three domains, a Nucleotide-binding domain, a Leucine-rich Repeat domain and a Pyrin domain, which form the basis for the abbreviation. There are 14 known NLRPs designated 1–14 which bind with several DAMPs and PAMPs to form inflammasomes (multiprotein complexes involved in inflammation). NLRP3 binds to S100A8 and S100A9, LMW hyaluronans, ATP, uric acid and mitochondrial DNA.

Interleukin Receptors (IL-IRs): IL-IRs bind interleukins released in response to injury or infection. They are widely distributed in numerous cell types in humans and activation has powerful pro-inflammatory effects.

P2X7: The P2X7 receptor is a member of the purinergic receptor family. Purinergic (P) receptors are membrane bound receptors present in almost all cells and bind to purines. P1 and P2 receptors bind adenosine and ATP, respectively. P2 receptors are further subdivided into five subclasses of which P2X is one. There are seven known P2X receptors designated P2X1-7. P2X7 has been implicated in cardiac inflammatory responses to high concentrations of extracellular ATP.

7

Roots and Branches

The ultimate purpose of the heart is to assist the transport of essential materials to and from specific locations around the body. Oxygen is delivered from the lungs to all tissues and carbon dioxide, the product of its metabolism, is taken back to the lungs to be exhaled. Nourishment from food is carried from the gut to all tissues and waste products are transported to the kidneys for excretion. Hormones, signalling molecules produced by endocrine glands to regulate the function of many organs, are circulated in the bloodstream to reach their target cells. Blood is the medium in which all of these are transported. Blood itself is far more than a passive transport medium, but rather a liquid organ with many unique functions, including acting as a surveillance system and first line of defence to maintain the integrity of blood vessels. All this transport takes place within a system of arteries and veins, referred to as the circulation, which carry blood from the heart to all regions of the body and return it for recirculation. Far from being a simple set of passive tubes, the circulatory system is highly evolved with distinctive features that are exquisitely suited to the pulsatile nature of the heart and on which the heart depends for its normal function.

Evolution of the Circulation

Around 800 million years ago, multicellular organisms called metazoa evolved from single-celled protozoa. The most primitive forms still extant include sponges, coral and jellyfish and these develop from two germ layers, ectoderm and endoderm. They are known as diploblasts (having just two germ layers) and derive oxygen and nourishment directly by diffusion from sea water through their skin and body cavities. Then between 600 and 700 million ago, a third germ layer, mesoderm, emerged to form triploblasts, which developed more complex animal forms. Triploblasts such as small flatworms are able to derive sufficient nourishment through diffusion and have no circulation (Figure 7.1(a)), however as animals became larger and more complex, their energy needs exceeded what could be met by simple diffusion and internal circulatory systems began to evolve.

Coelomic Circulation

The earliest circulatory system to evolve (Figure 7.1(b)) consisted of an internal body cavity called a coelom filled with coelomic fluid, which developed between the endoderm and ectoderm layers [1]. Coelomic fluid is produced by mesothelial cells derived from the mesoderm and has important functions in nutrient transport and immunity. Coelomic circulatory systems do not have a pump, instead fluid is circulated by the action of motile cilia. The coelomic cavity is lined with specialised mesothelial cells which acquired the ability to contract. These cells also have specialised hair-like extensions on the luminal facing surface. Movement of mesothelial cells and their cilia creates a current in the coelomic fluid which provides a means of transport of nutrients and gases. The coelomic cavity also acts as a cushion against shocks and may have originally evolved as a hydrostatic skeleton, which can be helpful in the movement of soft bodied animals. Coelomic circulations are found in some extant molluscs, annelids (e.g., segmented worms) and arthropods (e.g., insects and marine crustaceans). The chordates, which include vertebrates, are also described as coelomic because they develop a coelom during embryonic development. Coelomic circulation is far more efficient than simple diffusion and allows animals to develop larger and more complex bodies. However, convective fluid movement is relatively slow and poorly directed and most coelomates are segmented, with each segment having a separate coelomic cavity, providing local nutrition and gas exchange. These restrictions were overcome with the evolution of blood-based vascular systems which transport fluid throughout the length of the animal.

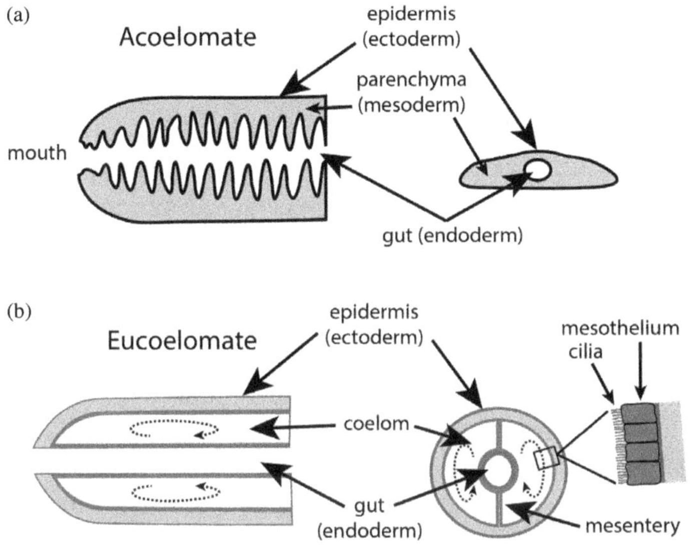

Figure 7.1: (a) Acoelomates rely on passive diffusion of gases and nutrients which is only possible over short distances. This is sufficient for some small animals such as flatworms. Nutrient uptake and gas exchange take place by passive diffusion through the skin which is lined with ectodermal cells and through the gut which is lined with endodermal cells. Many acoelomates are microscopic in size and larger species have flat bodies which reduces diffusion distances. Large species also have branched guts which increase the surface area and improves nutrient absorption. Parenchyma between the skin and gut is composed of mesoderm and lacks coelom spaces found in more evolved species and for this reason these species are known as acoelomates. (b) Eucoelomates are the earliest organisms to develop a circulatory system. They developed spaces within the parenchyma which coalesced to form a cavity called a coelom containing coelomic fluid. Diffusion of gases and nutrients occurs through the endoderm lining the gut into the coelomic fluid. Eucoelomates have not evolved pumping systems, rather they have specialised (mesothelial) cells which line the coelomic cavity. These cells have acquired the ability to contract and have hair-like extensions called cilia on the lumen facing side. This arrangement creates a current within the coelomic fluid and thereby distributes material absorbed from the gut. The latter is supported by mesentery composed of structural mesoderm within the coelomic cavity. In some animals, the body wall can also contract which further assists movement of coelomic fluid. Coelomic cavities are usually compartmentalised so that circulation is confined to local sections of the animal, in contrast with blood circulation which transports fluid throughout the body.

Blood-based Vascular Systems

Blood-based vascular systems are thought to have evolved around 600 million years ago. These consisted of spaces containing blood that were continuous throughout the body. Such spaces include vessels, simple contiguous gaps between cells called sinuses and cavities in which blood and extracellular fluid mix freely, called hemocoels, together with some form of pump mechanism. The latter was provided by (i) contracting myoepithelial cells derived from the mesoderm which generated peristaltic contraction waves to propel fluid and (ii) specialised muscular pumps in the form of tubular hearts (discussed in more detail in Chapter 2). Blood-based vascular systems gave several advantages by providing efficient transport of gases and nourishment, clotting factors, immune defences as well as improved regulation of heat and ultrafiltration of waste by perfusing kidneys. Blood perfusion of tissues also afforded protrusion of the head or foot which assisted movement and penile erection. Around 500 million years ago, blood vascular systems evolved a lining of endothelial cells derived from mesothelium which added a barrier function and a means for targeting the immune response against invading organisms. Circulatory systems have evolved broadly in two forms, open and closed systems. All vertebrates have closed systems, while invertebrates have evolved open and closed systems.

Open Circulatory Systems

Open circulations are found in arthropods such as insects and crustaceans and in some molluscs such as clams, snails and slugs. *Drosophila melanogaster* (fruit fly) exemplifies this arrangement (see Figure 2.6 and related text in Chapter 2). *Drosophila* has a heart tube which beats at 200–300 per minute circulating a fluid called haemolymph in an open circulation, meaning it lacks direct connections between arteries and veins. Haemolymph like blood carries oxygen, but it does so using haemocyanin, a protein containing copper rather than iron which is found in haemoglobin. Haemocyanin is colourless when deoxygenated but turns blue when it binds oxygen, accounting for the typical blue colour of circulating haemolymph. Haemolymph is pumped by a heart tube through the aorta directly into tissue spaces in the body. There are no veins; rather blood re-enters the heart directly from body tissues through a series of openings, called ostia. Lungs have not evolved in fruit flies. Air enters through tiny pores in the abdomen and passes through multiple channels where it exchanges oxygen and carbon dioxide with haemolymph. *Drosophila* evolved a highly efficient system of oxygenation by periodically reversing the direction of output from the heart to pump haemolymph into the abdominal space which allows gas exchange [2]. In some crustaceans like the horseshoe crab (*Limulus polyphemus*), a hemocoel forms part of the circulation. This consists of a large cavity in which haemolymph and extracellular fluid mix freely. Horseshoe crabs are of considerable interest to biologists because their haemolymph contains a single cell type, called amebocytes, which have immune and coagulant properties. These functions are also present in human blood but are carried separately in neutrophils and platelets. Amebocytes provide a highly effective innate immunity which has sustained the horseshoe crab for 200 million years.

Other crustaceans such as lobsters have an open circulation similar to fruit flies unlike vertebrates. Haemolymph is pumped by a single heart beating around 70 times per minute into two aortae. The anterior aorta supplies the head. The posterior aorta delivers haemolymph to the tail region and gives off a sternal artery which supplies the front of the body. Having no capillary network, tissue perfusion is achieved by haemolymph directly entering tissue spaces where gas exchange and nutrient delivery occur. Haemolymph in tissue spaces then collects in a central sinus which carries it back towards the heart. On the return route, it passes through gills which extract oxygen from the surrounding water. Lobsters use their tails to maintain a current of fresh water over the gills and intermittently reverse the flow to clear debris from the gill surfaces. Haemolymph is then delivered to the pericardial sac in which the heart is suspended. It enters the heart during diastole through openings in the ventricle, which are then closed by the ventricular contraction during systole [3]. Open circulations provide a more effective transport system than simple diffusion; they deliver oxygen and nutrients much more efficiently to all parts of the body and provide an integrated immune defence system.

Closed Circulatory Systems

Molluscs are invertebrates which evolved from *bilateria* (animals formed with bilateral symmetry). *Bilateria* separated as *protostomia* from *deuterostomia* around 600 million years ago. The vertebrates evolved from *deuterostomia* and invertebrates from *protostomia*. Molluscs evolved from the latter around 500 million years ago and gastropods (e.g., snails and mussels) were the first to have a two-chambered heart, with a heart rate of around 20 beats/min (see Chapter 2, Figure 2.8(a) and related text for more details). Gastropods also have an open circulation, which, like most open systems, works at low pressure, in this case, at 20–40 mmHg. Open circulations also require a larger fluid volume than closed systems. Typically, haemolymph and interstitial fluid which mix freely account for 50% of body weight.

Fossils of early shelled cephalopods have been found in Cambrian rocks dating to 520 million years. Cephalopods such as octopus and squid are also molluscs but have evolved closed circulatory systems. The earliest fossils of octopus ancestors are found in carboniferous rocks dating them to around 300 million years. Extant octopi are highly sophisticated animals with a large brain and separate nerve centres in the arms which give them independent motor control. They have three hearts (Figure 7.2(a)), a systemic heart which pumps haemolymph around arteries and two separate hearts called brachial hearts which direct it through the gills. The systemic heart generates a pressure of 50–70 mmHg in the arteries to perfuse and oxygenate the tissues of the body. Deoxygenated haemolymph then flows in veins to the gills for reoxygenation. During its course through the arteries and veins, haemolymph pressure falls and would not be sufficient to perfuse the gills. Octopi have evolved two additional brachial hearts for this purpose which pump at 20 mmHg, which is ideal for gill perfusion. This is the first known circulation with separates systemic and pulmonary pressures. However, it is achieved in a single circuit using hearts in series, unlike mammals, which have two circuits in parallel.

Cephalopods are more active and have higher energy needs than other molluscs. Their higher systemic pressure drives more efficient tissue perfusion and gas exchange which matches this. Gas exchange in gills, like lungs, takes place across thin-walled vessels which would be liable to rupture or leak if exposed to high transmural pressures. The two brachial hearts which generate pressure of around 20 mmHg provide an ideal solution for this. Octopi have a closed circulation meaning that the arteries and veins are connected by a network of small capillaries which ramify the tissues. In addition, all their blood vessels are lined with a layer of endothelial cells. These arrangements offer significant advantages which have allowed octopi to develop large and complex bodies. This is also reflected in the much lower volume of interstitial fluid and blood, which in octopi is around 20% of body weight compared with 50% in gastropods.

Vertebrates

Vertebrates have a closed circulatory system driven by a multi-chambered heart. Their blood contains cells, which in mammals, have no nucleus and are filled with haemoglobin for oxygen transport. The entire circulation is lined by endothelial cells which play active roles in diffusion of gases, transport of nutrients, haemostasis (maintaining blood in a liquid state) and immunity. Circulatory plans in vertebrates vary greatly depending on their energy requirements. For aquatic animals, the buoyant effect of water greatly reduces the effect of gravity. Terrestrial animals must withstand gravity, which increases their energy needs and standing upright increases it further. In addition, maintaining a warm body temperature requires energy. The ability to meet these increasing demands required evolutionary modifications of the heart and circulation. Hearts of land-based vertebrates have evolved the ability to pump blood at higher pressures. For example, in humans, blood pressure is normally 130/75 mmHg compared to 20/10 in fish and 40/20 mmHg in amphibians such as frogs. All vertebrates have a central pump; however, their chamber arrangements differ depending on energy requirements. Thus, fish hearts consist of two chambers, while amphibians have three and mammals and birds have four. The structure of these is discussed in more detail in Chapter 2. Their circulatory plans also vary depending on their energy needs and are discussed here.

(i) Fish have the simplest circulation plan among vertebrates. It consists of a single circulation (Figure 7.2(b)) driven by a two-chambered heart (see also Chapter 2, Figure 2.8(c)). Blood returns from the body through a vein which empties into a single atrium. From there it is propelled into the ventricle and ejected into the aorta which carries it directly to the gills which have a large surface area where gas exchange

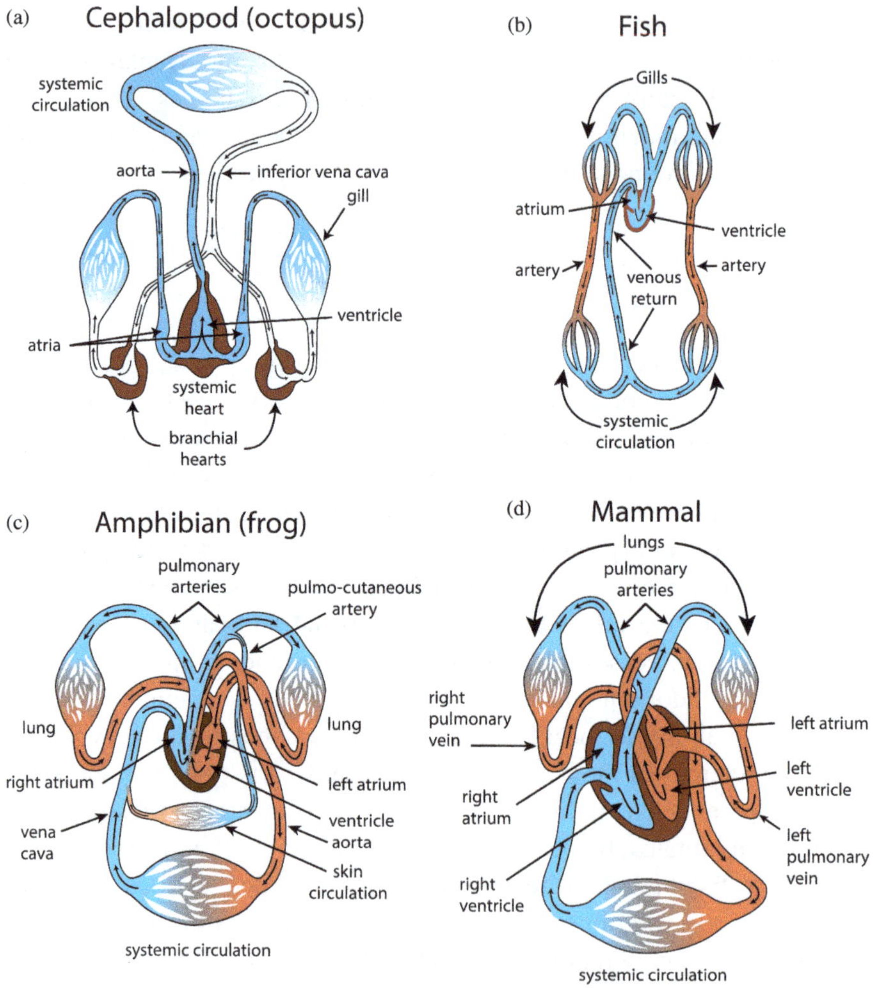

Figure 7.2: Diagrammatic illustration of closed circulatory plans which have evolved to suit the energy needs and lifestyles of different species. (a) Cephalopods are invertebrate molluscs and species such as octopi have a closed circulation. A systemic heart receives oxygenated haemolymph (which is blue in colour) from gills on each side of the body through left and right atria. The atria eject into a single ventricle which pumps haemolymph at a pressure of 70 mmHg into the aorta and through a network of vessels through the body tissues. Deoxygenated haemolymph loses its blue colour and drains into the vena cava which divides into two branches returning haemolymph to the gills. At this stage, haemolymph pressure has dissipated and would not be sufficient to perfuse the gills. Instead, two branchial hearts pump haemolymph at 20 mmHg through the gills and back to the systemic heart. (b) Fish have a two-chambered heart and a closed circulation

takes place and blood is oxygenated. Gill surface area increases with body mass and is larger in the most active predators, ranging from around 750 mm^2/g in striped marlin to 1,800 mm^2/g in some tuna [4]. Gas exchange takes place between blood circulating in the capillaries of gills and the water in which the animal is swimming. To achieve this efficiently, the walls of gill capillaries must be very thin, typically

---

Figure 7.2: (*see figure on facing page*) (see also Figure 2.8(c), Chapter 2). The heart receives deoxygenated blood (shown as blue) from the tissues of the body into an atrium which ejects it into the ventricle. Blood is then pumped at a pressure of 20 mmHg into the aorta. As cold-blooded animals in an aquatic environment, their metabolic rate is relatively low and this level of pressure is suitable for their energy needs. It is also ideal for perfusing the gills which receive blood directly from the aorta. Oxygenated blood (shown as red) then continues through arteries to perfuse the body tissues where it releases its oxygen, becomes deoxygenated again and is returned to the heart. (c) Amphibians evolved to live on land while retaining the ability for aquatic life and their circulations have evolved to match this. Frogs have a three-chambered heart consisting of a single ventricle and a left and right atrium. The single ventricle receives deoxygenated blood (shown as blue) from the right atrium and oxygenated blood (shown as red) from the left atrium. The ventricle has an incomplete interventricular septum and a spiral baffle in its outflow tract which directs blood from the right atrium to the pulmonary artery and from the left to the aorta. This arrangement results in functional separation of the pulmonary and systemic circulations (see also Figure 2.9, Chapter 2). Deoxygenated blood is delivered to the lungs via the left and right pulmonary arteries where it is oxygenated and is then returned to the left atrium via the left and right pulmonary veins. Oxygenated blood is pumped to the tissues of the body via the aorta and returns to the heart by the vena cava. Amphibians such as frogs maintain a moist covering of the skin which facilitates gas exchange. The pulmonary artery gives off a pulmonary-cutaneous branch which supplies deoxygenated blood to skin capillary vessels, which is oxygenated directly through the skin. Although not as efficient as lungs, this can provide sufficient oxygen to sustain animals during hibernation in water. (d) Mammals have a closed circulation and a four-chambered heart. The latter consists of left and right atria and ventricles and results in complete anatomical separation of the pulmonary and systemic circulations. This was a necessary evolutionary step because unlike amphibians, mammals are endothermic and therefore have much higher energy needs. This in turn requires a higher systemic blood pressure, which would be harmful to lung capillaries. Separation allows the left and right ventricles to pump blood into the aorta and pulmonary artery at around 120/70 mmHg and 25/15 mmHg, respectively. Deoxygenated blood (shown as blue) reaches the right atrium and from there is pumped into the right ventricle and in turn via the pulmonary arteries to the lungs. There it is oxygenated before returning to the left atrium via the pulmonary veins. The left atrium then pumps it into the left ventricle which in turn pumps into the aorta for delivery to the body tissues.

around 2–11 μm [4], but strong enough to withstand the blood pressure within them. In practice, this means that blood pressure in gas exchanging capillaries must be kept low. Fish are cold blooded vertebrates and the effects of gravity on them are limited. Consequently, their energy needs are also low compared to mammals, which is why they can also function with a lower blood pressure of around 20/10 mmHg. This also explains why fish hearts can pump blood directly to the gills without harming gill capillaries. To further protect the gills from the pulsatile action of the heart, fish have an elastic arterial cone, called the conus arteriosus, between the ventricle and aorta which buffers the systolic pulsation (see also Chapter 2, Figure 2.8(c)). From the gills, oxygenated blood then travels around the body where it gives up oxygen and captures carbon dioxide before returning to the heart for recirculation.

(ii) Amphibians moved from an exclusively aquatic to a terrestrial environment around 350 million years ago. They abandoned gills and became air breathing animals relying on lungs for gas exchange. Many amphibians such as frogs maintain a moist skin through which gas exchange can also occur. To facilitate this, they have also evolved a unique circulatory plan (Figure 7.2(c)). They have a three-chambered heart consisting of two atria and a single ventricle (see Chapter 2, Figure 2.9). Deoxygenated blood from the tissues drains into one atrium, while the other receives oxygenated blood from the lungs. Both atria drain into the single ventricle. This has a spiral baffle in the outflow tract which effectively directs deoxygenated blood from the tissues to the pulmonary artery and oxygenated blood from the lungs to the aorta. This arrangement provides a high degree of functional separation of pulmonary and systemic blood flows, but unlike the anatomical separation in mammals. In addition to supplying the lungs, the pulmonary artery gives off branches to the skin, which is also used for gas exchange. Oxygenated blood from the skin returns to the heart via the systemic veins and so is mixed with deoxygenated blood from the tissues. During normal activities, around 25% of pulmonary blood flow reaches the skin and lungs are the main source of oxygen. During diving and hibernation in cold water, the pulmonary arteries are actively constricted, diverting more deoxygenated blood to the skin [5] where it receives sufficient oxygen to meet the low energy deeds (and therefore low oxygen consumption) during prolonged hibernation. Amphibians, having a single ventricle, means that blood pressure is the same in pulmonary and systemic arteries. This is possible because blood pressure is low in amphibians at around 40/20 mmHg which is suitable for lung capillaries.

(iii) Reptiles occupy a transitional position in cardiovascular evolution between the single circulation in fishes and the double circulation in mammals. Some reptiles such as crocodiles and alligators have four-chambered hearts and a completely divided pulmonary and systemic circulations (see Chapter 2, Figure 2.10), whereas snakes, turtles, lizards and monitors have a single ventricle that is partly divided by an incomplete interventricular septum. The latter have some mixing of venous and arterial blood, but this is minimised by the incomplete septum which directs oxygenated blood to the aortas and venous blood to the pulmonary artery. Varanid lizards are terrestrial, highly active animals with a high metabolic rate and matching energy requirements. They can meet this by virtue of being air breathers and having almost complete functional separation of the pulmonary and systemic circulations. In contrast, cold-blooded diving reptiles have much lower energy needs. Typically their oxygen consumption is around five times less than warm blooded mammals [6]. When diving, they can divert blood from the pulmonary to the systemic circuit by constricting the pulmonary arteries. This increases systemic blood flow and nutrient delivery. They can also rely on anaerobic metabolism to generate energy to a much greater degree than mammals, due to their greater tolerance of lactic acidosis, a by-product of glycolysis (a process of generating energy from glucose that does not require oxygen).

Crocodiles are of particular interest in that they have fully developed four-chambered hearts and yet are well adapted to intense activity while diving. There is no mixing of venous and arterial blood within the heart; however, they have evolved a unique method for doing so outside the heart (see Figure 2.10, Chapter 2). The right ventricle gives off the left aorta and main pulmonary artery while the right aorta arises from the left ventricle. During low level activity, the right ventricle pumps blood to the main pulmonary artery and the left ventricle ejects into the right aorta. The foramen of Panizza (a hole between the roots of the left and right aortae) directs flow from the right to the left aorta. Under these conditions, the circulations are effectively separate. When diving, a special valve (known as a cog wheel valve) located at the origin of the pulmonary valve closes in response to neurological control. As a result, the outflow to the main pulmonary artery is blocked and the right ventricle ejects exclusively into the left aorta. This arrangement facilitates shunting of blood from the pulmonary to the systemic circuits during diving that is functionally similar to reptiles with three-chambered hearts, but by a unique mechanism. There is now evidence that crocodiles evolved this ability secondarily, having evolved from formerly terrestrial-based and possibly warm-blooded dinosaur ancestors (archosaurs) that had complete separation of the pulmonary and systemic circulations [6].

(iv) Mammals are endothermic meaning they can maintain a warm body temperature independent of the environment. Like all endothermic species, which includes some non-mammals such as birds, this allows them to remain active in cold or warm weather, but at the cost of maintaining a higher metabolic rate, requiring a larger cardiac output and higher blood pressure. This is made possible by a four-chambered heart and completely separate pulmonary and systemic circulations (Figure 7.2(d)) which allows them to function at different pressures. On average, mammals have a blood pressure around 130/70 mmHg, whereas pulmonary pressure in mammals is around 25/15. High systemic pressures ensure rapid delivery of nutrients and oxygen to peripheral tissues, while the lower pulmonary pressures are ideal for gas exchange in lungs. The latter takes place in tiny lung air sacs (called alveoli) of which there are estimated to be over 400 million in humans [7]. Gas exchange occurs between alveolar air and blood contained in surrounding capillaries through extremely thin tissue gaps of around 0.1–0.2 μm. This allows efficient gas transfer, but also requires that capillary blood pressure is low to protect these delicate structures. For this reason, pulmonary pressure in air breathing mammals must be lower than in the systemic circulation. The evolution of a four-chambered heart pumping dual circulations in parallel made this possible. An important requirement of dual circulations is that flow in each must be perfectly matched. This is made possible by careful regulation of the output of each ventricle. Synchronous control of the heart rhythm and rate ensures that the ventricles contract simultaneously. In addition, the Starling response (discussed in Chapter 3) provides independent responsiveness of each ventricle to stretch and therefore load, allowing each to respond to variations in output by the other. An additional requirement of the lower pressure in the pulmonary circulation is that the lungs must be anatomically close to the heart to avoid dissipation of pressure through long vessels. Hence co-location of the heart and lungs in the chest surrounded by a protective bony cage was an important evolutionary advantage.

Birds are reptiles, but like mammals they are also endothermic. They have even higher energy needs to meet the demands of flying. They do this by maintaining a higher blood pressure than mammals of around 160/50 mmHg with some as high as 200/70 mmHg. Avian heart rate varies with body size, ranging from 85–335 per minute in swans to 520–1,260 per minute in hummingbirds [8]. This provides a cardiac output up to six times greater than mammals. The high blood pressure in birds also assists withstanding high gravitational forces experienced during hunting flight. To achieve this, birds have also evolved a four-chambered heart and separate pulmonary and systemic circulations. However, the evolution of avian and mammal hearts occurred separately. Although both are vertebrates, mammals branched away from early vertebrates around 300 million years ago, while birds evolved from dinosaurs around 175 million years ago. Whether dinosaurs had already evolved a four-chambered heart is of particular interest because it relates to their metabolic requirements and whether they were also endothermic. Soft tissues are only rarely seen in the fossil record, however a well-preserved fossil of an herbivorous dinosaur from the late Cretaceous period (around 72–66 million years ago) was discovered in 2000 which allowed 3D reconstruction of the thorax. This revealed very good evidence of a four-chambered heart and a single aorta arising from the left ventricle [9]. Thus, birds and at least late-stage dinosaurs evolved four-chambered hearts. This occurred independently from that of mammals and is an example of evolutionary convergence driven by similar evolutionary pressures.

Thus, evolution has led to an extraordinarily diverse range of cardiovascular systems. Circulations can be open or closed and pumped by one, two-, three- and four-chambered hearts. What is so remarkable is how evolutionary pressures and natural selection have ensured that each circulatory plan perfectly suits the environment and needs of the animal it supports; there is no superior model. The circulatory performance of birds is many times greater than that of molluscs, however such a level of performance would be as wasteful for molluscs as theirs would be insufficient for birds. Our success as humans in spreading across the earth probably relates more to our brains and the adaptability it affords than to any other aspect of our physiology. Our cardiovascular system cannot be said to be superior to that of any other animal's. Rather it is simply exquisitely suited to our needs.

Anatomy of the Circulation

The human circulatory system (Figure 7.3), as in all mammals, consists of a closed circuit of arteries, capillaries and veins. The heart pumps blood into the arteries at high pressure which propels it to the capillaries in the tissues where nutrient delivery and gas exchange occur and from there it is collected in veins and returned to the heart for recycling. This arrangement works in the systemic and pulmonary circulations, although pressure is around five times higher in the aorta than the pulmonary arteries. Detailed descriptions of the circulation can be found in many excellent anatomical textbooks. Here an overview is provided to assist later discussion of the functions and roles of its component parts.

The heart ejects into the pulmonary and systemic circulations simultaneously. The aorta receives oxygenated blood to be circulated through the organs and tissues of the body. It arises from the base of the heart as the ascending aorta, which is about 3 cm in diameter and travels upwards for about 5 cm where it then forms the aortic arch, which curves posteriorly (towards the back) to become the descending (towards the feet) aorta. The descending aorta divides into left and right common iliac arteries which supply the pelvis and legs. The first branches of the ascending aorta are the left and right coronary arteries which arise close to its origin. It then

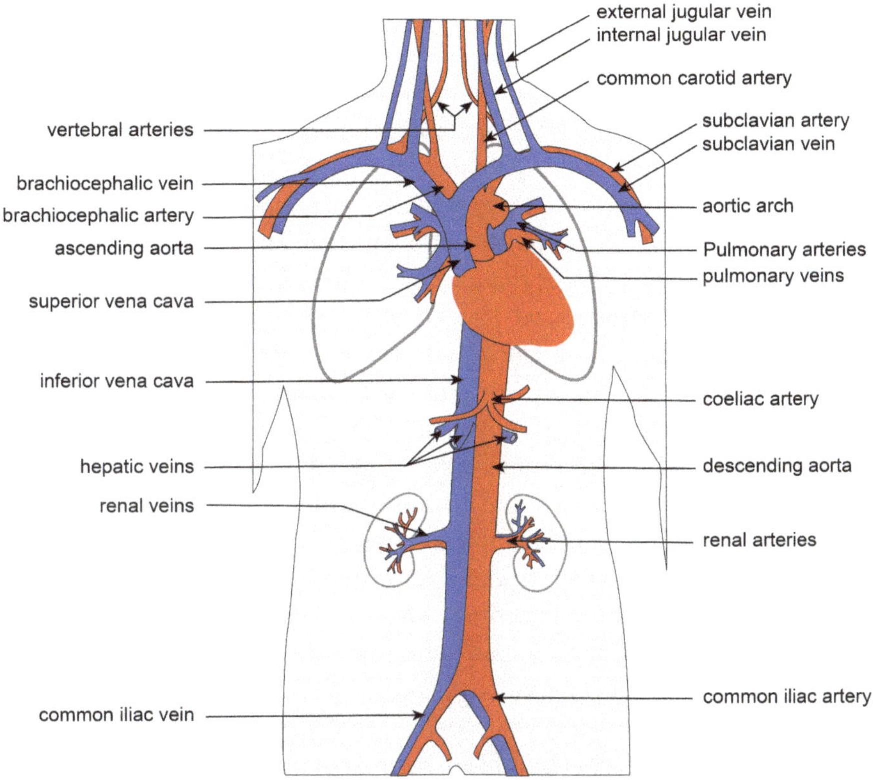

Figure 7.3: Diagrammatic illustration of the circulation in mammals (in this example human). The left ventricle ejects into the ascending aorta which arches rearwards and then downwards as the descending aorta behind the heart giving off branches which supply the arms, head, abdominal organs, pelvis and legs. Blood drains from the legs into the common iliac veins which join to form the inferior vena cava, which then travels upwards, receiving blood from the abdominal organs and continues to join the right atrium into which it drains. Blood from the arms, head and chest drains into the brachiocephalic veins which join to form the superior vena cava which also drains into the right atrium. The main pulmonary artery arises from the right ventricle and divides into left and right branches to supply each lung. Veins draining the lungs accompany them to return oxygenated blood to the left atrium from whence it is pumped into the left ventricle for recirculation.

travels cranially (towards the head) and gives off the brachiocephalic (brachial comes from the Latin for arm and cephalic from the Greek for head) artery which soon divides into the right common carotid and right subclavian arteries. The former supplies the head including the brain, while the latter supplies the right arm. At the apex of the aortic arch, the left common carotid and left subclavian arteries arise separately supplying the head and left arm. On its way inferiorly (towards the feet), the descending aorta gives off numerous small but crucial branches, including many small arteries which supply the spinal cord. Its first major branch is the coeliac artery which supplies the spleen, stomach, pancreas, liver and upper part of the intestine. This region of the circulation is known as the splanchnic vascular bed. It has a major role in digestion, metabolism and absorption of nutrients and as discussed later, it also plays an important role as a reservoir in the circulation. Unlike other arteries, the coeliac is not accompanied by a coeliac vein draining blood from tissues it supplies. Blood from the spleen, stomach, pancreas and intestines (sometimes referred to as the pre-portal organs) drains into the portal vein which travels to the liver where it branches into small vessels which ramify that organ, supporting the liver's role in digestion and metabolism. The descending aorta next gives rise to the superior and inferior mesenteric arteries which supply the lower parts of the intestines. The left and right renal arteries arise from the descending aorta about two-thirds of its length from the arch, supplying the left and right kidneys.

The common iliac veins drain deoxygenated blood from the legs and pelvis. They join to form the inferior vena cava (vena is Latin for vein and cava derives from the Latin for "hollow" at a time when the circulation of blood was unknown). The inferior vena cava is a large vessel which varies considerably in size between inspiration and expiration, being smaller on inspiration as blood is "sucked" into the right atrium. It runs upwards towards the heart, receiving branches from renal and mesenteric veins. As discussed above, the portal vein carries blood from the stomach and upper intestines to the liver, where nutrients absorbed are metabolised. Three large hepatic veins and several smaller ones connected to the inferior vena cava receive blood from the liver. The inferior vena cava continues upwards and joins the right atrium into which it drains. Deoxygenated blood from the upper body returns separately to the heart. Blood from the arms drains into the left and right subclavian veins. As these large veins continue medially (towards the centre of the body), they receive blood from the head via the left and right internal and external jugular veins to become the left and right brachiocephalic veins. They then curve medially and inferiorly and join about 7 cm above the heart to form the superior vena cava. This is a large vessel, 2.4 cm in diameter, which then continues to the heart, draining into the right atrium.

The pulmonary artery arises at the outflow of the right ventricle above the pulmonary valve. It is about 3 cm in diameter and runs upwards and backwards for about 5 cm where it divides to form the left and right pulmonary arteries which carry deoxygenated blood to the lungs. The right pulmonary artery is a little longer than the left and travels horizontally to the right lung where it divides into two main branches which further subdivide to ramify in the upper and lower lobes of the right lung. The left pulmonary artery also runs horizontally to the left lung where it divides into two main and then smaller branches to supply the upper and lower lobes. Branches of the left and right pulmonary arteries continue to divide into ever smaller vessels, eventually forming capillaries which surround the pulmonary alveoli where gas exchange takes place. Blood thus oxygenated is collected in small venules which join to form veins. These also join forming ever larger veins, eventually forming two pulmonary veins draining each lung. Although some variation on the anatomy of the pulmonary veins occurs, in most people the four pulmonary veins are connected separately to the left atrium. Oxygenated blood then flows from the left atrium into the left ventricle and from there is recirculated to the organs and tissues of the body.

Many other important structures in the upper and lower body are supplied by smaller arteries and veins, which are not included here for simplicity. Detailed descriptions of these can be found in many excellent anatomical textbooks.

Pulsatile Pumping

The heart pumps cyclically, meaning that blood is ejected into the aorta intermittently. This was ideally suited to the open circulations that initially evolved. However, for reasons discussed below, it has significant costs when used in closed circulations. Despite these, implementing a constant flow pump would have required a completely different evolutionary pathway and may not have been possible. This is because relaxation between heart beats is essential to allow perfusion of the muscle which drives the heart. Oxygen and nutrients are delivered in blood which is pumped through the coronary arteries and although the momentum for this flow is generated during systole, relatively little perfusion of the heart muscle can occur during this phase. This is because tension generated within the walls of the heart during systole inhibits blood flow in the coronary vessels which ramify them. It is when the muscle relaxes that most perfusion takes place. While this arrangement is very effective in supplying the energy to sustain cardiac performance, it also has adverse effects on the circulation which required important evolutionary adaptations.

Pulsatile ejection means that blood entering the circulation is also pulsatile so that pressure and flow increase and decrease with each heartbeat. This has significant disadvantages because (i) energy is required to generate pressure and much of this is lost as pressure falls at the end of the pulse cycle and (ii) the drag effect of the blood vessel lining on flow is greater with pulsatile than constant flow. The evolutionary solution for this is both simple and exquisite; the development of blood vessels that are elastic and distensible. In all closed circulatory systems, the large vessels that receive the cardiac output have evolved these properties. The aorta is such a vessel and is expanded by the left ventricular ejection during systole and then recoils elastically during diastole once the aortic valve closes. This has the effect of smoothing the pulsatile character of pressure and flow. For example, in humans the ratio of maximum aortic blood flow during systole to the mean (the average during systole and diastole) flow rate decreases from six in the aortic arch to two in the leg arteries as a result of this smoothing action [10] and decreases further more distally so that flow is almost constant in the small capillaries in which gas exchange and nutrients delivery occur. It also acts as a transient energy storage mechanism. Energy is expended in distending the aorta and then released as it recoils. About 80% of that energy is recovered as the aorta recoils and serves to maintain blood pressure during diastole. In contrast, a pulsatile pump operating in a rigid tube system would generate a much greater pulse pressure and flow during systole but flow would collapse to low levels between ejections and the energy required to eject would be much greater. The elastic property of arteries was first described in 1881 by Charles Roy in a series of elegant experiments [11] and have been confirmed in many species. Thus, elasticity of the arteries is a major element for controlling the behaviour of the circulation and enhancing its performance and efficiency.

Non-linear Elasticity

Arteries that behaved as simple elastic tubes would not be suitable for the circulation. This is because elastic materials when stretched do so in proportion to the force applied in a simple progressive manner, described as a linear relationship. In the context of the aorta, this would result in progressive dilatation as pressure increased with the risk of rupture. In practice, the elasticity of arteries is non-linear, meaning that as more force is applied, they become stiffer and less elastic with the result that the aorta dilates less at higher than lower pressures for the same increment in pressure. This non-linear elasticity is crucial for a distensible tube receiving the pulsatile output from the heart by allowing it to distend and recoil repeatedly in a stable manner. It is also essential for protecting the aorta from aneurysm (permanent dilatation) and rupture. The degree of elasticity of a material is the amount of stretch produced by a given force. It is expressed as the modulus which is measured as the resistance to stretch for a given force and so is inverse to the degree of "stretchiness". Since arteries are stiffer and less elastic at higher pressures, the elastic modulus is also greater at higher than lower pressures.

This increase in elastic modulus with increasing pressure has been observed in the aortae of a wide range of vertebrate and non-vertebrate species, including squid, lobster, toad, snake, tuna, shark, rat and whale. In each species, the modulus increases sharply at higher pressures. The relationship between pressure and modulus differs greatly between species (Figure 7.4, left panel (a)); however when normalised for the physiological blood pressure of each species, they are very similar (Figure 7.4, left panel (b)) [12]. Even more remarkable is the fact that the elastic modulus in each species is almost the same at its average *in vivo* blood pressure. Thus, despite blood pressures varying from 20/10 mmHg to 130/80 mmHg among lobster, squid, shark and rat, they all have an elastic modulus of the aorta which is around 0.4 MPa. This also means that the aorta in each species has the same ability to damp and smooth pressure and flow pulses

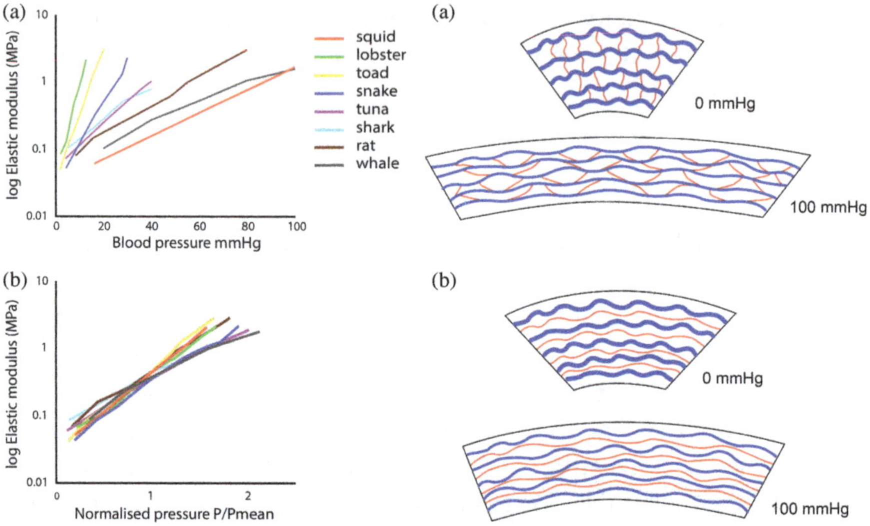

Figure 7.4: Left panel: (a) Aortic elastic modulus in several species measured at varying pressures. The results for elastic modulus are shown here as a log function, which converts the curve to a linear plot. The curves for these species differ, however the elastic modulus of the aorta increases markedly at higher pressures in all of them. (b) The results shown in (a) are replotted here showing the relationship between elastic modulus and pressure normalised to the physiological pressure of each species. In this case, the curves are closely overlapping, indicating that the elastic modulus of these aortae at the physiological pressure is similar in each species. Right panel: Diagrammatic illustration of the microscopic structure of the aortic wall in whales, showing elastin lamellae (darker lines) and collagen (lighter lines). (a) In the aortic arch, elastin lamellae are arranged circumferentially and at low pressure they have a wavy appearance. Collagen fibres are distributed between the elastin lamellae and arranged close to a radial orientation. As pressure is increased, the elastin is stretched and the collagen sandwiched between lamellae becomes more tangential. As this occurs, more of the load on the aortic wall is borne by collagen. Since collagen is much stiffer than elastin, the elasticity of the aorta is greatly reduced. (b) In the descending aorta, both elastin and collagen are arranged circumferentially so that even at low pressures the load is borne by collagen. As a result, the descending aorta has relatively little elasticity compared to the arch. These examples illustrate the most marked differences in aortic elasticity observed between aortic sections found in aorta of whales. In many other mammals, a more graded difference is found and is likely to reflect structures intermediate to those illustrated in (a) and (b) [12].

generated by the heart and suggests that the aorta has evolved in each species to match the ejection performance of the heart. This in turn suggests that it must have been a vital evolutionary advantage.

Structure of Arteries

The distensible behaviour of arteries is dependent on the presence of elastic and rigid elements in the wall. At a microscopic level, arterial walls are composed of an inner, middle and outer layer, called intima, media and adventitia, respectively. The media is the thickest and contains circumferential layers (called lamellae) of elastic tissue interspersed with collagen and smooth muscle cells. Electron micrographic studies indicate that elastin and collagen are both connected to smooth muscle cells which act as a link between them [13]. The number of lamellae varies with the thickness of the wall and in turn with the size of the species; for example, from 6 in the toad to 820 in the whale. In vertebrates, the elastic layer is made of elastin, an elastic protein which is responsible for the elastic quality of tissues, such as skin as well as and many others throughout the body. In contrast, collagen is a structural protein which has a high tensile strength but very little elasticity and is abundant in tendons and strong ligaments. The characteristic non-linear elastic behaviour of the aorta is due to the combined action of elastin and collagen which act as a composite material in the arterial wall. Microscopic examination of arteries studied at varying pressures has shown elastin to have a wavy appearance at low pressure which then straightens as pressure is increased. As this happens, the load on the arterial wall is also increasingly transferred to collagen, explaining the increasing stiffness as load increases.

The distensibility of arteries varies with their size and proximity to the heart, larger and proximal vessel being more compliant. This means that the upper part of the aorta, particularly the arch, expands most, an arrangement which is optimal for absorbing the energy of the left ventricular pressure wave and releasing it during diastole. In addition, it allows damping and smoothing of the pressure wave to occur early in the cycle, affording the best protection of the distal vessels. This gradation of compliance between the proximal and distal aorta is found in all species but appears to be greatest in the whale. In large whales, the diameter of the arch can be over three times that of the descending aorta and contain almost all the compliance of the arterial system, so that the distal aorta can be up to 30 times stiffer than the arch. The unusually large size of the aortic arch in whales helps maintain pressure during the long diastolic intervals when heart rate can slow from around 25–30 beats/min to 4–8 beats/min during dives [14]. The structural basis for this proximal to distal transition in aortic compliance is related to the content and structural arrangement of elastin and collagen. A much higher (up to sixfold) ratio of elastin to collagen in the aortic arch than the descending aorta of fin whales contributes to its greater compliance [15]. In addition, however, the arrangement of elastin and collagen at a microscopic level is important. In the proximal aorta, while elastin is arranged in circumferential lamellae around the vessel wall, collagen fibres are aligned radially (Figure 7.4, right panel (a)). Therefore, in the arch, elastin takes almost all the load during early stretch, the collagen only engaging as elastin is stretched further. This arrangement means that at physiological blood pressure the arch is compliant. In contrast, in the distal aorta, collagen and elastin are both arranged circumferentially (Figure 7.4, right panel (b)), and therefore collagen carries the load earlier, rendering the vessel stiffer [12]. This description accounts for the extreme differences in compliance at the top and bottom of the aorta and one can readily envisage gradations in structure to account for a gradual transition along its length.

Invertebrate Arteries

Matching of the elastic modulus of the aorta to the ventricular pressure pulse occurs across many species including vertebrates and invertebrates, and yet elastin is only present in the former and not the latter. This means that some alternative elastic mechanism must have evolved in invertebrates and recent research has confirmed this. Octopi also have highly distensible elastic arteries and have elastic fibres in the wall of the aorta. Although these have properties similar to elastin, they are provided by a separate and unique protein. Other invertebrates also have elastic arteries, which are based on the presence of elastic proteins called microfibrils. These provide the non-linear elastic properties of the aortae in the low-pressure circulations found in many invertebrates. Microfibrils are also crucial for the correct building of elastin fibres in vertebrate arteries by forming a scaffold on which elastin is assembled. They are also incorporated into construction of mature elastin fibres.

Microfibrils and Elastic Fibre Assembly

Microfibrils are important components of the extracellular matrix of all tissues and are vital to the elastic property of tissues such as skin, lung and blood vessels. They are ancient and highly conserved structures in biology. For example, microfibrils are the main source of the elastic properties of tissues in simple invertebrates such as jellyfish. Microfibrils also play a key role in the elastic properties of vertebrates because they are essential for producing the elastin fibres that provide this function. Microfibrils (Figure 7.5, upper panel) are multi-molecular assemblies of the glycoprotein fibrillin formed into fine strands of around 20 nm in diameter. Fibrillin is widely present in animals from *cnidaria* (e.g., jellyfish) to mammals and is likely to have evolved at least 750 million years ago. *Cnidaria* and molluscs have just one fibrillin gene whereas most mammals have three (named fibrillin 1, 2 and 3), which evolved by gene duplication. Fibrillin 1 is the main constituent of arteries. The assembly of multiple fibrillin molecules into microfibrils is a carefully regulated process [16]. Electron microscopic examination shows fibrillin microfibrils to be roughly cylindrical in shape and composed of eight fibrillin molecules in cross section. Microfibrils are around 20 nm in diameter and have a beaded appearance along the length of the microfibril, repeating at 50–60 nm which increases to about 80 nm when stretched. In invertebrates with low blood pressures, microfibrils form a major component of the structure of arterial walls, conferring a non-linear elasticity property to them.

Elastin fibres (Figure 7.5, bottom panel) are composed of multiple elastin fibrils joined end to end and wound together. They can be arranged as single fibres or branched and in adults can be as large as 2–14 μm. Elastin fibrils are about 120 nm in diameter and are assembled as polymers of tropoelastin. Tropoelastin is a protein with a molecular weight of 70 kDa. It is synthesised by many cells including the smooth muscle cells in arterial walls. Tropoelastin evolution coincided with the appearance of closed circulations in vertebrates and therefore much later than fibrillin. It has been found in all vertebrate aortae with the exception of primitive jawless fishes such as lampreys and hagfish. However, microfibrils also form a crucial element of mature elastin fibres.

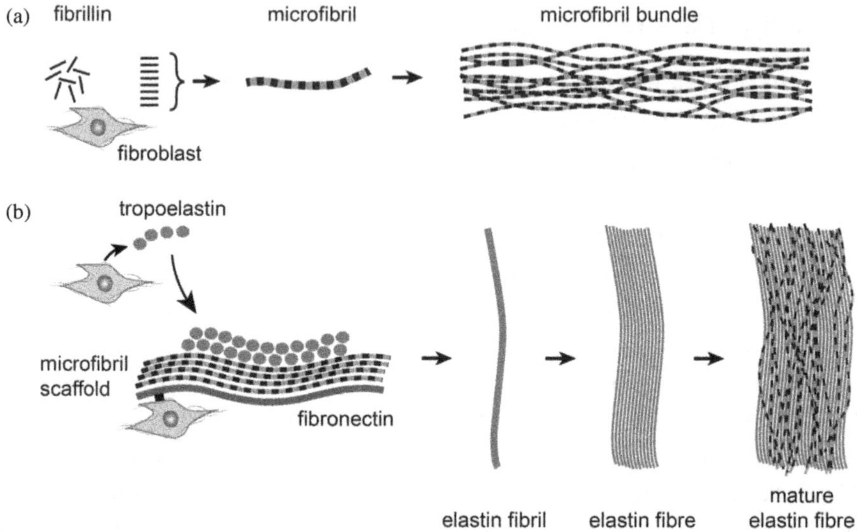

Figure 7.5: Diagrammatic illustration of microfibril and elastin formation (not drawn to scale). (a) Fibrillin is a glycoprotein secreted by fibroblasts and is assembled into microfibrils by a highly regulated process in the presence of fibroblasts. Microfibrils have a diameter of around 20 nm and are formed into bundles which provide the elastic properties of many tissues, including the arteries of invertebrates that have low blood pressures. (b) Working model of elastin fibre formation proposed by Yanagisawa and Davis [17]. Tropoelastin is a precursor of elastin secreted by elastogenic cells. It is assembled into elastin fibrils in a highly organised process. This involves fibrillin microfibrils which form a scaffold connected to fibronectin and fibroblasts on which tropoelastin polymerisation takes place to form fibrils of elastin. Elastin fibrils are then crosslinked and wound together to form elastin fibres. Mature elastin fibres are then completed by the addition of microfibrils which are bound to the exterior of elastin fibres and in gaps between elastin fibrils. Elastin fibrils are much larger than fibrillin microfibrils with a diameter of around 120 nm and mature elastin fibres can be a large as 14 μm in diameter. Elastin fibres provide the elastic properties of arteries in almost all vertebrates whose closed circulations work at higher pressures than invertebrates.

The assembly of elastin fibrils from tropoelastin is a carefully regulated and organised process involving a number of other proteins and enzymes [17]. In simple terms, elastin fibril assembly always takes place in the presence of microfibrils which form a scaffold required for correct deposition of tropoelastin molecules. Tropoelastin molecules are added along the length of microfibril scaffolds which assist their binding end to end and cross linking to form the fibril. Elastin fibrils are then cross linked to microfibrils and collagen in varying amounts depending on the tissue in which they are formed, and individual elastin fibrils are wound to form mature elastin fibres. The importance of microfibrils for elastin assembly is indicated by conditions like Marfan syndrome which is due to a genetic mutation in fibrillin 1 (the main isoform present in human arteries) and causes dilatation of the aorta (aneurysm) and increased risk of aortic rupture. Elastin fibres are highly insoluble and very stable with a measured half-life of around 74 years [18].

Thus, the aorta and arteries which transport blood from the heart to the tissues are constructed to exquisitely match the output of the heart. They act as a brief inter-beat energy storage mechanism which helps to maintain pressure and flow during diastole. They damp and smooth pressure and flow pulses in the circulation which reduces wall stress on blood vessels and improves nutrient and gas transfer across capillaries in the tissues. The elasticity of arteries which provides this function is present in all animals. It was made possible by the evolution of fibrillin microfibrils in ancient invertebrates which have low circulation pressures but required the additional evolution of tropoelastin and its polymerisation to form elastin fibrils to match the higher pressures of vertebrate arteries.

Resistance Arteries

The large elastic arteries which transport blood from the heart are sometimes referred to as conducting vessels but as we have seen their function goes far beyond this. It is nevertheless a useful label to distinguish them from the more distal resistance arteries which play a very important role in regulating blood flow (discussed in more detail in Chapter 5). This is facilitated by a gradual change in the structure of arteries with increasing distance from the heart. As discussed above, the walls of all arteries are composed of three layers, intima, media and adventitia. The intima consists of a layer of endothelial cells which line the lumen. The adventitia is the outer layer and consists of collagen and elastin. The media is sandwiched between these two and is the thickest of the three layers. It is also the site at which most changes are observed in arteries further from the heart. As the arteries divide into smaller vessels, the media contains an increasing number of smooth muscle cells, which gradually predominate its content. The media of intermediate-sized arteries such as the radial and ulnar artery in the forearm contains some elastin but is predominantly composed of smooth muscle cells and this gradually transitions to the smaller resistance arteries (called arterioles) in which the media is solely composed of smooth muscle cells.

Muscular Arteries

Smooth muscle cells in arterial walls confer the ability of these vessels to contract and relax and this is essential for their functions in regulating blood flow and pressure. Smooth muscle cells are spindle shaped and like cardiac and skeletal muscle they contain the contractile proteins actin and myosin. However, they differ in many other respects. Actin and myosin are arranged in regular tightly packed sarcomeres in skeletal and cardiac muscle (discussed in detail in Chapter 3), whereas the actin and myosin filaments are less densely packed in smooth muscle cells and are not arranged along the long axis of the cell, but rather at an angle to it in both directions, forming a network of filaments throughout the cytoplasm. Actin myosin filaments are connected end to end forming continuous fibrils which are then attached to the cell membrane at the end of each span. In this way, the cell as a whole shortens and lengthens with contraction and relaxation. Unlike skeletal muscle, smooth muscle is involuntary, meaning it is not regulated by conscious control, but rather by autonomic nerve stimulation and by hormones and other substances in the blood stream. A more detailed discussion of smooth muscle cell function and regulation is presented in Chapter 8.

Blood Pressure Regulation

In closed circulations, blood pressure is a function of the force of ventricular contraction and the resistance to flow through it. That resistance is determined by many factors such as the viscosity of blood, the frictional effect of vessel walls on blood flow and the total luminal area of the vascular tree. The latter is especially important because it is actively regulated to maintain a normal blood pressure. At the end of the arterial circulation, blood flow continues in capillaries. These are the smallest vessels in the circulation, 5–10 μm in diameter, and are composed of a single layer of endothelial cells. They are therefore delicate structures and would be incapable of withstanding the force of arterial blood pressure. The arterioles which feed capillaries are known as resistance vessels because they actively regulate flow through capillary beds. Collectively they also represent the main resistance against which the heart pumps and are therefore the main determinants of blood pressure. The ability to contract and relax smooth muscle cells in the walls of these arteries allows the total luminal diameter of resistance arteries to be regulated, and this in turn provides an important mechanism for controlling blood pressure. The level of blood pressure is continuously monitored by sensors called baroreceptors located in aortic arch and the carotid arteries. These function as mechanical sensors which detect stretch in these elastic vessels and provide a continuous output of electrical impulses which are transmitted to the brain stem via the vagus nerve from the aortic arch and the ninth cranial nerves from the carotid arteries. An increase in arterial stretch due to elevated pressure increases the impulse frequency and in turn activates the autonomic nervous control of the heart and resistance arteries. This acts to reduce cardiac output and relaxes the smooth muscle cells in the resistance arteries. Both effects lead to a reduction in blood pressure. The opposite occurs in response to reductions on blood pressure. This provides a mechanism for rapid, short-term regulation of blood pressure and is known as the baroreceptor reflex.

Avoiding excessive increases and decreases in blood pressure is important for maintaining a stable circulation. Hypertension is harmful because it increases stress on arteries due to excessive wall stretch and shear forces on the endothelial lining. Over time this can cause the development of arterial diseases, most notably atherosclerosis and loss of arterial elasticity. These are important causes of heart attacks and strokes. Hypertension also increases the risk of vessel rupture and haemorrhage in the brain, another cause of stroke. Baroreflexes provide a central mechanism by the autonomic nervous system for controlling blood pressure in the circulation as a whole. In addition, as discussed in Chapter 5, autonomic control of arteries regulates the distribution of cardiac output to meet increased metabolic demand of tissues and organs at a local level.

Blood flow through the microcirculation is also regulated at a local level by two intrinsic properties of arteries:

(i) Flow mediated dilatation: As discussed in Chapter 5, the endothelial lining of blood vessels is sensitive to increases in blood flow and responds by producing nitric oxide which results in relaxation of smooth muscle in the arterial wall causing dilatation. This provides an important mechanism for facilitating increased blood flow locally while at the same time limiting stress on the arterial wall.
(ii) Myogenic tone: Smooth muscle cells differ from cardiac and skeletal muscle in that they have an intrinsic ability to maintain a constant level of contraction. This is referred to as myogenic tone. The level of myogenic tone in arteries is independent of the autonomic nervous system; however, it is responsive to blood pressure. In fact, by stretching smooth muscle cells in the arterial wall, blood pressure activates signalling pathways which cause the contraction responsible for myogenic tone. Myogenic tone then sets a level of contraction which determines the arterial lumen diameter. It is against this baseline that vasodilatation or vasoconstriction occurs in response to neural, hormonal and other local stimuli. The sensitivity of small arteries to luminal pressure was first observed in 1903 by William Bayliss [19] and is now known to be a vital mechanism for maintaining flow constant during variations in pressure. The capacity to constrict in response to increased pressure also lowers pressure distally in the thin-walled capillary vessels at which tissue perfusion occurs and thereby protects them from harmful effects of excessive pressure. This is important for protecting the capillary bed in all tissues but especially so in the cerebral circulation where it provides a specialised system for selective diffusion between the blood and tissues of the brain, known as the blood-brain barrier. The regulation of contraction and relaxation of resistance arteries by external and intrinsic mechanisms is discussed in more detail in Chapter 8.

Hypertension (sustained high levels of blood pressure) is a major risk factor for cardiovascular disease, including heart attacks, stroke and heart failure. It would be difficult to overestimate its importance and significance.

Indeed, hypertension has emerged as the greatest risk factor for disease burden globally during the early years of the 21st century. In 1990, the leading risks for the global disease burden were childhood underweight, household air pollution due to solid fuels burning in the home and tobacco smoking, which collectively accounted for 20% of total risk. Substantial improvements in childhood health and survival have transformed this so that in 2010, hypertension, followed by tobacco smoking and household air pollution were the leading causes of disease globally [20]. Hypertension accounts for almost half of all deaths from cardiovascular disease globally and is a growing burden in developing nations. For example, an estimated 30% of adults in south Asia have high blood pressure of whom only 1 in 7 are controlled with treatment [21]. Treatments for reducing blood pressure have improved hugely during the past half century so that it can now be well controlled in most people. Nevertheless, the underlying causes as to why and how hypertension develops in some people remain poorly understood. This is now the subject of intense research and hopefully will lead to new ways to prevent it in the future. A summary of vascular changes known to occur in hypertension is given here. More detailed accounts of areas of current research interest can be found in reviews [22–25].

Vascular remodelling

A characteristic feature of hypertension is a resetting of the baroreflex which maintains pressure at a higher level. The reasons why this occurs are unclear. Over time however, the sustained increase in blood pressure is associated with changes in the structure of arteries and an increased arterial vascular resistance. This is known as vascular remodelling and involves changes in smooth muscle cells in the arterial wall as well as other cellular components, the endothelial lining and the content of collagen and elastin. Remodelling occurs in both large conductive arteries and small resistance arteries with some differences in each.

Conductive arteries

In large conductive vessels, the most prominent change is an increase in wall thickness. This is mostly due to an increase in the medial layer secondary to hypertrophy of smooth muscle cells. The effect of this is to render these arteries stiffer and less elastic which limits aortic distension during systole resulting in a reduced ability to absorb the energy of left ventricular ejection. This in turn causes a greater systolic pressure rise (known as pulse pressure) with each heartbeat (Figure 7.6(a)). In addition, the absence of aortic recoil during diastole impairs diastolic blood flow. This also increases the load against which the left ventricle has to eject and contributes to the development of ventricular hypertrophy and ultimately failure.

Resistance arteries

Small resistance arteries also undergo hypertrophy in hypertension. This matters particularly because the vascular resistance in small arteries and arterioles (lumen diameter 100–350 μm and <100 μm, respectively) accounts for almost 50% of total peripheral resistance. Consequently, they are major determinants of blood pressure, which is sensitive to even small changes in their lumen diameter. Essential hypertension, meaning no primary cause is known, is the most frequent form of high blood pressure. In these cases, remodelling of resistance arteries takes the form of an increase in medial thickness and a reduction in lumen diameter [26], known as eutrophic remodelling (Figure 7.6(b)). A reduction in lumen diameter is highly significant because vascular resistance is related to the fourth power of the lumen radius, meaning that even small changes in radius and lumen size have a large effect on resistance. One possibility is that this remodelling is a protective response to hypertension as a means to mitigate the effect of elevated pressure on the distal circulation; however, this remains to be clarified.

In secondary hypertension due to kidney disease or other causes, resistance arteries undergo remodelling in which smooth muscle cells in the media increase in size (hypertrophy) and number (hyperplasia). This results in an increased thickness and cross-sectional area of the medial layer as well as a reduction in lumen diameter and is known as hypertrophic remodelling (Figure 7.6(b)) resulting in an increase in vascular resistance and increased blood pressure. Again, it remains unclear whether this is a secondary response to protect the distal circulation from elevated pressure or part of the primary cause of it. It has also been suggested that

(a)

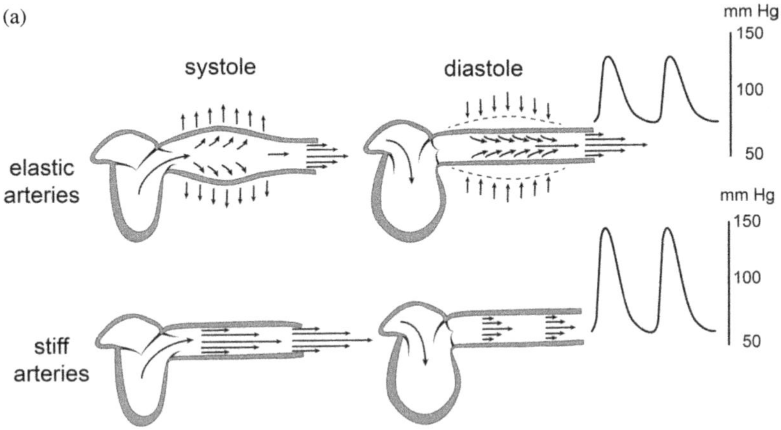

(b)

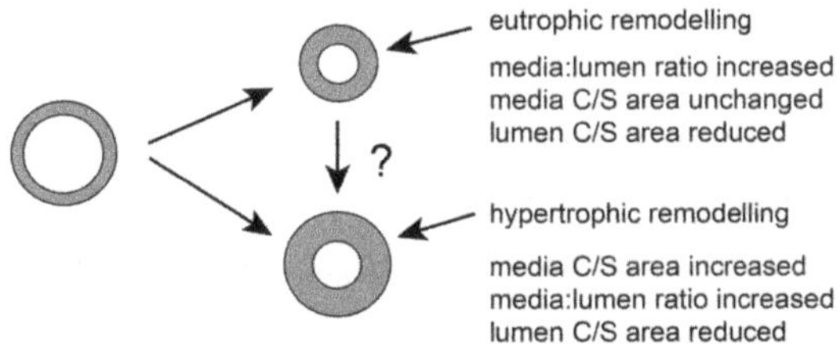

Figure 7.6: (a) Diagrammatic illustration of the effect of arterial stiffness on haemodynamics. The normal elastic properties of the large proximal arteries allow them to distend during systole (top left), absorbing some energy of left ventricular ejection and limiting the pressure generated. During diastole (top right), this distension recoils, releasing stored energy and assisting blood flow. Increased stiffness in the large arteries prevents distension during systole (bottom left). This has the effect of increasing pressure during left ventricular ejection and accelerating flow during systole. The absence of elastic recoil during diastole (bottom right) results in reduced flow and a greater fall in pressure during diastole. The effect on blood pressure is an increase in pulse pressure (shown on the right), increased stiffness results in higher systolic and lower diastolic pressure. After Briet *et al.* (2012) [36]. (b) Resistance artery remodelling in hypertension. Eutrophic remodelling occurs in essential hypertension and consists of an increased thickness of the medial layer. The vessel lumen and external diameters are reduced with the effect that the medial cross-sectional area remains unchanged. Hypertrophic remodelling occurs in secondary hypertension due to kidney disease or other causes and consists of growth of the media layer, resulting in an increase in its cross-sectional area. This is due to an increase in the size and number of smooth muscle cells. It is unclear whether eutrophic remodelling can progress to hypertrophic remodelling over time. After Intengan and Shiffrin (2000) [37].

eutrophic remodelling may be an early response to hypertension which may later progress to hypertrophic remodelling. However, at present this also remains to be established.

Resistance arteries play a crucial role in regulating blood pressure due to their ability to relax and contract in response to local and remote stimuli. These effects are mediated by smooth muscle cells in the media layer and are abnormal in hypertension. The capacity of resistance arteries to relax (and therefore dilate and reduce resistance to flow) can be measured by the flow response at a given pressure. The flow at maximal dilatation (also referred to as minimal resistance) is an indication of the vasodilator reserve, meaning the ability to increase flow above the resting level. The minimal resistance to blood flow at maximal vasodilatation in the forearms of hypertensive patients was first shown to be increased [27] by Folkow in 1958. This abnormality has since been confirmed in many studies which have also clarified that it is caused by a structural abnormality in the resistance vessels. Given the magnitude of the global health burden of hypertension, understanding the processes which bring about these vascular changes has to be one of the greatest challenges facing modern medicine. Despite great improvement in the treatments available for hypertension over the past decades, all are geared to lowering blood pressure by modifying aspects of cardiovascular physiology. All require life-long treatment and, importantly, in essential hypertension which accounts for most cases, none target the underlying cause. It seems not unrealistic to hope that new knowledge will emerge in the next decades to clarify the underlying causes of hypertension and that this will lead to more effective measures that can be implemented at its onset to prevent its progression and the consequent structural abnormalities.

Capillary filtration and transport

Beyond the smallest arterioles, blood flow continues in capillary networks which ramify tissues. This is the site at which nutrient delivery and waste product removal take place. Capillaries are small vessels, 5–10 μ in diameter, consisting of a single layer of endothelial cells. Transport of molecules to and from blood to the tissues takes place by four principal mechanisms, some of which are tightly regulated (Figure 7.7(a)).

(i) Small molecules like oxygen and carbon dioxide simply diffuse through the endothelial cell membrane. This is also the case for small lipid soluble molecules such as alcohol and some drugs and is dependent on concentrations on each side of the membranes involved.

(ii) Cell membranes are composed of two lipid layers facing the extracellular and intracellular environment. Water soluble molecules are not amenable to diffusion through these. Instead, transport takes place between endothelial cells. This is also an important route for inflammatory cell migration from blood to tissues and is a crucial aspect of inflammatory and immune responses to injury and infection. It is an actively regulated process assisted by the expression of adhesion molecules on the surface endothelial cells (see Figure 6.5, Chapter 6 and related text for more details). This is the case in most tissues of the body, with the exception of the brain where transport of molecules between blood and tissues is limited by what is known as the blood brain barrier, discussed below.

(iii) Cell membranes are also impermeable to proteins. Transport takes place via receptors which bind specific proteins to the endothelial cell surface and facilitate delivery to the tissues.

(iv) Some molecules which are critical to tissue survival have specific transport mechanisms that are tightly regulated. Glucose, which is an essential energy source for cells, is such an example. Glucose transporters are a large family of proteins which facilitate the transport of glucose across cell membranes and are found in all animals. Glucose transporter 1 (GLUT1) is present in humans, bound to the cell membranes of endothelial cells where it facilitates the uptake of glucose to tissues. GLUT1 is also present on the surface of red blood cells, which act as temporary storage sites for glucose to maintain blood glucose levels constant, in this case acting as a bidirectional transporter.

Blood brain barrier

Transport of molecules and cells from blood and the brain is much more tightly regulated than in other parts of the body. To achieve this, endothelial cells of the cerebral circulation form a more restrictive barrier, known as the blood brain barrier, which is crucial for normal brain function.

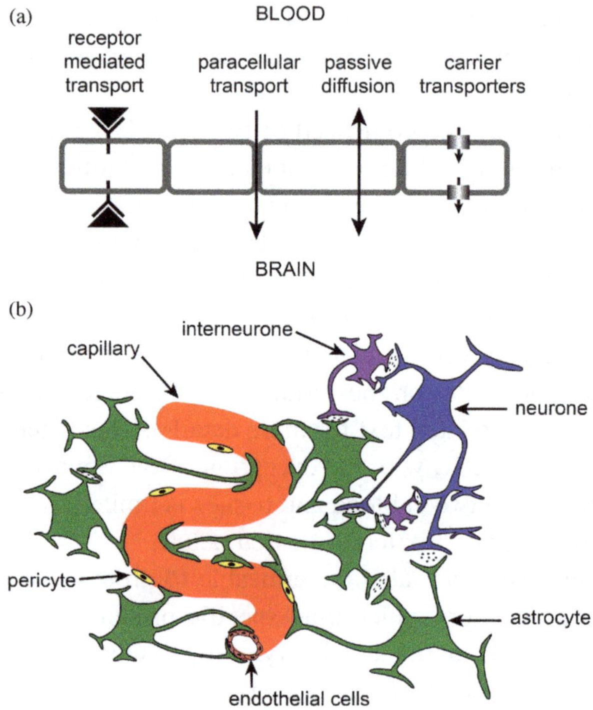

Figure 7.7: (a) Capillary filtration and transport. Transport of molecules from blood to tissues takes place by several mechanisms. Passive diffusion: gases like oxygen and carbon dioxide and small lipid soluble molecules such as alcohol are freely diffusible through cell membranes dependent on local concentration gradients. Paracellular transport: water soluble molecules cannot diffuse through cell membranes; transport between blood and tissues takes place between endothelial cells. Inflammatory cells also migrate into tissues by this route. Receptor mediated transport: large molecules such as proteins are bound to specific receptors, transported through the endothelial cell cytoplasm and released from the cell surface facing tissues. Carrier transporters: molecules which are in high demand by tissues such as glucose have dedicated transporters. Uptake of glucose to tissues is done by a tightly regulated membrane bound transporter. Blood brain barrier: capillary transport in the brain is more restricted than in other parts of the circulation. Brain capillary endothelial cells have fewer receptors and large protein uptake is restricted. Endothelial cells are much more tightly joined in brain capillaries which restricts paracellular transport. Together these limit the movement of many toxic molecules including infections and inflammatory cells from entering the brain. (b) Neurovascular unit: neurovascular units regulate local blood flow in the brain to match energy demands. Neurones are responsible for most of the brain's activity and are the biggest energy users. Neurones (blue) interact with astrocytes (green) by neurotransmission, signalling their energy needs. Astrocytes in turn are in contact with capillary endothelial cells and transmit this information. Endothelial cells respond by producing nitric oxide (when demand increases) which

Diffusion of gases and small lipid soluble molecules is similar to other parts of the circulation. However individual endothelial cells are much more tightly joined together by multiple gap junctions than in other capillary networks and this has the effect of limiting the transport of water-soluble molecules and inflammatory cells. Gap junctions are areas of very close contact between adjacent cells and play an important role in cell interaction in many tissues including the heart (see Figure 4.5 and related text in Chapter 4). In addition, expression of adhesion molecules (which are involved in attracting and binding inflammatory cells) on endothelial cells is reduced compared to other tissues. Receptors which bind specific proteins are fewer and limited to those essential for normal cerebral function such as water and amino acids. It is for this reason that most infections do not cross the blood brain barrier, although there are some exceptions to this such as syphilis and toxoplasmosis. Specific transporters such as GLUT1 are abundant on endothelial cells of the cerebral circulation. This is because the brain almost exclusively uses glucose for energy production. In addition, brain function depends on constant electrical neurotransmission, which has a high energy usage. Thus, although the brain weighs just 2% of body weight, it receives 15% of the cardiac output, accounts for consumption of 20% of total body oxygen and 25% of total blood glucose.

Neurovascular unit

Capillary endothelial cells in the brain are supported by other brain cells, forming neurovascular units which regulates blood flow in relation to energy needs (Figure 7.7(b)). Neurones are the brain cells which carry signals within the brain and to and from it to the rest of the body and account for most of the brain's energy use. Neurones also form part of the neurovascular units which regulate blood flow. They do this through contact with second brain cell type called astrocytes. Astrocytes in turn are in

Figure 7.7: (*see figure on facing page*) causes an increase in local blood flow by dilating resistance arteries. Nitric oxide also depolarises endothelial cells and this is transmitted upstream by intercellular conduction and causes vasodilatation of proximal arterioles. Astrocytes are also in contact with pericytes, which regulate capillary growth and repair.

contact with the endothelial cells of capillaries (Figure 7.7(b)). Neurones convey signals to astrocytes about their energy needs and in turn astrocytes communicate this to endothelial cells, which in turn can regulate local blood flow by modifying the tone in local resistance arteries. They do this by producing nitric oxide which acts as a vasodilator and by electrical communication with adjacent endothelial cells and smooth muscle cells (for more details of flow regulation, see Figure 5.2 and related text in Chapter 5). Pericytes are closely attached to endothelial cells and are important for growth and repair as well as maintaining normal blood brain barrier function.

Thus, while we may be inclined to think of the heart as the active centre of the circulation, its normal function is only possible in association with a large number of perfectly matched mechanisms in the arteries which allow them to receive its output and deliver it to the tissues with exquisite efficiency. Each section of the delivery side of the circulation has its unique function from expanding energy storing proximal vessels to small arteries which can self-regulate their response to pressure and flow and capillary networks that ensure appropriate filtration and transport of molecules to each tissue to match its needs. The heart works at the centre of this complex system with consummate efficiency. The return side of the circulation collects blood from the tissues and returns it to the heart and as we might expect, this too turns out to be more than just a delivery system.

The Circulation's Reservoir

Water is a vital resource for life and to ensure its supply during variations in rainfall and drought we build reservoirs to retain a large capacity in reserve. An adequate blood volume is similarly essential for the circulation or more specifically, an adequate volume of blood returning to the heart for recirculation. We can readily envisage that the availability of such a reserve supply of blood could be lifesaving in the event of a major blood loss through haemorrhage. However, the circulation in mammals is closed and the blood volume is confined to it, so that under normal circumstances such a reserve might not immediately appear to be an important requirement. In fact, a blood reservoir forms a vital part of the circulation and is constantly used to adjust and regulate the performance of the circulation.

As discussed in Chapter 3, cardiac muscle function is sensitive to stretch, meaning that the degree of ventricular filling is an important determinant of the force of contraction. Known as the Frank–Starling relationship, this is especially the case for the right ventricle, which operates at lower pressures than the left and has a thinner wall that is more compliant (distensible) than the left ventricle. In fact, the right ventricle is exquisitely sensitive to filling pressure, meaning that small changes in filling pressure have a large effect on cardiac output. For example, under conditions in which aortic pressure is maintained constant within the physiological range, an increase in right atrial pressure of 1 mmHg can result an increase in stroke volume of 50 mL/kg body weight [28]. The implication from this is that there has to be a mechanism to regulate right atrial pressure at a constant level to maintain stability of cardiac performance. The importance of this becomes clearer, considering the fact that standing causes pooling of around 1 L of blood in the lower body [29], especially in humans. Almost all of this pooling occurs in the veins and would lead to a marked reduction in blood flowing back to the heart without some compensatory mechanism. In addition, regulation of body temperature involves substantial increases of blood flow to the skin as a means of shedding heat. This increases the volume of blood in skin which has to be accompanied by a compensatory reduction in the blood volume elsewhere. The importance of such a mechanism can be seen in a condition called postural hypotension, in which this mechanism is deficient or absent. Standing for such people can cause fainting due to a sharp drop in cardiac output and blood pressure. This can also occur following space travel of more than a few days. Here the absence of gravity in space removes the normal posture effects on blood pooling and as a result the normal compensatory mechanisms lose their efficiency probably due to lack of use. On return to earth, astronauts frequently experience postural hypotension for a few days before returning to normal.

Blood pressure in veins is much lower than in arteries and their walls are much thinner and more distensible. It is for this reason that veins can hold as much as 75% of the systemic blood volume. It is also why they are able to store a volume of blood which can be added to or removed from the circulating blood volume to compensate for redistributions. In mammals including man, this reservoir is provided by the splanchnic circulation, which comprises blood flow to the spleen, stomach, pancreas, intestines and liver (Figure 7.8(a)). As discussed in Chapter 5, exercise involves an increase in blood flow to the active muscles and a redistribution of the circulating blood volume is necessary to compensate for this. The splanchnic circulation plays a major part in this. For example, in men undertaking supine cycling (and therefore avoiding the pooling effect of upright posture), splanchnic blood volume decreases by around 40% [30]. The displaced blood flows back to the heart, adding to right atrial filling. During intense exercise, heart rate may increase up 180 beats/min and cardiac output by up to fivefold due to sympathetic stimulation. To allow this, the heart must receive an equal volume from the tissues. After an interval of a few minutes exercise, the increased cardiac output also results in an increased venous return, reaching a new equilibrium in which return matches output. However, in the short-term, venous return is inadequate for this and without an additional input volume, right atrial pressure would fall, causing a reduction in right ventricular contraction and stroke volume, hindering the response to exercise. Therefore, the ability to redistribute blood stored in the splanchnic circulation to increase venous return within seconds of the onset of exercise provides an essential mechanism to maintain right atrial pressure until that new equilibrium is established.

Exercise is just one example in which this mechanism plays a vital role. A similar response occurs following a substantial haemorrhage. The loss of a significant blood volume results in a reduction in venous return and right atrial pressure which would in turn lead to a fall in cardiac output and pressure. Here also, blood is mobilised from the splanchnic bed as a compensatory response to maintain venous return and pressure. Following acute haemorrhage of 17% of the blood volume (around twice the amount donated in a single sitting) in humans, redistribution of blood from the splanchnic bed amounted to around half the loss [31]. The increased blood flow to skin to increase heat loss in response to increased body temperature similarly involves redistribution of blood from the splanchnic bed. The opposite is also true. For example, a sudden rise in blood pressure tends to increase the return of venous blood to the heart, which would further increase cardiac output and pressure. This effect can be mitigated by increasing the amount of blood stored in the splanchnic bed. The splanchnic bed therefore acts as a reservoir of blood with the ability to increase or decrease its capacity in response to circulatory needs. Almost all of this storage capacity resides in the splanchnic veins and the ability to vary its volume is by constricting or dilating these vessels. There are two mechanisms for doing this: (i) passive and (ii) active venoconstriction and venodilatation (Figure 7.8).

Passive changes in venous capacity

Like all blood vessels, veins are distended by the pressure of blood within them; if pressure increases, they distend and in turn undergo elastic recoil if pressure falls. The pressure in any organ is related to its blood flow and the vascular resistance to that flow. Therefore, when blood flow to an organ is reduced and the resistance is unchanged, then the pressure within it also falls. This means that the distending pressure on the veins is less, and they undergo a passive elastic recoil, which reduces their volume capacity. The immediate consequence of this is that blood is passively displaced from the organ into the venous drainage system. During exercise in humans, blood flow to the splanchnic bed falls depending on the severity of exercise. At maximal exercise capacity, this reduction can be up to 25% of resting value [32]. The passive shift of blood to venous return which results from this is a crucial mechanism for maintaining right atrial pressure and therefore cardiac output during exercise. The increased skin blood flow that occurs in response to heat stress is similarly accompanied by reduced splanchnic blood flow and the redistributed splanchnic blood that results compensates for blood diverted to the skin. Thus, passive changes in splanchnic vascular capacity provide an important mechanism for regulating the circulation.

Active changes in venous capacity

Active mobilisation of the splanchnic reservoir can occur by two mechanisms.

(a) Bundles of sympathetic nerves are present in the adventitial (outer) layer of the splanchnic and hepatic arteries. Stimulation of these nerves causes constriction which reduces splanchnic arterial inflow.

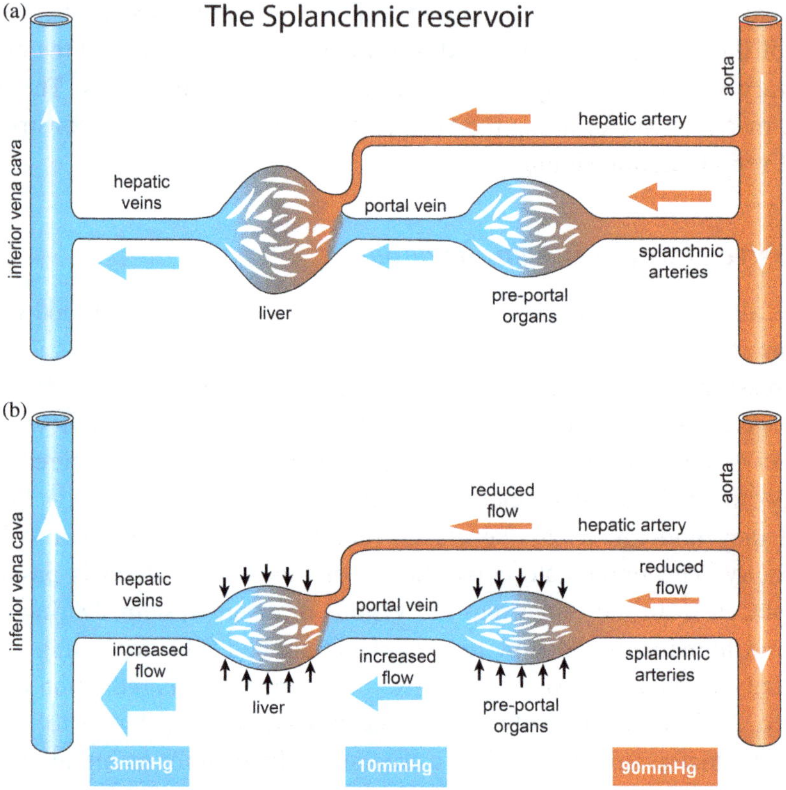

Figure 7.8: Diagrammatic illustration of the splanchnic reservoir. (a) Anatomy of the splanchnic reservoir. The liver and pre-portal organs (spleen, stomach, pancreas, and intestines) receive oxygenated blood (shown in red) at high pressure (~120/80 mmHg) via the hepatic and splanchnic arteries. Deoxygenated blood (shown in blue) from the pre-portal organs is carried by the portal vein at lower pressure (~10 mmHg) to the liver where absorbed food is metabolised. The vessels of the liver and pre-portal organs comprise the splanchnic vascular bed which normally contain around 25% of the total blood volume. This is mostly in the veins of these organs which are thin walled and elastic making them highly distensible depending on the pressure within them. This is the basis for their role as a reservoir in the circulation. Reduction in the luminal diameter of splanchnic veins also reduces their capacity to hold blood. When this happens, as much as 65% splanchnic blood can be shifted into systemic circulation. This ability to rapidly mobilise blood from the splanchnic bed to the systemic circulation is a crucial mechanism for regulating right atrial pressure. Blood can be mobilised from the splanchnic bed by two mechanisms. (b) Mobilisation of the splanchnic reservoir. Passive mobilisation results from reduced pressure and flow (red arrows) in the hepatic and splanchnic arteries. This lowers the distending pressure in the splanchnic veins, which then undergo elastic recoil (black arrows).

This in turn reduces the distention pressure of splanchnic veins and decreases their volume capacity. As a result, blood is diverted into the systemic circulation via the inferior vena cava circulation [33].

(b) The portal vein (Figure 7.8(a)) is a large vessel (8–10 mm in diameter in adult humans) which carries blood to the liver from the other splanchnic organs (stomach, pancreas, spleen and intestines) and accounts for around 20% of the cardiac output [34]. Although the portal vein wall is thinner and more compliant than that of arteries, it is similarly composed of three layers, intima, media and adventitia. The media consists of two layers of smooth muscle: an inner circumferential layer and an outer longitudinal layer [35]. Bundles of sympathetic nerve fibres are also present in the adventitial and medial layers in close contact with smooth muscle cells. Active constriction of the portal vein and its tributaries would reduce the splanchnic volume capacity and transfer blood into the systemic circulation. Evidence that this can occur is suggested by the response to blood loss. In healthy humans, controlled haemorrhage of 17% of total blood volume over 25–35 min had little change on blood pressure, heart rate, cardiac output or splanchnic blood flow, but reduced splanchnic volume by 40%. As a result, the blood displaced from the splanchnic circulation replaced around half the haemorrhage volume [31] and thereby helped to maintain central haemodynamics. Such a reduction in splanchnic volume in the absence of a change in splanchnic inflow suggests that this effect was mediated actively by reflex induced venoconstriction rather than passively by elastic recoil of veins.

Figure 7.8: *(see figure on facing page)* This in turn reduces their volume capacity to retain blood, causing an increase in venous outflow (blue arrows) via the hepatic veins to the inferior vena cava. Active mobilisation of splanchnic blood occurs when sympathetic stimulation induces constriction of splanchnic arteries and veins. This has the same effect of lowering splanchnic volume capacity (black arrows) and increasing venous outflow (blue arrows) to the inferior vena cava and increases right atrial pressure. Both mechanisms are known to occur within seconds as a rapid adaptation to changing physiological conditions such as exercise and standing from a lying position.

Thus, the splanchnic circulation has an important reservoir function in maintaining stability of systemic haemodynamics and this is especially so in humans adapted to upright posture. This works by the transfer of blood from the splanchnic bed to the systemic circulation in response to a reduction in splanchnic inflow due either to lower blood pressure or arterial constriction by reflex activation of the sympathetic splanchnic nerves. Transfer of blood from the splanchnic bed to the systemic circulation can also occur due to constriction of splanchnic veins induced by reflex sympathetic stimulation. Together, these changes provide for rapid circulatory adjustments during changes in posture and exercise.

Conclusions

The heart provides the power to drive blood through the circulation and its regular beating is a constant reminder of its presence within us. Our awareness of it gives it a special place in human consciousness and as a signal of our emotions. However, from a physiological perspective, the heart is part of the circulation as a whole and is closely integrated with it. It evolved to work at higher pressures than our ancestors to support the increased energy demands of terrestrial life and upright posture. At the same time, the arteries and veins evolved to precisely match its changing characteristics. This is especially evident in the elastic properties of the arteries as they adapted to higher pressures generated by the heart as it evolved. Far from being passive conduits of blood, arteries and veins are intimately involved in regulating the circulation. Cells within the walls of vessels have evolved mechanisms to regulate blood pressure and flow by sensing changes in them and modifying vessel calibre in response. They also act as pressure sensors for the autonomic nervous system, providing the input for reflex control of the circulation. Changes in posture and energy demands require rapid adjustments of the distribution of blood flow. The heart provides the increased cardiac output for this; however, redistribution of blood flow is also a vital component in meeting energy demands and is provided by selective dilatation and constriction in different vascular beds so as to prioritise regions of highest demand. At the same time, the total circulating blood volume is modulated by releasing blood from the splanchnic reservoir. Thus, the heart and blood vessels are a fully integrated system in which every action of one is matched by the other and many of their combined functions are co-regulated by the autonomic nervous system. There is still much to learn about the function of blood vessels and this has become a pressing issue as high blood pressure has emerged as the biggest risk factor for early death in the world.

References

[1] Monahan-Earley R, Dvorak AM, Aird WC. Evolutionary origins of the blood vascular system and endothelium. *Journal of Thrombosis and Haemostasis: JTH*. 2013;11(Suppl 1):46–66.

[2] Wasserthal LT. Periodic heartbeat reversals cause cardiogenic inspiration and expiration with coupled spiracle leakage in resting blowflies, *Calliphora vicina. The Journal of Experimental Biology*. 2014;217(Pt 9):1543–1554.

[3] Bettex DA, Pretre R, Chassot PG. Is our heart a well-designed pump? The heart along animal evolution. *European Heart Journal*. 2014;35(34):2322–2332.

[4] Wegner NC, Sepulveda CA, Bull KB, Graham JB. Gill morphometrics in relation to gas transfer and ram ventilation in high-energy demand teleosts: Scombrids and billfishes. *Journal of Morphology*. 2010;271(1):36–49.

[5] Moalli R, Meyers RS, Jackson DC, Millard RW. Skin circulation of the frog, *Rana catesbeiana*: Distribution and dynamics. *Respiration Physiology*. 1980;40(2):137–148.

[6] Seymour RS, Bennett-Stamper CL, Johnston SD, Carrier DR, Grigg GC. Evidence for endothermic ancestors of crocodiles at the stem of archosaur evolution. *Physiological and Biochemical Zoology: PBZ*. 2004;77(6):1051–1067.

[7] Ochs M, Nyengaard JR, Jung A, Knudsen L, Voigt M, Wahlers T, *et al.* The number of alveoli in the human lung. *American Journal of Respiratory and Critical Care Medicine*. 2004;169(1):120–124.

[8] Machida N, Aohagi Y. Electrocardiography, heart rates, and heart weights of free-living birds. *Journal of Zoo and Wildlife Medicine: Official Publication of the American Association of Zoo Veterinarians*. 2001;32(1):47–54.

[9] Fisher PE, Russell DA, Stoskopf MK, Barrick RE, Hammer M, Kuzmitz AA. Cardiovascular evidence for an intermediate or higher metabolic rate in an ornithischian dinosaur. *Science*. 2000;288(5465):503–505.

[10] Milnor WR. *Hemodynamics*. Baltimore: Williams & Wilkins; 1982.

[11] Roy CS. The elastic properties of the arterial wall. *The Journal of Physiology*. 1881;3(2):125–159.

[12] Shadwick RE. Mechanical design in arteries. *The Journal of Experimental Biology*. 1999;202(Pt 23):3305–3313.

[13] Clark JM, Glagov S. Transmural organization of the arterial media. The lamellar unit revisited. *Arteriosclerosis (Dallas, TX)*. 1985;5(1):19–34.

[14] Goldbogen JA, Cade DE, Calambokidis J, Czapanskiy MF, Fahlbusch J, Friedlaender AS, *et al.* Extreme bradycardia and tachycardia in the world's largest animal. *Proceedings of the National Academy of Sciences of the United States of America*. 2019;116(50):25329–25332.

[15] Gosline JM, Shadwick RE. The mechanical properties of fin whale arteries are explained by novel connective tissue designs. *The Journal of Experimental Biology*. 1996;199(Pt 4):985–997.

[16] Godwin ARF, Singh M, Lockhart-Cairns MP, Alanazi YF, Cain SA, Baldock C. The role of fibrillin and microfibril binding proteins in elastin and elastic fibre assembly. *Matrix Biology: Journal of the International Society for Matrix Biology*. 2019;84:17–30.

[17] Yanagisawa H, Davis EC. Unraveling the mechanism of elastic fiber assembly: The roles of short fibulins. *The International Journal of Biochemistry & Cell Biology*. 2010;42(7):1084–1093.

[18] Campbell E, Pierce J, Endicott S, Shapiro S. Evaluation of extracellular matrix turnover. Methods and results for normal human lung parenchymal elastin. *Chest*. 1991;99(3 Suppl):49s.

[19] Bayliss WM. On the local reactions of the arterial wall to changes of internal pressure. *The Journal of Physiology*. 1902;28(3):220–231.

[20] Lim SS, Vos T, Flaxman AD, Danaei G, Shibuya K, Adair-Rohani H, *et al.* A comparative risk assessment of burden of disease and injury attributable to 67 risk factors and risk factor clusters in 21 regions, 1990–2010: A systematic analysis for the Global Burden of Disease Study 2010. *Lancet (London, England)*. 2012;380(9859):2224–2260.

[21] Prabhakaran D, Jeemon P, Ghosh S, Shivashankar R, Ajay VS, Kondal D, *et al.* Prevalence and incidence of hypertension: Results from a representative

cohort of over 16,000 adults in three cities of South Asia. *Indian Heart Journal*. 2017;69(4):434–441.

[22] Tykocki NR, Boerman EM, Jackson WF. Smooth muscle ion channels and regulation of vascular tone in resistance arteries and arterioles. *Comprehensive Physiology*. 2017;7(2):485–581.

[23] Brown IAM, Diederich L, Good ME, DeLalio LJ, Murphy SA, Cortese-Krott MM, *et al.* Vascular smooth muscle remodeling in conductive and resistance arteries in hypertension. *Arteriosclerosis, Thrombosis, and Vascular Biology*. 2018;38(9):1969–1985.

[24] Walsh MP, Cole WC. The role of actin filament dynamics in the myogenic response of cerebral resistance arteries. *Journal of Cerebral Blood Flow and Metabolism: Official Journal of the International Society of Cerebral Blood Flow and Metabolism*. 2013;33(1):1–12.

[25] Folkow B. Physiological aspects of primary hypertension. *Physiological Reviews*. 1982;62(2):347–504.

[26] Laurent S, Boutouyrie P. The structural factor of hypertension: Large and small artery alterations. *Circulation Research*. 2015;116(6):1007–1021.

[27] Folkow B, Grimby G, Thulesius O. Adaptive structural changes of the vascular walls in hypertension and their relation to the control of the peripheral resistance. *Acta Physiologica Scandinavica*. 1958;44(3–4):255–272.

[28] Herndon CW, Sagawa K. Combined effects of aortic and right atrial pressures on aortic flow. *American Journal of Physiology*. 1969;217(1):65–72.

[29] Rothe CF. Physiology of venous return. An unappreciated boost to the heart. *Archives of Internal Medicine*. 1986;146(5):977–982.

[30] Bradley SE, Childs AW, Combes B, Cournand A, Wade OL, Wheeler HO. The effect of exercise on the splanchnic blood flow and splanchnic blood volume in normal man. *Clinical Science*. 1956;15(3):457–463.

[31] Price HL, Deutsch S, Marshall BE, Stephen GW, Behar MG, Neufeld GR. Hemodynamic and metabolic effects of hemorrhage in man, with particular reference to the splanchnic circulation. *Circulation Research*. 1966;18(5):469–474.

[32] Rowell LB. Human cardiovascular adjustments to exercise and thermal stress. *Physiological Reviews*. 1974;54(1):75–159.

[33] Brooksby GA, Donald DE. Release of blood from the splanchnic circulation in dogs. *Circulation Research*. 1972;31(1):105–118.

[34] Greenway CV, Lister GE. Capacitance effects and blood reservoir function in the splanchnic vascular bed during non-hypotensive haemorrhage and

blood volume expansion in anaesthetized cats. *The Journal of Physiology*. 1974;237(2):279–294.

[35] Holman ME, Kasby CB, Suthers MB, Wilson JA. Some properties of the smooth muscle of rabbit portal vein. *The Journal of Physiology*. 1968;196(1):111–132.

[36] Briet M, Boutouyrie P, Laurent S, London GM. Arterial stiffness and pulse pressure in CKD and ESRD. *Kidney International*. 2012;82(4):388–400.

[37] Intengan HD, Schiffrin EL. Structure and mechanical properties of resistance arteries in hypertension: Role of adhesion molecules and extracellular matrix determinants. *Hypertension (Dallas, TX: 1979)*. 2000;36(3):312–318.

8

Intelligent Tubes

Introduction

Tubes have played an important role in human civilisation for thousands of years and not least in the field of health. Access to clean water and its safe disposal was a primary factor in the location and size of early human settlements. Gravity-fed aqueducts were used in the ancient world in Greece, Egypt, Rome and Mesoamerica and remain an important method for controlling water supplies today. Tubes have the advantage of enclosing the liquids they deliver, thereby protecting them from contamination as in the case of water for human consumption and in preventing contamination of the environment by toxic liquids such as sewage and oil. Tubes are therefore ideal conduits for liquids and given the importance of water supplies and its disposal for human health, they also emerged early in its management. The Sumerians are credited with introducing clay pipes for sanitation around 4000 BCE. Lead pipes were widely used in Roman sanitation systems and they continued to be used up to and during much of the 20th century. Lead poisoning has been known for over 2,000 years, described by the Greek physician Nicander, in the second century BCE and Nero's physician, Dioscorides in the first century CE [1]. Lead continued to be used in plumbing systems because of its inert properties and was not widely replaced by copper and durable plastic pipes until the 20th century. Interestingly, even as late as the 18th and 19th centuries, tubes made from bored tree trunks were also used in sanitation, an example of which was recently displayed at the Weston library in Oxford.

Monitoring the conditions in which tubes operate is widely used [2]. Engine temperature and the integrity of brake systems are constantly monitored in vehicles. In large scale systems, monitoring may include the temperature, pressure and physical stresses to which pipes are exposed. For this purpose, appropriate sensors are placed in strategic locations in the system, the output of which can be used to decide servicing schedules, urgent part replacement or even complete shut-down of the system. Modern aircraft jet engine sensors transmit such information not only to the pilot, but also by satellite links to the engine manufacturer, providing real time data that can be used to make recommendations about engine maintenance.

Intelligent Tubes in the Circulation

At one level, blood vessels may appear to be simple conduits for the distribution of constituents of blood such as oxygen, nutrients, hormones, waste products and heat around the body. In reality however, they undertake a wide range of functions, some of which we have encountered in previous chapters. In order to maintain normal levels of blood pressure and flow around the body, it is necessary that blood pumped by the heart is distributed in a highly regulated way. This is achieved by controlling the calibre of blood vessels in every organ of the body to ensure that flow matches local tissue needs. This is possible due to the presence of muscle in the walls of blood vessels, which can contract and relax to regulate the vessel's calibre. The body's response to exercise or changes in posture, for example, is only possible as a result of prompt alterations in the calibre of specific vessels. Those which feed the area of increased need dilate to facilitate extra blood flow, while others supplying areas of less urgent need constrict to limit flow, thereby prioritising energy delivery where demand is highest. In addition, splanchnic veins, which function as the reservoir of the circulation (see Figure 7.8 and related text in Chapter 7), constrict to expand the circulating blood volume. All of this requires appropriate regulation of arteries and veins in different parts of the circulation. This regulation is undertaken by the autonomic nervous system, circulating hormones and by the ability of blood vessels to sense and respond to local blood pressure and flow. In addition to their role in regulating haemodynamics, blood vessels are also responsible for many other functions. As discussed in Chapter 6, healing of infarcted myocardium is heavily dependent of the ability of blood vessels within the injured area to respond in multiple complex ways. They produce signalling molecules which orchestrate activation of the immune system and attract immune cells to the injured tissue. Cells derived from small vessels in the injured area assist in the growth of new blood vessels to supply the healed tissue. The muscle cells which generate force to control the calibre of blood vessels are normally dedicated to that function, but in the presence of injury are able to revert to more primitive forms, which can proliferate and produce proteins to replace damaged tissue. Yet a further function of blood vessels is to provide a mechanism which maintains blood in a liquid state by preventing clot formation. This is partly due to the anticoagulant properties of the endothelial cells which line all blood vessels. Thus blood vessels, far from being simple conduits, are part of a highly complex organ capable of sensing and responding to the large number of remote and local signals which regulate numerous functions of the body.

Discussing all of these is beyond the scope of this work. Instead, this chapter will focus on the ability of blood vessels to regulate pressure and flow within the circulation. The tissue needs for oxygen and other nutrients vary continuously and independently in different parts of the body and consequently blood flow to each organ and tissue has to be individually regulated. Blood vessels contain a layer of muscle which is capable of contracting and relaxing and thereby modulating their lumen diameter. This provides a mechanism by which blood pressure and flow can be regulated and in a continuous and dynamic way. For this to work effectively, it has to be responsive to variations in blood pressure and flow as well as to the metabolic needs of tissues in real-time. In turn, this implies that there must be mechanisms for sensing physiological conditions in the circulation, to which the blood vessels can respond. There are in fact several such mechanisms which can broadly be classified into two groups, intrinsic and extrinsic. Blood vessel regulation in which the sensing and response mechanisms are integral to the vessel itself are intrinsic while vessel responses to conditions sensed remotely are extrinsic.

In general, intrinsic mechanisms are involved in regulating local pressure and flow, while extrinsic mechanisms regulate conditions in the circulation more widely. All of these functions are carried out by the layers of smooth muscle (named after its microscopic appearance) in the walls of blood vessels. To do so, smooth muscle relies on its remarkable ability to sense conditions in the circulation and respond in a way that maintains normal haemodynamics as well as protecting the vessels themselves from excessive stress.

Myogenic Tone

William Bayliss (1860–1924) was a British physiologist who made several important discoveries, including secretin, a hormone produced by the intestine which regulates pancreatic secretion. In a series of experiments published in 1902, he also showed that blood vessels constrict and dilate in response to increases and decreases in blood pressure, respectively, and that this occurs in the absence of any external stimulation [3]. Having been aware that contraction of many muscles is sensitive to stretch (for example, the increased force of contraction by heart muscle in response to increased filling), he proposed that the muscle layer in blood vessels is similarly sensitive to stretch and as a result maintains a constant level of contraction (also called tone) commensurate with the prevailing blood pressure. He further deduced that this tone is reduced when blood pressure is lowered and increased when blood pressure is elevated. This property is referred to as myogenic tone and means that blood vessels regulate their degree of contraction and relaxation in relation to the pressure within them and do so independently of external factors. Myogenic tone requires that vessels must have a mechanism for sensing the pressure of blood within them and to use this as a signal to regulate their behaviour.

Smooth Muscle

The property of motion is characteristic of animals, allowing them to move within their environment and became possible with the evolution of muscle. The motive power of muscle is generated by myocytes which form the contractile unit in all muscles. Mammals have three types of muscles: skeletal, cardiac (discussed in more detail in Chapters 3) and smooth. Cardiac and skeletal muscles are described as striated muscles due to the regular pattern of sarcomeres observed on light microscopy. This is due to the fact that in cardiac and skeletal muscles the contractile elements actin and myosin are organised into regular and tightly packed sarcomeres (see Figure 3.1 and related text in Chapter 3). The absence of such a pattern led to the description of smooth muscle. Smooth muscle is found in organs and tissues where motility is not regulated under conscious control such as intestines, bladder and blood vessels. Whereas skeletal muscle responds to signals from the central nervous system under conscious control, smooth muscle responds to the autonomic nervous system. The molecular motors in all muscle types consist of actin and myosin and despite the absence of sarcomeres, smooth muscle cells (SMC) contain large amounts of both. There are however important differences in the structure and physiology of striated and smooth muscles.

Contractile Function

Skeletal muscles are capable of contracting and relaxing rapidly and can sustain contraction when necessary as in steadily holding an object. Rapid rhythmic contraction and relaxation are characteristic features of cardiac muscle and the basis of the heart's ability to function as an intermittent pump. In contrast, smooth muscles generate contractions for prolonged periods to control vessel calibre against the expansile effect of blood pressure and does so independent of any external stimulation. In organs where smooth muscle generates movement, such as in the peristaltic contraction waves in intestines, this is many times slower than in striated muscle. In addition, smooth muscle can shorten to a much greater extent than striated muscle. Thus, some SMCs can shorten to 25% of their resting length.

Length Adaptability

Smooth and striated muscles also differ in how their contractile behaviour responds to changes in muscle length. Cardiac and skeletal muscles have an optimal resting length for force generation and if stretched beyond that, contractile function declines. In contrast, smooth muscle can adapt to increased length in a way that allows it to maintain contractile function [4]. In some cases, SMCs retain the ability to generate force when stretched to several times their original length, and even if stretched beyond that, they can recover with passive restoration of length [5]. This length adaptability of smooth muscle is essential for hollow organs such as the stomach and bladder which undergo large volume changes. It is also important for some blood vessels. The circulation has an important reservoir function which is present in the abdominal splanchnic circulation (discussed in greater detail in Chapter 7). This property depends on the ability of the splanchnic veins to accommodate a substantial part (around 25%) of the total blood volume and expel it into the circulation when needed as during changes in posture, exercise or haemorrhage. This is possible due to the ability of the smooth muscle they contain, to contract and relax when stretched over greater lengths than would be possible for cardiac or skeletal muscle.

Smooth Muscle Structure

Cell structure

Smooth and striated muscles have evolved different structural and physiological properties which allow them to undertake their distinct roles. Typically, SMCs are arranged in layers around tubular structures such as arteries and intestines. In the case of blood vessels, smooth muscle cells are arranged in two layers, an inner circular layer surrounding the lumen and an outer layer along the length of the vessel (Figure 8.1). Pressure is many times higher in arteries than veins and the distribution of smooth muscle reflects this. In arteries, the inner circular layer is much thicker than the longitudinal layer and is also more developed than in veins, which allows them to contract against a greater resistance. In other organs too, the structure of smooth muscle is well suited to their function. For example, intestines have a much more developed longitudinal layer consistent with their function in propelling contents along their length and bladder smooth muscle is arranged in several differently aligned layers which is ideal for achieving large concentric volume reduction.

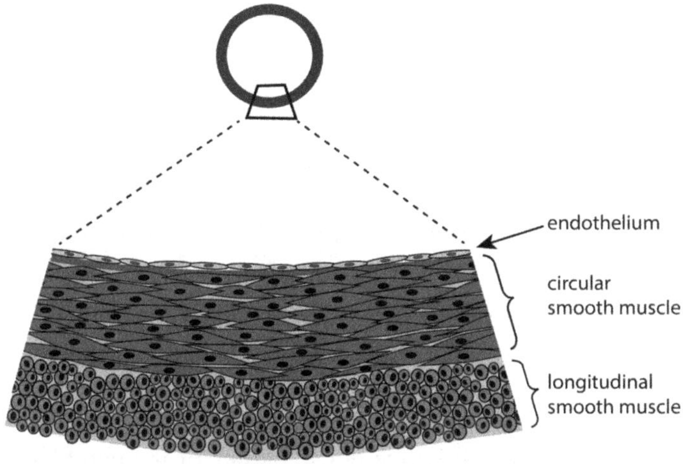

Figure 8.1: Diagrammatic illustration of a blood vessel in cross section. All blood vessels are lined with a thin layer of endothelial cells. Spindle-shaped smooth muscle cells (SMCs) are arranged in two layers, an inner circular layer and an outer longitudinal layer. In arteries, the inner circular layer is thicker than in veins, reflecting the higher luminal pressure which they must withstand. SMCs are attached side to side which allows them to contract in parallel, greatly increasing the capacity for force generation by these layers.

At a microscopic level too, SMCs differ in their size, structure and composition compared to striated muscle. Typically, SMCs are spindle shaped, around 200 μm long with a maximal width of 2–8 μm. Skeletal myocytes are huge by comparison, on average 3 cm long in humans (but can be up to 30 cm) and from 10 to 100 μm in diameter. Cardiac myocytes are similar in length to SMCs but thicker, being 10–25 μm in diameter and more cylindrical in shape. SMCs contain about 75% of the contractile elements, actin and myosin, present in striated muscle (by volume fraction) and sarcomeres are much more densely packed in striated muscle. This is well suited to the highly developed and dedicated function of the latter. Under normal conditions, vascular SMCs are dedicated to contractile function, regulating vessel tone. However, they can also fulfil a range of other functions when necessary. In response to injury, for example, they can dedifferentiate into more primitive forms and re-enter the cycle of cell division to proliferate. They then migrate through tissue layers and secrete extracellular proteins involved in tissue repair including collagen, elastin and proteoglycans. Once the repair process is complete, SMCs revert to their mature specialised contractile role.

Smooth muscle cytoskeleton

In contrast to the regular sarcomeres in striated muscle, smooth muscle cells contain a strong cytoskeleton consisting of an interconnected lattice of filaments throughout the cell and attached to the cell membrane. The lattice contains both contractile and non-contractile elements joined to anchor points known as dense bodies in the cytoplasm and cell membrane (Figure 8.2(a)). Intermediate filaments form the non-contractile elements of the lattice and actin–myosin filaments generate contractile force in the cell. Both are linked to the lattice dense bodies of the cytoplasm and cell membrane.

Intermediate filaments

Intermediate filaments (IFs) are structural elements present in vertebrate and many invertebrate cells and are formed from elongated structural proteins. At least 70 of these structural proteins are known, which provide the physical architecture of cells by polymerising to form IFs. IFs in skin, hair and nails are formed from keratin, whereas IFs in SMCs are composed of vimentin and desmin. Vimentin (named from the Latin, *vimentum* for flexible rod or stick) is more abundant in large arteries and desmin (named after the Greek word *δεσμός* meaning link) in smaller arteries. Assembly of IFs is similar in all tissues due to the similarity of their structural proteins. Initially two molecules align parallel and bind to form a coiled coil (like the strands in a rope) dimer. Dimers then align antiparallel to each other and bind to form tetramers known as unit filament lengths (UFLs). UFLs then join end to end and side to side to form complete IFs [6]. Intermediate filaments are closely associated with cytoplasmic and membrane dense bodies, forming a cell-wide lattice structure (Figure 8.2(a)). The cytoskeleton of smooth muscle provides a much less compact and dense contractile apparatus than striated muscle sarcomeres, however as discussed below it compensates for this in being much more flexible and adaptable to changing physical conditions.

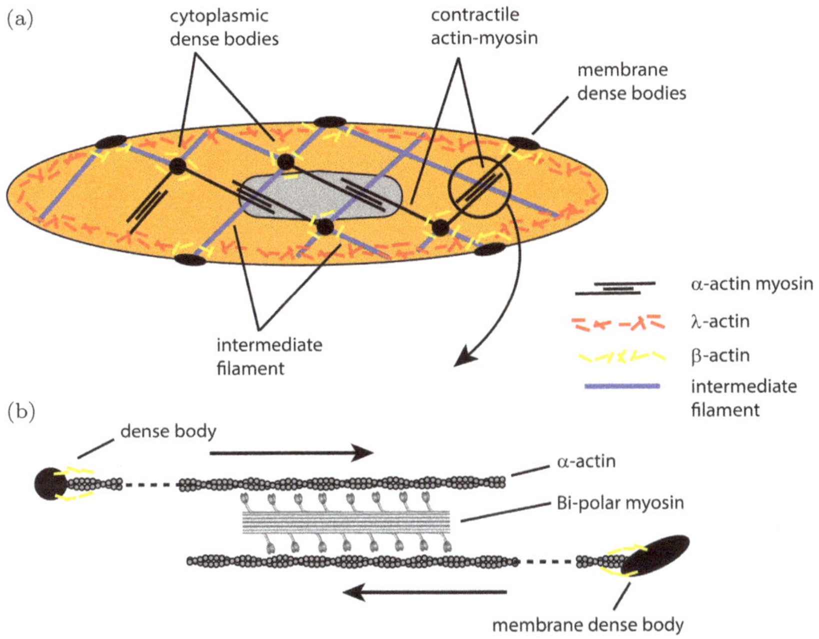

Figure 8.2: (a) Diagrammatic illustration of the cytoskeleton and contractile elements of a smooth muscle cell. The cytoskeleton consists of a lattice of intermediate filaments and actin–myosin contractile elements. Cytoplasmic and membrane dense bodies composed of α-actinin (black circles/ellipses) act as anchor points for the lattice. A cortical layer of non-muscle λ-actin is present inside the cell membrane and bundles of non-muscle β-actin are also present around the cytoplasmic and membrane dense bodies. These non-muscle actins are involved in binding intermediate filaments and contractile actin to dense bodies (β-actin) and to the cell membrane (λ-actin). The cytoskeletal lattice transmits contraction of the actin–myosin filaments through entire cell and cell membrane and cell-to-cell attachments ensure coordinated contraction of smooth muscle layers. (b) Diagrammatic illustration of an α-actin and myosin contractile unit. Bi-polar myosin interacts with actin filaments on each side of the filament resulting in a sliding action which pulls the actin ends closer together, shortening the element. α-actin is shown connected to dense bodies via β-actin filaments.

Actin–myosin filaments

Actin–myosin filaments of the contractile units are also connected to the cytoplasmic and membrane dense bodies of the cytoskeleton. In this way, the cytoskeleton transmits force generated by contraction through the cell. Although the cytoskeleton does not contract itself, recent evidence indicates that it can indirectly affect contractile function [7]. In studies of genetically modified SMCs in which intermittent filaments were absent, active force generation and the maximal rate of contraction were substantially reduced [7]. In addition, passive stretch induced less stress in these cells, indicating reduced stiffness. Such findings indicate that the cytoskeleton can adapt in a way that enhances force generation. Three isoforms of actin are present in SMCs. Only α-actin forms part of the contractile machinery and is known as muscle actin. β-Actin and λ-actin form part of the cytoskeleton and are referred to as non-muscle actin. β-Actin is present in close association with cytoplasmic and membrane dense bodies and is involved in anchoring IFs and contractile actin. λ-Actin is present closely associated with the cell membrane throughout the cell (Figure 8.2(a)).

Dense bodies

Dense bodies (Figure 8.2(a)) provide anchor points for the intermediate filaments and contractile elements of SMCs. Cytoplasmic and membrane dense bodies are composed of α-actinin. α-Actinin and α-actin are distinct proteins but are functionally related because they bind together forming the attachments of filaments to dense bodies of the cytoskeleton. The molecular structure of α-actinin consists of an actin binding domain at one end with a flexible neck region connecting it to a rod-like domain. The neck region allows for variability in the angular orientation between α-actinin and bound actin [8]. α-Actinin, dystrophin and α- and β-spectrin are members of a large family of proteins known as spectrins which are present as structural proteins in single cell organisms, invertebrates, birds and mammals. α-Actinin binds to actin in all muscle types, forming the Z junctions in striated muscle (see Figure 3.1 in Chapter 3) and dense bodies in smooth muscle [9]. α-Actinin is assembled as an antiparallel dimer, meaning two molecules of α-actinin join together to form a rod-like structure, but aligned in opposite directions. Each has an actin binding site at one end, so that the dimer can bind an actin filament at each end. In this way, α-actinin in the Z-band of striated muscle links adjacent sarcomeres. α-Actinin in the dense bodies of smooth muscle is closely associated with β-actin, which provides a molecular anchor for contractile α-actin and intermediate filaments [10]. Originally dense bodies were thought to act simply as fixed anchor points within the cytoskeleton, however recent studies of their molecular structure indicate that they have more complex dynamic roles.

A Dynamic Cytoskeleton

As mentioned above, smooth muscle needs to be flexible to function properly in distensible organs such as intestine, bladder and some blood vessels. This also means that its contractile elements must be capable of adapting to changes in length. This is possible in part due to the flexibility built into the cytoskeleton. Dense bodies were originally thought to be fixed anchor points but recent studies using high power 3-D reconstruction electron microscopy have led to a reappraisal of their structure. By capturing images of successive ultra-thin layers through cytoplasmic and membrane dense bodies, it has been possible to trace their dimensions more accurately. Such studies have shown that dense bodies in association with intermediate filaments are present as cable-like structures in the cell and that these can straighten and slacken in response to change in length [11, 12]. These findings have led to the concept that the cytoskeleton is not simply a fixed lattice within SMCs, but rather a mechanically dynamic structural framework that can stretch and recoil allowing the cell to change length while maintaining passive tension. In addition, by providing a dynamic lattice, to which the contractile elements are attached and exert force, the cytoskeleton also provides a means of transmitting force at different cell lengths. This arrangement is ideally suited to smooth muscles which function under varying conditions of length.

In addition to being a structural protein of the cytoskeleton, α-actinin also has an important role in binding adjacent cells. Membrane bound proteins such as integrins and intercellular adhesion molecules (ICAMs) are important structural junctions between adjacent cells including SMCs. By binding to the intracellular domains of these molecules, α-actinin links the cytoskeleton of adjacent cells, ensuring strong mechanical connections and coordinated contraction in SMCs' layers [13]. Importantly too, these connections between adjacent SMCs are mostly side to side creating a parallel array of contracting cells. In contrast, striated and cardiac muscle cells are mostly connected end to end. This is one reason why SMCs can generate force equal to or greater than striated muscle, despite having less contractile protein per cell.

α-Actinin also binds to several signalling proteins and ion channels which regulate SMC function, including calcium (Ca^{2+}) [14] and potassium (K^+) [15] ion channels and adenosine receptors [16]. The importance of this is exemplified by the role of α-actinin in regulating both striated and smooth muscle L-type Ca^{2+} ion channels which are key regulators of contraction in smooth muscle. Recent evidence [14] has shown that α-actinin is involved in locating these ion channels in the cell membrane as follows. L-type Ca^{2+} channels are formed in vesicles within the cell cytoplasm, which coalesce, promoting clustering of these channels. They are then transported by mechanisms involving the cytoskeleton to the cell surface where α-actinin binds to the pore forming domain of the channel. This has two important effects; first, it organises these ion channels in clusters at specific sites on the cell membrane and second, it restricts L-type Ca^{2+} channel activity until located on the cell membrane because they are inactive until bound to α-actinin located there [14]. This tells us that α-actinin is not simply structural elements of cells, but rather provides a sophisticated system for managing the transport, location and function of major regulating components of cells.

Smooth Muscle Contraction

Contraction in SMCs is generated by the same contractile proteins, actin and myosin as in skeletal muscle. However, the concentration ratio of myosin to actin in SMCs is about half that in striated muscle and actin thin filaments are longer in smooth muscle [17]. The distribution of actin and myosin at a molecular level in SMCs and the mechanisms of interaction between them are also quite different from that in striated muscle (Figures 8.2 and 8.3). Actin and myosin filaments which form the contractile system in SMCs are structurally integrated with the cytoskeleton through their connections to the cytoplasmic and membrane dense bodies. In this way, cytoplasmic dense bodies are anchor points between consecutive intermediate filaments of the cytoskeleton as well as contractile actin. In addition, adjacent cells are joined at areas corresponding to membrane dense bodies forming strong attachments between adjacent cells. Therefore, actin–myosin filaments are

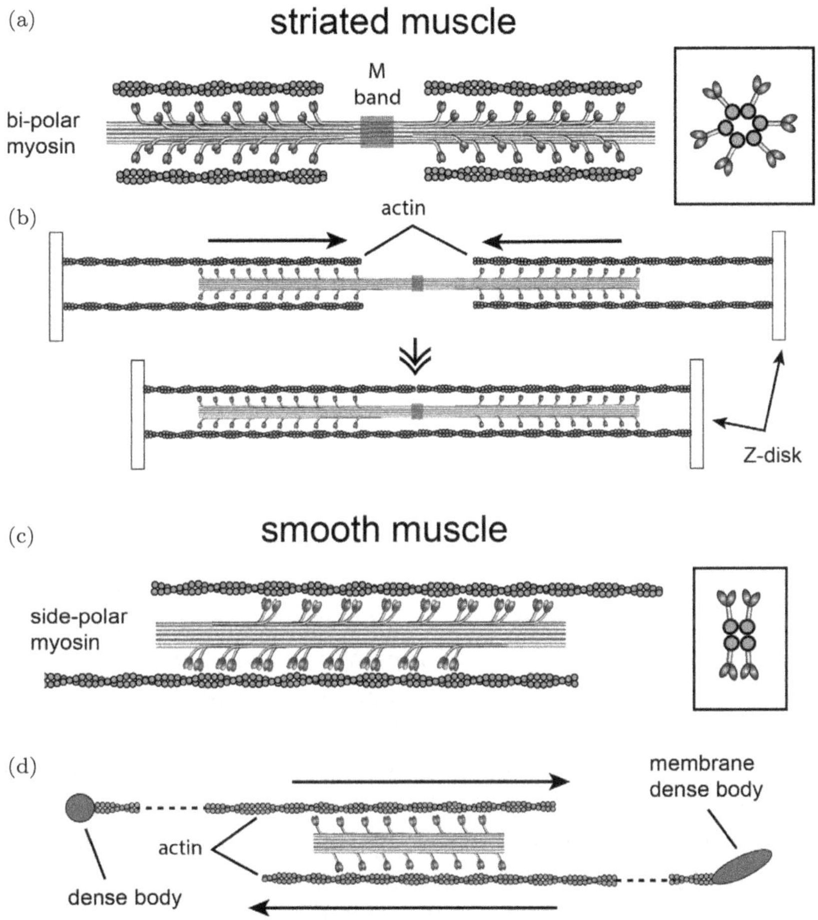

Figure 8.3: (a) Illustration of the bi-polar configuration of striated muscle myosin. Myosin thick filaments are formed by parallel binding of the tails of myosin dimers with heads protruding from the filament in a helical array around cylindrical-shaped filament backbone (see inset on the right). Each contractile element of sarcomeres contains two thick filaments oriented in opposite directions with their ends overlapping anti-parallel at the M-band. (b) Contraction of striated muscle sarcomeres is due to actin–myosin interaction which pulls the actin filaments towards the centre of the sarcomere and drawing the sarcomere Z disks closer together. Shortening is limited by the length of actin and its limited ability to overlap at the centre of the sarcomere (double arrow). (c) Illustration of the side-polar configuration of smooth muscle myosin. On each side of the thick filament, myosin dimers are aligned anti-parallel with those on the other side, forming a filament rectangular in cross section (inset right) with myosin heads projecting on each side and oriented in opposite directions. (d) The thick filaments are sandwiched between actin filaments on each side. As a result, the thin filaments are always separated and can overlap without hindering each other. This, together with the presence of longer thin filaments in smooth muscle, allows SMCs to shorten more than striated myocytes.

connected longitudinally within each cell and also between adjacent cells. In this way, SMCs contract in parallel, contributing to a strong contracting layer of smooth muscle [9].

Actin–myosin configuration in SMCs

Contraction of SMCs results from an energy using interaction between thin actin and thick myosin filaments as in striated muscle. The thin filaments are similar to those in striated muscle consisting of two strands of polymerised actin arranged in a double helix (the origins and formation of actin are discussed in Chapter 3). SMCs produce three forms of actin called α, β and λ, and despite their similar amino acid sequences, they undertake different functions. Vascular SMC muscle actin filaments are composed of α-actin (contractile λ-actin is present in other SMCs such as intestines) and the cytoskeleton contains non-muscle β-actin and λ-actins [18]. The actin filaments in SMCs are longer than in striated muscle and are aligned on each side of the thick myosin filaments (Figure 8.2(b)). SMC actin filaments also contain tropomyosin arranged in a similar way as seen in striated muscle (see Figure 3.4 and related text in Chapter 3). Two tropomyosin molecules form elongated coiled-coil protein dimers which assemble in the grove between strands of actin filaments. As discussed in Chapter 3, tropomyosin plays a key role in regulating actin–myosin interaction in striated muscle by controlling access to the binding sites on actin to which myosin binds through the action of troponin. However, troponin is not present in SMC actin, activation of which depends on a different mechanism (discussed below).

At least 37 classes of myosin have been described which fulfil different motor functions in biology. Class II myosin is present in smooth and striated muscles and although the amino acid sequences of each are very similar, they have different biochemical properties and they assemble into thick filaments differently. In both muscle types, myosin II is formed of two individual myosin molecules, known as heavy chains, each having a head, neck and tail section. The tails of each are wrapped around each other while the heads which contain the ATP and actin binding sites are separate. The neck region of each also contains two additional protein structures known as light chains, a regulatory light chain and an essential light chain.

The wrapped tails of each myosin dimer are attached to many others to form the backbone of thick myosin filaments while the heads protrude from the side at an angle controlled by the neck region (Figure 8.4).

Bi-polar and side-polar myosin

A striking feature of smooth muscle is the extent to which it can shorten. In some cases, SMCs can contract to around 25% of their resting length, which is much greater than striated muscle. One of the reasons for this is the way actin and myosin filaments are arranged in smooth muscle and how this differs from striated muscle. In striated muscle, the backbone of myosin filaments is cylindrical in cross section and myosin heads protrude in a helical array around the circumference [19, 20] (Figure 8.3(a)). The thick filaments in each half of the sarcomere are oriented in opposite directions with the ends of filaments overlapping antiparallel at the M-band (Figure 8.3(a)). This configuration is described as bi-polar reflecting the opposite polarity of myosin heads at end of the sarcomere. Through their interaction with actin, the myosin filaments move in opposite directions towards the Z disk at each end of the sarcomere (Figure 8.3(b)). Consequently, the movement of myosin in each sarcomere is limited to the length of the actin filaments and in their ability to overlap at the centre of the sarcomere.

In smooth muscle, the assembly of myosin results in a different thick filament structure, described as side-polar myosin [19, 20] (Figure 8.3(c)). Here, thick myosin filaments are square in cross section rather than cylindrical and the heads protrude on each side of it rather than in a helical array. The heads are arranged with the same polarity along the entire length of one side of thick filament and with the opposite polarity on the other side. Unlike striated muscle myosin thick filaments, which overlap and bind antiparallel at the ends of the filaments (corresponding to the M-band), smooth muscle myosin filaments overlap and bind antiparallel along the entire length of the filament. This has important consequences for the way actin and myosin interact. In striated muscle, the maximal contraction of sarcomeres is limited by the length of actin filaments in each half, being around 1.6 μm. In smooth muscle, actin filaments are attached to dense bodies in the cytoplasm and at the cell membrane and are much longer than in striated muscle, being 3–10 μm, with an average length of 4.5 μm [21]. The basic contractile apparatus in SMCs consists of a thick myosin filament sandwiched between two thin actin filaments extending in opposite directions. The side-polar configuration of myosin allows it to traffic along the length of both thin filaments simultaneously and their alignment on opposite sides of the thick filaments allows them to overlap as contraction progresses without hindering each other (Figures 8.3(c) and (d)). With a myosin length of 1.6 μm [21] and thin filaments on average 3–6 times longer than in striated muscle, smooth muscle contractile units are capable of much greater shortening than striated muscle sarcomeres. This helps explain why smooth muscle cells can contract to less than 30% of resting length.

Actin–myosin interaction in SMCs

Interaction between actin and myosin is triggered by different mechanisms in smooth and striated muscles. However, the interaction itself is very similar in both. This is described in detail in Figures 3.4 and 3.5 and related text in Chapter 3 and will be summarised here. Interaction between the myosin heads and the myosin binding sites on actin utilises energy derived from the breakdown of ATP to ADP and is a repeating cycle of four steps: (1) detached, (2) weakly bound, (3) strongly bound and (4) stroke phase.

1. Myosin heads have a site which binds to actin and one which binds to ATP. Binding to ATP occurs at the end of the power stroke and changes the configuration of the myosin head forcing it to detach from actin.
2. Once detached the head hydrolyses the ATP to ADP and a phosphate group (P) and undergoes a further change in configuration which brings it close to actin again.
3. The myosin head then binds weakly to actin and releases the P product of ATP hydrolysis.
4. This results in myosin binding strongly to actin and undergoing a configuration change which pulls the thin filament along the thick filament. This is usually referred to as the power stroke. The myosin head then

releases the ADP product of ATP hydrolysis leaving the ATP site vacant. This returns the head to state (1) and the process can be repeated.

Repetition of the cycle many times occurs on both sides of the bipolar myosin filament drawing the thin filaments on each side towards the thick filament.

Activation of smooth muscle contraction

Vascular smooth muscle is required to contract and relax in response to a variety of physiological demands such as exercise and changes in posture as well as contracting in response to stretch (myogenic tone). These functions are regulated by the autonomic nervous system, hormones and autacoids (locally released regulating substances), which act as agonists stimulating specific receptors on the cell membranes of SMCs. The ability of smooth muscle to sense and react to stretch is possible because SMCs have also evolved mechano-sensing receptors that are activated when the cell membrane is stretched. SMC responses are implemented using several intracellular signalling pathways and although the originating stimuli act through different receptors on the cell surface, they often act via the same signalling pathways to regulate and balance the contraction and relaxation and hence the lumen diameter in blood vessels.

Signalling Pathways Regulating Contraction

Contraction of all muscle types is initiated by a rise in the level of intracellular calcium $[Ca^{2+}]_i$. As we saw in Chapter 3, in striated muscle a rise in $[Ca^{2+}]_i$ activates troponin which in turn releases tropomyosin from its tight position in the actin filament groove. However, troponin is not present in smooth muscle. Instead, increased $[Ca^{2+}]_i$ activates enzymatic phosphorylation of myosin regulatory light chains (MRLCs) which enables contraction (Figure 8.4). This begins with the increased $[Ca^{2+}]_i$ binding to calmodulin (CAM), a ubiquitous and avid calcium binding protein which rapidly combines with Ca^{2+} forming Ca^{2+}/CAM. Ca^{2+}/CAM binds to the enzyme myosin light chain kinase (MLCK) which activates it. MLCK then phosphorylates its target, MRLC. At low levels of $[Ca^{2+}]_i$, MLCK is

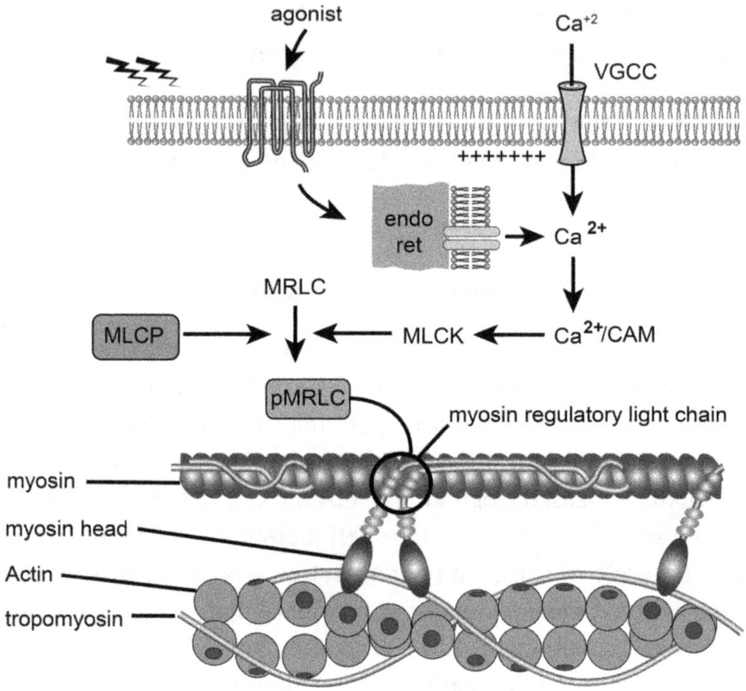

Figure 8.4: Diagrammatic illustration of actin–myosin interaction in smooth muscle. Elevation of intracellular calcium $[Ca^{2+}]_i$ is triggered by agonist stimulation or stretch of the cell membrane which cause an influx of extracellular Ca^{2+} via voltage-gated calcium channels (VGCCs) and release of Ca^{2+} stored in the endoplasmic reticulum. Raised $[Ca^{2+}]_i$ binds to calmodulin forming Ca^{2+}/CAM, which binds to myosin light chain kinase (MLCK), activating it. Activated MLCK then phosphorylates myosin regulatory light chains (MRLCs), which enables interaction with actin and contraction. Actin–myosin contractile filaments are illustrated in the lower part of the figure. Actin is shown as a polymer of actin monomers forming a double helix with tropomyosin attached to each groove (note that troponin is not present in SMCs). Myosin is shown with two heads projecting towards actin on the thin filament with myosin regulatory light chains (MRLCs) and essential light chains attached to the neck domains.

Notes:

Ca^{2+}/CAM	Ca^{2+} bound to calmodulin
CAM	Calmodulin
endo ret	Endoplasmic reticulum
MLCK	Myosin light chain kinase
MLCP	Myosin light chain phosphatase
MRLC	Myosin regulatory light chain
VGCC	Voltage-gated calcium channel

inactive because its active site for MRLCs is inaccessible. Binding to Ca^{2+}/CAM induces a conformational change in MLCK, which exposes its active site and enables it to phosphorylate MRLCs [22]. This activates ATPase on myosin heads, enabling contraction. The degree of smooth muscle myosin phosphorylation is also regulated by a second enzyme, myosin light chain phosphatase (MLCP), which dephosphorylates and deactivates it (Figure 8.4). In this way, contraction is controlled by positive and negative factors; MRLCs are phosphorylated in response to elevated $[Ca^{2+}]_i$ and dephosphorylated by the action of MLCP. In addition, both of these enzymes are subject to further modulation, which fine tunes the degree of contraction and relaxation.

Unlike the contraction of cardiac and skeletal muscles which is rapidly activated and relaxed by the on/off switch-like action of troponin/tropomyosin, smooth muscle contraction and relaxation are regulated by enzymatic phosphorylation and dephosphorylation of myosin, which are slower processes. As a result, contraction and relaxation are also slower in onset and offset in smooth muscle. This also allows for contraction to be maintained for as long as myosin remains phosphorylated.

Multiple Interacting Controls

In order to regulate the circulation correctly, blood vessels receive signals from several sources and must be able to respond to each of them. An outline of these is given here and a summary of them is illustrated in Figure 8.5. A more detailed account of the signalling pathways involved in each will follow below. The autonomic nervous system provides a central mechanism for regulating the circulation as a whole and for this blood vessels are richly endowed with autonomic nerve endings. These work by releasing noradrenaline which targets G-protein coupled receptors (GPCR) on the cell membrane. SMCs contain many GPCRs which act as specific receptors for several other agonists. Some of these are produced locally (called autacoids) while others are secreted remotely and circulate in the blood stream as hormones. Examples of the former include vasoconstricting agonists such as endothelin, thrombin, sphingosine and thromboxane A2 and also vasodilating agonists like prostacyclin. Angiotensin II, adrenaline and oestrogen are

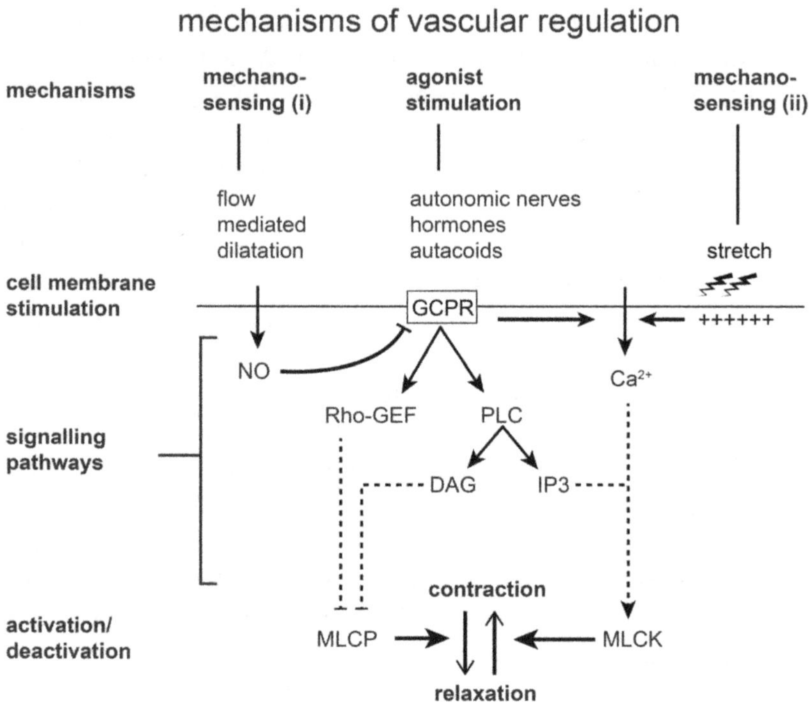

Figure 8.5: Summary illustration of vascular regulation. Blood vessels are controlled by several local and remote regulators. Mechano-sensors in endothelial cells and smooth muscle cells are sensitive to blood flow and pressure, respectively. Increased flow induces vasodilatation (flow mediated dilatation) while increased pressure causes vasoconstriction, both acting as local control mechanisms independent of remote regulation. The autonomic nervous system provides central control of the circulation as a whole via nerve endings that release noradrenaline on specific GPCRs on the cell membrane. In addition, SMCs contain many such GPCRs which act as specific targets for vasoactive agonists produced locally and remotely. All of these stimuli activate intracellular signalling pathways leading to contraction or relaxation.

Notes:

DAG	Diacylglycerol
GPCR	G-protein coupled receptor
IP3	Inositol trisphosphate
MLCK	Myosin light chain kinase
MLCP	Myosin light chain phosphatase
Rho	Ras homologue
Rho-GEF	Guanine nucleotide enhancing factor

examples of circulating hormones. In addition, blood vessels can sense and respond to conditions in the circulation. This is possible because blood vessels have evolved two mechano-sensors that are sensitive to blood flow and pressure. (i) Flow mediated dilatation operates at the level of endothelial cells which line all blood vessels. Endothelial cells are sensitive to the frictional effect of blood flow and when blood flow increases, they release nitric oxide which diffuses into the vessel wall and causes smooth muscle relaxation and dilatation of the vessel. (ii) Blood vessels are also sensitive to blood pressure due to the presence of stretch sensitive receptors on SMC membranes so that stretching due to an increase in pressure causes them to contract and this is the basis of myogenic tone.

Each of these mechanisms activates a specific receptor on the cell membrane which in turn activates an intracellular pathway leading to contraction or relaxation. In the following section, these are discussed in more detail under the headings of agonist stimulation and mechanosensing. Although the source and mechanism of stimulation may vary, often the intracellular signalling pathways they activate are shared. In the following, signalling pathways are discussed in relation to the type of stimulation which activates them (Figure 8.6).

Vasoactive agonist stimulation

The membranes of SMCs contain receptors which bind and respond to vasoactive molecules. These receptors are G-protein coupled receptors (GPCRs) which detect extracellular stimuli and signal their presence to the cell interior. Several GPCR subtypes are known, most of which mediate contraction in response to agonists such as endothelin, noradrenaline, angiotensin II and phorbol ester. Some result in relaxation, for example, in response to prostacyclin. The latter stimulates GPCRs linked to G-proteins and causes muscle relaxation via increased production of cAMP. However, most GPCRs are linked to intracellular pathways that initiate or modulate contraction in smooth muscle, e.g., GPCR subtypes Gq-11 and Gq-12.

Phenylephrine and noradrenaline are α-adrenergic agonists which stimulate GPCRs linked to Gq11, and which in turn activates the membrane bound

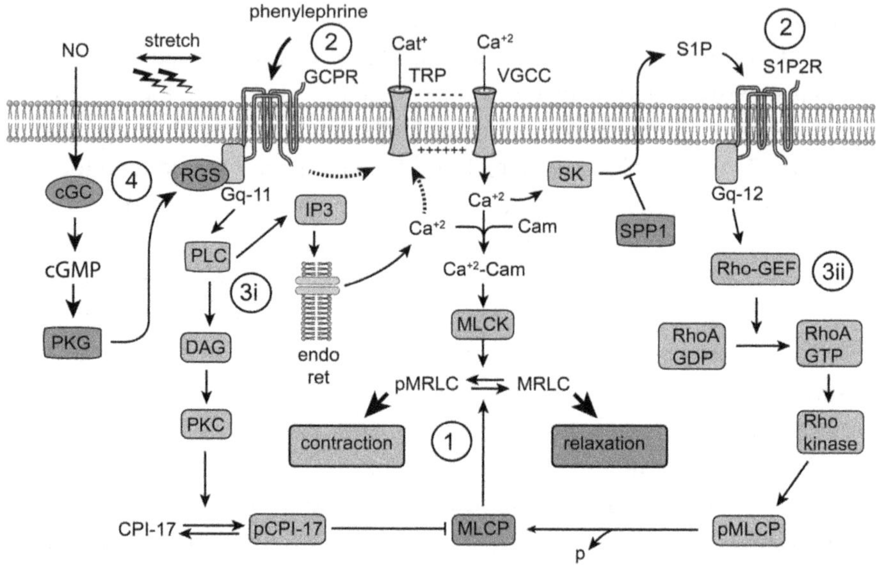

Figure 8.6: Illustration of synergistic pathways leading to smooth muscle contraction. (1) Smooth muscle contraction is enabled by phosphorylation of the MRLCs by MLCK while relaxation occurs due to its dephosphorylation by MLCP. (2) Stimulation of GPCR by agonists such as phenylephrine or S1P or stretching the cell membrane causes contraction of smooth muscle. In each case, this process is mediated by shared downstream signalling pathways. The common trigger for contraction is membrane depolarisation and raised intracellular calcium $[Ca^{2+}]_i$, the latter due to an influx of extracellular Ca^{2+} via TRP or VGCCs and/or IP3 mediated release of Ca^{2+} stored in the ER. Increased $[Ca^{2+}]_i$ promotes calmodulin binding to Ca^{2+}, to form Ca^{2+}/CAM, which activates MLCK. MLCK then phosphorylates MRLCs, enabling contraction. (3) Agonist stimulation of GPCRs also potentiates contraction by inhibiting relaxation. This can occur by two mechanisms. (i) Activation of the G-protein Gq-11 activates PLC. PLC produces IP3 and DAG from PIP2, a phospholipid component of the cell membrane. IP3 stimulates release of Ca^{2+} from the ER while DAG activates PKC. PKC phosphorylates CPI-17 to pCPI-17, which is a potent inhibitor of MLCP, thereby inhibiting relaxation. (ii) Stimulation of GPCRs linked to Gq-12 by agonists, such as S1P, thrombin or thromboxane A2, causes activation of Rho kinase by recruitment of Rho-GEF which accelerates formation of RhoA GTP from RhoA GDP. RhoA GTP binds to Rho kinase, activating it. Rho kinase then phosphorylates MLCP inactivating it and thereby inhibiting relaxation. (4) Nitric oxide (NO) produced by endothelial cells in response to shear stress diffuses into SMCs where it activates cGC to produce cyclic GMP, which in turn activates PKG. PKG phosphorylates proteins which regulate G-proteins, in this case Gq-11, activating it causing Gq-11 inhibition.

Figure 8.6: *(see figure on facing page)*

Notes:

AT1R	Angiotensin II receptor 1
Ca^{2+}/CAM	Ca^{2+} bound to calmodulin
CAM	Calmodulin
Cat^{+}	Cation
cGC	Guanylate cyclase
cGMP	Cyclic GMP
CPI 17	C-kinase potentiated protein phosphatase-1 inhibitor
DAG	Diacylglycerol
endo ret	Endoplasmic reticulum
GPCR	G-protein coupled receptor
Gq-11	GPCR subtype
Gq-12	GPCR subtype
IP3	inositol trisphosphate
MLCK	Myosin light chain kinase
MLCP	Myosin light chain phosphatase
MRLC	Myosin regulatory light chain
NO	Nitric oxide
pCPI 17	Phosphorylated CPI 17
PKC	Protein kinase C
PKG	Protein kinase G
PLC	Phospholipase
pMRLC	Phosphorylated myosin regulatory light chain
RGS	Regulator of G-protein signalling
Rho	Ras homologue
Rho-GEF	Guanine nucleotide enhancing factor
S1P	Sphingosine phosphate
S1P2R	Sphingosine receptor 2
SK	Sphingosine kinase
SPP1	Sphingosine phosphohydrolase 1
TRP	Transient receptor potential
VGCC	Voltage-gated calcium channel

enzyme, phospholipase C (PLC) (Figure 8.6). PLC breaks down a component of cell membranes, phosphatidylinositol 4,5-bisphosphate (PIP2), into diacylglycerol (DAG) and inositol trisphosphate (IP3). IP3 contributes to elevated $[Ca^{2+}]_i$ by releasing Ca^{2+} stored within the cell in the endoplasmic reticulum (ER). This is accompanied by depolarisation of the cell membrane and activation of voltage-gated calcium ion channels (VGCCs), resulting in an influx of extracellular Ca^{2+}, which allows replenishment of ER Ca^{2+} stores and maintenance of the stimulus response [23].

Production of DAG from PIP2 by the action of the enzyme phosphokinase C (PKC) enhances the contractile response by increasing sensitivity to Ca^{2+}. DAG activates PKC which phosphorylates CPI-17 to pICP-17, a 17 kDa protein and potent inhibitor of myosin light chain phosphatase (MLCP) (Figure 8.6). Since myosin regulatory light chains (MRLCs) are dephosphorylated by MLCP, its inhibition reduces the rate of MRLC dephosphorylation by MLCP and thereby potentiates contraction. PKC is increasingly recognised as an important contractile mechanism in smooth muscle independent of increases in $[Ca^{2+}]_i$ and is discussed further below.

Injury can lead to severe blood loss and one way for limiting this is by vasoconstricting the vessels which supply the area involved. Stimulation of GPCR subtype Gq12 results in vasoconstriction in response to injury via a separate signalling pathway. These GPCRs mediate contraction by activation of the Rho signalling pathway (Rho stands for RAS homologue, named after genes, first identified in rats which developed a viral induced sarcoma [a type of tumour]) (Figure 8.6). GPCRs, Gq12, are targeted by thrombin, thromboxane A2 and sphingosine-1-phosphate [24], all of which are produced locally in response to tissue injury, resulting in activation of the Rho kinase pathway. Thrombin is produced directly during clot formation, while thromboxane A2 and sphingosine are both released from activated platelets. The higher concentrations present have a role in defence and repair. For example, vasoconstriction, as one component of the response to injury, would serve to limit any associated haemorrhage. Stimulation of GPCRs linked to Gq-12 activates Rho kinase, which phosphorylates MLCP. This inhibits MLCP, slowing the rate of dephosphorylation of MRLCs which potentiates contraction. This process is illustrated in Figure 8.6 by the example of sphingosine. In the presence of Ca^{2+}, the enzyme sphingosine kinase (SK) is activated and translocated across the cell membrane where it phosphorylates locally produced sphingosine to form sphingosine-1-phosphate (S1P). S1P then binds to and activates S1P receptors which in turn activate Rho-GEF and Rho kinase via formation of RhoA GTP.

Yet a further mechanism for modulating smooth muscle contraction involves a family of proteins which regulate G-proteins (RGS) (Figure 8.6). G-proteins such as Gq-11 are activated by binding to GTP and deactivated by dephosphorylation of this to GDP. RGS proteins accelerate the rate of G-protein-GTP dephosphorylation, which has the effect of greatly reducing the duration of G-protein activation and hence its downstream effects. Since Gq-11 activation potentiates contraction by the PLC pathway, the effect of RGS proteins is to reverse this, promoting relaxation. As discussed previously (see Figure 5.2 and related text in Chapter 5), nitric oxide (NO) is produced by endothelial cells in response to shear stress and diffuses into adjacent SMCs where it activates guanylate cyclase (cGC) to produce cyclic GMP (cGMP), which in turn activates phosphokinase G (PKG). PKG phosphorylates RGS proteins activating them. Activated RGS bound to Gq-11 inhibits its ability to potentiate contraction, resulting in a net relaxation (Figure 8.6).

Thus, vascular smooth muscle is controlled by several vasoactive agonists which regulate the tone and calibre of blood vessels by activating signalling pathways within SMCs to regulate contraction. Changing the level of $[Ca^{2+}]_i$ is the primary driver for contraction and this is modulated by several enzyme pathways that either enhance or inhibit the response to fine tune the degree of contraction. The ability of blood vessels to respond to different agonists is the basis by which the circulation is controlled by external factors such as the autonomic nervous system, circulating hormones and locally produced autacoids. In addition, arteries and veins have intrinsic mechanisms for regulating their lumen diameter by sensing and responding to local conditions.

Mechano-sensing in Blood Vessels

The ability of blood vessels to alter their lumen diameter in response to changes in blood pressure and flow is crucial for regulating blood pressure and the distribution of blood flow to meet tissue needs. This is possible because they have evolved mechanisms for sensing the flow and pressure and responding by contracting or relaxing the smooth muscle which lines their walls.

Flow mediated dilatation

The flow of blood creates a frictional effect on the lumen of blood vessels. Known as shear stress, the endothelial cells which line blood vessels can sense this and respond by causing the vessel to dilate. This process is known as flow mediated dilatation (FMD) and has been discussed previously (see Figure 5.2 and related text in Chapter 5) and will be summarised here. FMD has two important benefits; first by dilating in response to increased flow, the adverse effect of excessive shear stress is mitigated. Second, FMD facilitates increased flow to meet increased demand. FMD works by relaxing smooth muscle in the walls of these vessels in response to nitric oxide (NO) and prostacyclin, which are produced by endothelial cells in the presence of increased shear stress.

Stretch induced contraction — Myogenic tone

A second form of mechano-sensing in blood vessels is their ability to sense and respond to changes in blood pressure. Myogenic tone is crucial for regulating blood pressure and flow and disturbances in myogenic tone are likely to be implicated in the causes of hypertension. The mechanisms underlying myogenic tone have yet to be fully elucidated, however research in recent years has begun to shed light on how blood vessels are able to sense changes in pressure (or stretch) and how this is relayed to regulate contraction and relaxation of SMCs.

Sensing stretch

Recent evidence indicates that SMCs sense stretch via a cooperative activation of the angiotensin receptor (AT1R) and transient receptor potential (TRP) ion channels in the cell membrane [25] (Figure 8.7). As well as being responsive to the agonist angiotensin II, AT1Rs are also directly sensitive to stretch [26]. TRP ion channels are subtypes of ion channels present in the cell membranes of all cells including SMCs. The name

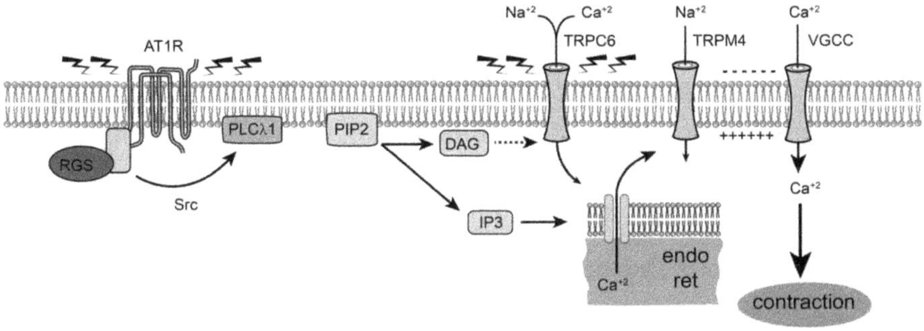

Figure 8.7: Illustration of pathways of stretch-mediated vasoconstriction. TRPM4 and TRPC6 are necessary for the myogenic response and function cooperatively with activation of AT1Rs in mediating contraction in response to stretch. AT1Rs are directly sensitive to stretch and are primary mechano-sensors in arteries. Stimulation of AT1Rs causes activation of an isoform of PLC, PLCλ1, that is co-localised in the cell membrane with AT1Rs, and TRPC6 and TRPM4. Stretch activation of AT1R activates PLCλ1 leading to the production of DAG and IP3 from phosphatidylinositol 4,5-bisphosphate. DAG augments cation influx through stretch activation of TRPC6 ion channels. This and IP3 stimulate release of Ca^{2+} from stores in the ER, which in turn opens TRPM4 ion channels causing an influx of Na^{+} and membrane depolarisation. VGCCs are activated by membrane depolarisation resulting in a much larger influx of Ca^{2+} and elevation of $[Ca^{2+}]_i$, which enables contraction [25].

Notes:	AT1R	Angiotensin II receptor 1
	DAG	Diacylglycerol
	endo ret	Endoplasmic reticulum
	IP3	Inositol trisphosphate
	PIP2	Phosphatidylinositol 4,5-bisphosphate
	PLC	Phospholipase C
	RGS	Regulator of G-protein signalling
	TRP	Transient receptor potential
	VGCC	Voltage-gated calcium channel

"transient receptor potential" comes from an observed transient change in membrane potential related to the channel when it was first discovered. Unlike most ion channels, TRP channels are not sensitive to membrane voltage changes, but instead are sensitive to a variety of other stimuli, including stretch, vibration, light and various chemicals, and have a key role in many sensing mechanisms such as vision, hearing, taste, smell and touch. Several subtypes of TRP ion channels are known to be present in

SMCs and recently TRPM4 and TRPC6 have been shown to be involved in mechano-sensing. Stretch sensing in SMCs depends on cooperative activation of AT1Rs, TRPM4 and TRPC6, which together cause membrane depolarisation and activation of a signal cascade resulting in contraction (Figure 8.7) [25].

Stretch-induced depolarisation

Smooth muscle cells like all cells have a negatively charged interior due to the action of ion pumps which maintain a stable intracellular osmotic and ionic environment (see Figure 4.1 and related text in Chapter 4). In 1984, it was clearly demonstrated that SMCs become depolarised when stretched [27], by an increase in luminal pressure [28]. This is due to an influx of Na^+ and Ca^{2+}, as follows.

Stimulation of AT1Rs in response to stretch causes activation of PLC (Figure 8.6). Several isoforms of PLC are present in SMCs, of which PLCλ1 is co-localised in the cell membrane with AT1Rs, TRPC6 and TRPM4. Stretch activation of AT1Rs activates PLCλ1 leading to the production of DAG and IP3 from PIP2. TRPC6 [29] and TRPM4 [25] are required for the myogenic response. DAG activates TRPC6 and IP3 stimulates release of Ca^{2+} from the endoplasmic reticulum (ER). This in turn activates TRPM4, resulting in an influx of cations via TRPC6 and TRPM4 causing membrane depolarisation. Voltage-gated calcium channels (VGCCs) are selective for calcium ions and open during membrane depolarisation resulting in an influx of extracellular Ca^{2+} [30], causing an increase in $[Ca^{2+}]_i$ (Figure 8.7). Calmodulin rapidly binds elevated Ca^{2+} forming Ca^{2+}/CAM, which activates MLCK, enabling it to phosphorylate MRLCs, which initiates contraction.

Thus the myogenic response in arteries is due to mechano-sensitive properties of SMCs. Stretching the cell membrane of SMCs causes depolarisation by cooperative action of AT1Rs and TRP channels. Cell depolarisation then opens VGCCs resulting in an influx of Ca^{2+} ions, which activates contraction via phosphorylation of MLCK. Again we see that the signalling pathway for this is shared with agonist stimulation (Figure 8.6). By the same token, a reduction in blood pressure reduces stretch activation of SMCs and thereby the degree of contraction in arteries. In this way, myogenic tone serves to provide a continuous mechanism for matching vessel calibre and tone to prevailing haemodynamics.

Evolution of Muscle

Despite the striking differences in the structure and function of striated and smooth muscles, they share a common evolutionary origin. Although the microscopic appearance of Z-bands in striated muscle and dense bodies in smooth muscle are quite different, they have the same function in binding actin and they are both composed of α-actinin. Further evidence for this is provided by evolutionary hybrids of actin–myosin. As discussed above, striated muscle is composed of bi-polar myosin and smooth muscle of side-polar myosin. Recent studies of invertebrate species have shown that this is not always the case.

Schistosoma mansoni is a worm and an important human pathogen. Around one centimetre in length, it infects around 200 million people around the world. The adult worms are mobile, containing circular and longitudinal layers of smooth muscle. Evidence that praziquantel, a drug used to treat schistosomiasis, works by targeting the worm's myosin regulatory light chain [31] stimulated research to understand its contractile apparatus. Surprisingly, this revealed that *Schistosoma* smooth muscle contains bi-polar myosin with heads arranged in a helical array, identical to that present in vertebrate striated muscle [32].

Further evidence for a common origin of striated and smooth muscle comes from the discovery that some jellyfish have striated muscle which is activated by phosphorylation of myosin. As discussed above, striated muscle contains troponin bound to actin and activation of troponin by a rise in intracellular calcium triggers contraction. Smooth muscle, on the other hand, lacks troponin. Instead contraction is activated by enzymatic phosphorylation of myosin. Remarkably, adult moon jellyfish, also called common jellyfish (*Aurelia aurita*), contain striated muscle which lacks troponin. Instead it contains myosin light chain kinase which phosphorylates myosin regulatory light chains in the same way that smooth muscle does [33]. These examples of hybrid elements of smooth and striated muscles in *Schistosoma* and jellyfish suggest that both forms of actin–myosin derived from a common ancestor. This is also supported by evidence that the core motor protein, myosin heavy chain, is present in unicellular organisms, such as sponges, which do not have fully formed muscle and evolved before multicellular organisms appeared around 500 million years ago. Thus both smooth and striated muscles are likely to have evolved by the addition of new proteins to an ancient contractile system [34].

Smooth Muscle Cells are Flexible in Structure and Function

Smooth muscle cells are highly flexible in their ability to adapt to prevailing conditions and needs by altering their function and structure. Many of these changes are regulated by proteins which bind to actin. Actin binding proteins regulate not just the onset and force of contraction, but also structural changes in SMCs. When subjected to prolonged stretch, SMCs can alter, becoming longer, extending their cytoskeleton and contractile elements. They also dedifferentiate into primitive cell types which can proliferate and transform into secretory cells to assist healing following injury. Many of these adaptations are regulated by actin and actin binding proteins, the most important of which are tropomyosin, caldesmon and calponin.

Actin Binding Proteins

Smooth muscle tropomyosin

Tropomyosin is bound to actin filaments in smooth muscle as it is in striated muscle and in both tropomyosin maintains the relaxed state by blocking interaction between actin and myosin. By the same token, movement of tropomyosin is required in both to enable contraction. In striated muscle, this is achieved by the action of troponin. However, troponin is not present in SMCs and tropomyosin is moved by an alternative mechanism to enable contraction. Studies using fluorescence resonance energy transfer (FRET, a technique which uses fluorescent probes to measure distances at the molecular level) have demonstrated that tropomyosin is moved during actin–myosin interaction in smooth muscle and that this is dependent on phosphorylation of myosin heads [35]. This difference in behaviour of tropomyosin in smooth and striated muscles is likely due to a subtle difference in the way tropomyosin binds to actin. Four genes encode for tropomyosin, which produce at least 40 isoforms based on alternative splicing. These regulate actin functions in different cell types. They all bind to actin, but in slightly different alignments. The isoform tropomyosin 6 predominates in vascular smooth muscle and has been shown to bind differently to actin than cardiac tropomyosin [36]. In SMCs, tropomyosin binding to actin allows a degree of access to the myosin binding sites on actin, such that once myosin heads are phosphorylated by MLCK, they can engage with actin in a way that moves tropomyosin further, giving full exposure to myosin binding sites on actin. In other words, the isoform of tropomyosin in smooth muscle binds to actin in such a way that phosphorylated myosin can engage with actin and dislodge tropomyosin sufficiently to completely expose the myosin binding sites on actin. Thus, tropomyosin regulates the contraction in both striated and smooth muscles, but in SMCs this is mediated by phosphorylation of myosin rather than by troponin.

Caldesmon

Caldesmon is a protein of 793 amino-acid and molecular weight of 87 kDa, which is present in thin filaments in the ratio of 1–14 actins. Like tropomyosin, caldesmon binds to actin and interacts with tropomyosin in two ways. Caldesmon is an important inhibitor of actomyosin ATPase and therefore of actin–myosin interaction. This effect is strongly enhanced by the presence of tropomyosin [37]. Second, the presence of caldesmon increases the affinity of tropomyosin binding to actin [38]. Thus in smooth muscle, contraction is actively inhibited by caldesmon. This inhibition is highly dependent on the level of intracellular calcium $[Ca^{2+}]_i$. In the presence of raised $[Ca^{2+}]_i$, calcium activated calmodulin (Ca^{2+}/CAM) binds to caldesmon and reverses its inhibitory action on actomyosin ATPase, which enables actin–myosin interaction. Therefore activation of actin–myosin interaction in smooth muscle is regulated by two Ca^{2+} dependent mechanisms; (i) Ca^{2+}/CAM activates MLCK, which then phosphorylates MRLCs enabling myosin interaction with actin and (ii) Ca^{2+}/CAM also binds to caldesmon reversing its inhibitory action on actin–myosin interaction [39]. Caldesmon can be seen as acting somewhat like troponin in striated muscle maintaining smooth muscle in the relaxed state and controlling activation in the presence of raised intracellular calcium.

Calponin and PKC mediated contraction

As discussed above, PKC is an important Ca^{2+} independent contractile mechanism in smooth muscle and recent evidence has shown that calponin has an important role in regulating this action of PKC. Phosphorylation of myosin by the enzyme MLCK follows an increase in $[Ca^{2+}]_i$ and is particularly involved in phasic smooth muscle contraction as in the intestines and initiating tonic contractions. PKC in contrast is important in maintaining smooth muscle tone [40]. PKC is a member of a family of protein kinase enzymes, which regulate cellular functions by phosphorylating serine and threonine amino acid residues of many proteins. One of these is CPI-17 which when phosphorylated, potentiates contraction by inhibiting myosin regulatory light chain phosphatase (Figure 8.6). Of the several isoforms of PKC, some are activated by a rise in $[Ca^{2+}]_i$. Others are sensitive to DAG, produced by PLC mediated breakdown of PIP2 following stimulation by agonists such as phenylephrine (Figure 8.6). All isoforms of PKC require initial phosphorylation to become fully active enzymes and one of these involves an autophosphorylation step (meaning the enzyme phosphorylates itself) and calponin has been shown to play a role in this [41].

Calponin is a 34 kDa protein present in smooth muscle, which also binds to actin. In its absence, smooth muscle contractions induced by α-adrenergic agonists such as phenylephrine or exposure to phorbol ester are greatly reduced [42]. Phosphorylation of PKC is necessary for its activation and calponin is essential for this to occur by facilitating its autophosphorylation [41]. In this way, calponin facilitates activation of PKC and as a result enhances phosphorylation of its target ICP-17, which potentiates contraction. In addition, calponin also binds to actin with some similarities to caldesmon in that it has an inhibitory effect on actin–myosin interaction. However, unlike caldesmon, its inhibitory effect is independent of Ca^{2+}/CAM and tropomyosin [43], but is reversed by phosphorylation by PKC resulting in contraction in coronary artery [44] and uterine [45] smooth muscle. Thus PKC is an important additional mechanism for smooth muscle contraction that is independent of Ca^{2+}. This fits well with the ability of smooth muscle to maintain long lasting contractions needed for myogenic tone.

Thus, smooth muscle contraction is regulated by a large number of mechanisms. Myosin light chain phosphorylation by MLCK triggered by raised $[Ca^{2+}]_i$ is the main stimulus for contraction. But this can be modulated by several additional enhancing and inhibiting signalling pathways. One might ask why are there so many? The answer lies in the wide range of conditions which blood vessels must respond and adapt to. Maintenance of a constant level of blood pressure and flow requires an ability to sustain a stable vessel calibre and hence the degree of smooth muscle tone. In contrast, meeting the increased energy needs of physiological changes such as exercise requires a more dynamic response. Yet again, blood vessels have several roles in responding to tissue injury, for example, by contracting vigorously to limit blood loss following trauma, promoting healing of myocardial infarcts (discussed in Chapter 6) or signalling to promote activation of the immune system. Blood vessels therefore undertake a more diverse range of functions than either the heart or skeletal muscle for which they have evolved the ability to respond to a wider range of stimuli. Although these are detected by multiple receptors and ion channels on the cell surface, the signalling pathways they activate are often shared.

Tropomyosin: A Master Actin Filament Regulator

In addition to its role as a key contractile protein in muscles, actin is essential for numerous other functions in all cell types. Broadly, these relate to its presence as part of the cytoskeleton or as a motor protein. As discussed in Chapter 3, actin and myosin function as molecular motors in many ways other than as muscles, by providing an essential transport system for components of cells as well as cell motility. These basic properties are essential for a wide range of higher level cell functions, including cell division and maturation and the ability of cells to migrate within tissues. As part of the cytoskeleton, actin is essential for cell metabolism, for example, by correctly locating the glucose transporter GLUT4 in cell membranes, and for maintaining the physical properties, e.g., stiffness of cells. At the molecular level, these different roles depend on the ability of actin to undertake numerous interactions with other actin binding proteins. And yet only three isoforms of actin are present in most cell types, for example, SMCs have α, β and λ actin. How then can actin undertake such a wide range of functions? Recent evidence has shown that tropomyosin is the key to this [46] by specifying the functional properties of actin [47]. The basis for this is that all actin filaments contain two tropomyosin polymers bound on each side of the actin double helix (Figure 8.4). Although mammalian cells have just three actin isoforms, they have over 40 tropomyosin isoforms and it is the isoform of tropomyosin bound to actin that determines its function. Tropomyosin does this because each isoform determines which actin binding protein can bind to actin. Binding of tropomyosin to actin occurs when actin filaments are being assembled. This process is regulated by a protein called formin and it is at this point that the function of actin is determined by the isoform of tropomyosin it binds [46]. Thus by varying the isoforms of tropomyosin it binds, actin is responsible for many cell functions other than contraction. Examples include maintaining the integrity of the immune system and platelet function [48].

Smooth Muscle Length Adaptability

Smooth muscle has a remarkable ability to function over a large length range. This is particularly the case in organs such as stomach and uterus, but also in capacitance blood vessels, notably of the splanchnic circulation. Length adaptability has been demonstrated by many studies and means that smooth muscle capacity for power generation actually increases when it is stretched. For example, when passively stretched to 150% of initial length and then allowed to adapt to the new length for 120 s, smooth muscles contracted more rapidly and with greater force [49]. This is unique to smooth muscle. The cytoskeleton of smooth muscle, through which force is transmitted, is much less tightly packed than striated muscle sarcomeres. This allows for greater flexibility and adaptability and this is the key to length adaptability of smooth muscle.

Dynamic actin polymerisation

Extension and rearrangement of the cytoskeleton and contractile elements of smooth muscle is the basis for its ability to adapt to increased length. This involves the formation of new actin filaments. As discussed in Chapter 3, actin filaments (f-actin) are formed by polymerisation of g-actin monomers. This is a dynamic process in smooth muscle, increasing with contraction and decreasing by depolymerisation with relaxation. Thus, experimental inhibition of actin polymerisation markedly impairs force generation, while inhibition of depolymerisation enhances it [50]. In addition, constriction of smooth muscle by stimulation with agonists such as noradrenaline increases the concentration of f-actin and reduces g-actin. This dynamic formation of actin is also required for the myogenic response of arteries; increasing the luminal pressure of arteries induces not only contraction, but also reduces the concentration of g-actin [51], indicating increased polymerisation into f-actin. Thus the ability to dynamically restructure the cytoskeleton, including its contractile components, provides an important mechanism for the length adaptability of smooth muscle. This involves the capacity to form new actin filaments from a pool of g-actin monomers when required.

Dynamic myosin polymerisation

The ability to form new contractile elements implies that myosin filament assembly in some degree must occur simultaneously with actin polymerisation. Evidence for this comes from the discovery that myosin is present in SMCs in three forms. As discussed above, myosin thick filaments are assembled from myosin monomers, consisting of a head, neck and tail domain with an essential light chain and a regulatory light chain attached to the neck. In the cytoplasm, myosin is present as dimers with the tail domain of each wrapped together in a coiled coil, while the two heads are separated with an essential light chain and a regulatory light chain attached to each neck domain (Figure 8.8). In addition to myosin present in contractile thick filaments, up to 30% may be present in an unassembled

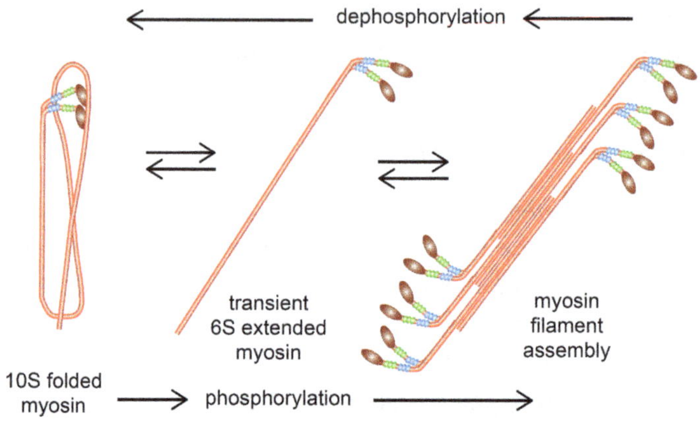

Figure 8.8: Illustration of dynamic phosphorylation induced myosin assembly in smooth muscle. Myosin is present in the cytoplasm as a dimer, consisting of two myosin monomers, of which the tails are wrapped in a coiled coil with the two heads (brown) separated and an essential light chain (blue) and a regulatory light chain (green) are attached to the neck domain of each. Myosin adopts a folded conformation (10S) when the RLCs are dephosphorylated. This causes the tail domain to fold at two sites in such a way that the second fold wraps the heads blocking their ATPase activity and inhibiting interaction with actin. Phosphorylation of the regulatory light chain by myosin light chain kinase unfolds the 10S conformation to an extended form, 6S myosin, which can be rapidly assembled into thick filaments. The transition from 10S myosin to assembled filaments is dynamic and reversible by phosphorylation and dephosphorylation of RLCs [52].

folded form, known as 10S myosin (10S refers to Svedberg units, a measure of how quickly it sediments in a liquid). An extended form known as 6S myosin is also present in smaller amounts, which can be rapidly assembled into filaments [52]. Folded 10S myosin is maintained in an inactive state because one of its two folded regions binds to the heads (Figure 8.8) in a way that blocks ATPase activity and hinders interaction with actin [53]. It also has a more globular structure which not only allows it to sediment more quickly than the extended form, but also facilitates its transport in the cytoplasm to regions of filament assembly. The availability of a pool of 10S myosin, which can be rapidly assembled into filaments, provides a mechanism for dynamic restructuring of the contractile system when required, for example, when blood vessels or other organs are distended.

Although 10S myosin is inactive in terms of interaction with actin, its regulatory light chains (MRLCs) are accessible to phosphorylation by MLCK and when phosphorylated causes unfolding of the molecule to adopt the extended 6S form of myosin which is immediately available for incorporation into myosin filaments. Thus activation of MLCK and phosphorylation of MRLCs not only initiates contraction of the actin–myosin filaments, but also actively unfolds inactive 10S myosin making it immediately available for assembly into new thick filaments. In other words, stimulation of smooth muscle not only initiates contraction but also promotes the formation of new contractile filaments. In studies in which smooth muscle is stretched and allowed to adapt to its new length, shortening velocity and power output increase (length adaptability) and this is accompanied by a reduction in the proportion of myosin present as monomers [49] as they are increasingly incorporated into assembled filaments. Thus, adapting to increased length is possible due to the active assembly of additional myosin filaments and this is reversible with restoration of length.

Length adaptability is yet another example of the flexibility of smooth muscle to perform many functions. Cardiac muscle astonishes us with its consistent and regular performance made possible by its highly organised and relatively fixed structure. Smooth muscle is equally remarkable for its functional adaptability, which is the basis for its ability to provide motor power in so many different organs and to meet different physiological and pathological challenges. The ability to restructure its cytoskeleton and contractile apparatus and to do so rapidly and reversibly in response to external stimulation by agonists or as part of the myogenic response is crucial for normal regulation of vascular tone and the circulation. Better understanding of these mechanisms and the extraordinary adaptability of smooth muscle holds great promise for finding better ways to treat and prevent hypertension which is the largest risk for ill health in the world today.

Conclusions

This book ostensibly concerns the heart and its durability, longevity and capacity to sustain our lives. However, those attributes risk being overblown without some acknowledgement of the physiological framework on which they depend. In fact, the heart has many allies in meeting the demands placed upon it. The vasculature is its most intimate and contiguous partner, distributing its output and returning it continuously. It is precisely because of the extraordinary adaptability and flexibility of blood vessels that the heart can function as it does and for so long; if blood vessels were simple rigid tubes, its life span would be greatly reduced.

The ability of blood vessels to sense pressure and flow and to respond continuously to changes in them not only protects the vessels themselves and ensures appropriate distribution of blood flow, but also protects the heart by moderating excessive swings in blood pressure and load on the heart. The ability of the endothelial cells which line all blood vessels to sense the rate of blood flow and to produce a signal which instructs the smooth muscle in the vessel wall to relax is truly elegant. So too is the sensitivity and contractile response of vascular smooth muscle cells to stretch and thereby changes in blood pressure. This allows them to modulate vessel diameter in a way that maintains a constant flow rate and protects the smallest capillaries from excessive pressures. The structure and function of smooth muscle are quite different from that of heart muscle; never the less they share a common evolutionary heritage. One might debate whether the word intelligent is appropriate in describing blood vessels, but they are undeniably smart tubes.

References

[1] Needleman H. Low level lead exposure: History and discovery. *Annals of Epidemiology*. 2009;19(4):235–238.

[2] Hedblom E. A fusion of advanced stainless steel and cloud technology is giving unprecedented insight into industrial and chemical processes 2019. Available from: https://www.eurekamagazine.co.uk/design-engineering-features/technology/a-fusion-of-advanced-stainless-steel-and-cloud-technology-is-giving-unprecedented-insight-into-industrial-and-chemical-processes/199669/. (Accessed on 10/3/2020).

[3] Bayliss WM. On the local reactions of the arterial wall to changes of internal pressure. *Journal of Physiology*. 1902;28(3):220–231.

[4] Harris DE, Warshaw DM. Length vs. active force relationship in single isolated smooth muscle cells. *American Journal of Physiology*. 1991;260(5 Pt 1):C1104–C1112.

[5] Uvelius B. Isometric and isotonic length-tension relations and variaitons in cell length in longitudinal smooth muscle from rabbit urinary bladder. *Acta Physiologica Scandinavica*. 1976;97(1):1–12.

[6] Kirmse R, Portet S, Mucke N, Aebi U, Herrmann H, Langowski J. A quantitative kinetic model for the in vitro assembly of intermediate filaments from tetrameric vimentin. *Journal of Biological Chemistry*. 2007;282(25):18563–18572.

[7] Sjuve R, Arner A, Li Z, Mies B, Paulin D, Schmittner M, *et al.* Mechanical alterations in smooth muscle from mice lacking desmin. *Journal of Muscle Research and Cell Motility*. 1998;19(4):415–429.

[8] Ribeiro Ede A, Jr., Pinotsis N, Ghisleni A, Salmazo A, Konarev PV, Kostan J, *et al.* The structure and regulation of human muscle alpha-actinin. *Cell*. 2014;159(6):1447–1460.

[9] Sweeney HL, Hammers DW. Muscle contraction. *Cold Spring Harbor Perspectives in Biology*. 2018;10(2):a030288.

[10] Chou RG, Stromer MH, Robson RM, Huiatt TW. Substructure of cytoplasmic dense bodies and changes in distribution of desmin and alpha-actinin in developing smooth muscle cells. *Cell Motility and the Cytoskeleton*. 1994;29(3):204–214.

[11] McGuffee LJ, Mercure J, Little SA. Three-dimensional structure of dense bodies in rabbit renal artery smooth muscle. *Anatomical Record*. 1991;229(4):499–504.

[12] Zhang J, Herrera AM, Pare PD, Seow CY. Dense-body aggregates as plastic structures supporting tension in smooth muscle cells. *American Journal of Physiology. Lung Cellular and Molecular Physiology*. 2010;299(5): L631–L638.

[13] Sjoblom B, Salmazo A, Djinovic-Carugo K. Alpha-actinin structure and regulation. *Cellular and Molecular Life Sciences*. 2008;65(17):2688–2701.

[14] Turner M, Anderson DE, Bartels P, Nieves-Cintron M, Coleman AM, Henderson PB, *et al.* α-Actinin-1 promotes activity of the L-type Ca^{2+} channel $Ca_v1.2$. *EMBO Journal*. 2020;39(5):e102622.

[15] Cukovic D, Lu GW, Wible B, Steele DF, Fedida D. A discrete amino terminal domain of Kv1.5 and Kv1.4 potassium channels interacts with the spectrin repeats of α-actinin-2. *FEBS Letters*. 2001;498(1):87–92.

[16] Burgueno J, Blake DJ, Benson MA, Tinsley CL, Esapa CT, Canela EI, *et al.* The adenosine A_{2A} receptor interacts with the actin-binding protein α-actinin. *Journal of Biological Chemistry*. 2003;278(39):37545–37552.

[17] Gabella G. Structural apparatus for force transmission in smooth muscles. *Physiological Reviews*. 1984;64(2):455–477.

[18] North AJ, Gimona M, Lando Z, Small JV. Actin isoform compartments in chicken gizzard smooth muscle cells. *Journal of Cell Science*. 1994;107 (Pt 3):445–455.

[19] Xu JQ, Harder BA, Uman P, Craig R. Myosin filament structure in vertebrate smooth muscle. *Journal of Cell Biology*. 1996;134(1):53–66.

[20] Craig R, Megerman J. Assembly of smooth muscle myosin into side-polar filaments. *Journal of Cell Biology*. 1977;75(3):990–996.

[21] Small JV, Herzog M, Barth M, Draeger A. Supercontracted state of vertebrate smooth muscle cell fragments reveals myofilament lengths. *Journal of Cell Biology*. 1990;111(6 Pt 1):2451–2461.

[22] Baumann F, Bauer MS, Rees M, Alexandrovich A, Gautel M, Pippig DA, *et al.* Increasing evidence of mechanical force as a functional regulator in smooth muscle myosin light chain kinase. *Elife*. 2017;6:e26473.

[23] Lee CH, Poburko D, Sahota P, Sandhu J, Ruehlmann DO, van Breemen C. The mechanism of phenylephrine-mediated $[Ca^{2+}]_i$ oscillations underlying tonic contraction in the rabbit inferior vena cava. *Journal of Physiology*. 2001;534(Pt 3):641–650.

[24] Yu OM, Brown JH. G protein-coupled receptor and RhoA-stimulated transcriptional responses: Links to inflammation, differentiation, and cell proliferation. *Molecular Pharmacology*. 2015;88(1):171–180.

[25] Gonzales AL, Yang Y, Sullivan MN, Sanders L, Dabertrand F, Hill-Eubanks DC, *et al.* A PLCγ1-dependent, force-sensitive signaling network in the myogenic constriction of cerebral arteries. *Science Signaling*. 2014;7(327):ra49.

[26] Pires PW, Ko EA, Pritchard HAT, Rudokas M, Yamasaki E, Earley S. The angiotensin II receptor type 1b is the primary sensor of intraluminal pressure in cerebral artery smooth muscle cells. *Journal of Physiology*. 2017;595(14):4735–4753.

[27] Harder DR. Pressure-dependent membrane depolarization in cat middle cerebral artery. *Circulation Research*. 1984;55(2):197–202.

[28] Davis MJ, Hill MA. Signaling mechanisms underlying the vascular myogenic response. *Physiological Reviews*. 1999;79(2):387–423.

[29] Welsh DG, Morielli AD, Nelson MT, Brayden JE. Transient receptor potential channels regulate myogenic tone of resistance arteries. *Circulation Research*. 2002;90(3):248–250.

[30] Knot HJ, Nelson MT. Regulation of arterial diameter and wall [Ca^{2+}] in cerebral arteries of rat by membrane potential and intravascular pressure. *Journal of Physiology*. 1998;508 (Pt 1):199–209.

[31] Gnanasekar M, Salunkhe AM, Mallia AK, He YX, Kalyanasundaram R. Praziquantel affects the regulatory myosin light chain of *Schistosoma mansoni*. *Antimicrobial Agents and Chemotherapy*. 2009;53(3):1054–1060.

[32] Sulbaran G, Alamo L, Pinto A, Marquez G, Mendez F, Padron R, *et al.* An invertebrate smooth muscle with striated muscle myosin filaments. *Proceedings of the National Academy of Sciences of the United States of America*. 2015;112(42):E5660–E5668.

[33] Tanaka H, Ishimaru S, Nagatsuka Y, Ohashi K. Smooth muscle-like Ca^{2+}-regulation of actin-myosin interaction in adult jellyfish striated muscle. *Scientific Reports*. 2018;8(1):7776.

[34] Steinmetz PR, Kraus JE, Larroux C, Hammel JU, Amon-Hassenzahl A, Houliston E, *et al.* Independent evolution of striated muscles in cnidarians and bilaterians. *Nature*. 2012;487(7406):231–234.

[35] Graceffa P. Phosphorylation of smooth muscle myosin heads regulates the head-induced movement of tropomyosin. *Journal of Biological Chemistry*. 2000;275(22):17143–17148.

[36] Lehman W, Morgan KG. Structure and dynamics of the actin-based smooth muscle contractile and cytoskeletal apparatus. *Journal of Muscle Research and Cell Motility*. 2012;33(6):461–469.

[37] Dabrowska R, Goch A, Galazkiewicz B, Osinska H. The influence of caldesmon on ATPase activity of the skeletal muscle actomyosin and bundling of actin filaments. *Biochimica et Biophysica Acta*. 1985;842(1):70–75.

[38] Yamashiro-Matsumura S, Matsumura F. Characterization of 83-kilodalton nonmuscle caldesmon from cultured rat cells: Stimulation of actin binding of nonmuscle tropomyosin and periodic localization along microfilaments like tropomyosin. *Journal of Cell Biology*. 1988;106(6): 1973–1983.

[39] Marston S, El-Mezgueldi M. Role of tropomyosin in the regulation of contraction in smooth muscle. *Advances in Experimental Medicine and Biology*. 2008;644:110–123.

[40] Horowitz A, Menice CB, Laporte R, Morgan KG. Mechanisms of smooth muscle contraction. *Physiological Reviews*. 1996;76(4):967–1003.

[41] Kim HR, Gallant C, Morgan KG. Regulation of PKC autophosphorylation by calponin in contractile vascular smooth muscle tissue. *BioMed Research International*. 2013;2013:358643.

[42] Je HD, Gangopadhyay SS, Ashworth TD, Morgan KG. Calponin is required for agonist-induced signal transduction — Evidence from an antisense approach in ferret smooth muscle. *Journal of Physiology*. 2001;537(Pt 2):567–577.

[43] Winder SJ, Walsh MP, Vasulka C, Johnson JD. Calponin-calmodulin interaction: Properties and effects on smooth and skeletal muscle actin binding and actomyosin ATPases. *Biochemistry*. 1993;32(48):13327–13333.

[44] Mino T, Yuasa U, Naka M, Tanaka T. Phosphorylation of calponin mediated by protein kinase C in association with contraction in porcine coronary artery. *Biochemical and Biophysical Research Communications*. 1995;208(1):397–404.

[45] Li L, Zhang Y, Zhou C. Phosphorylation of h1 calponin by PKC epsilon may contribute to facilitate the contraction of uterine myometrium in mice during pregnancy and labor. *Reproductive Biology and Endocrinology*. 2012;10:37.

[46] Gunning PW, Hardeman EC, Lappalainen P, Mulvihill DP. Tropomyosin — Master regulator of actin filament function in the cytoskeleton. *Journal of Cell Science*. 2015;128(16):2965–2974.

[47] Gateva G, Kremneva E, Reindl T, Kotila T, Kogan K, Gressin L, *et al.* Tropomyosin isoforms specify functionally distinct actin filament populations *in vitro*. *Current Biology*. 2017;27(5):705–713.

[48] Reinhard M, Halbrugge M, Scheer U, Wiegand C, Jockusch BM, Walter U. The 46/50 kDa phosphoprotein VASP purified from human platelets is a novel protein associated with actin filaments and focal contacts. *EMBO Journal*. 1992;11(6):2063–2070.

[49] Chitano P, Wang L, Tin GYY, Ikebe M, Pare PD, Seow CY. Smooth muscle function and myosin polymerization. *Journal of Cell Science*. 2017;130(15):2468–2480.

[50] Walsh MP, Cole WC. The role of actin filament dynamics in the myogenic response of cerebral resistance arteries. *Journal of Cerebral Blood Flow and Metabolism*. 2013;33(1):1–12.

[51] Moreno-Dominguez A, El-Yazbi AF, Zhu HL, Colinas O, Zhong XZ, Walsh EJ, *et al.* Cytoskeletal reorganization evoked by Rho-associated kinase- and protein kinase C-catalyzed phosphorylation of cofilin and heat shock protein 27, respectively, contributes to myogenic constriction of rat cerebral arteries. *Journal of Biological Chemistry*. 2014;289(30):20939–20952.

[52] Milton DL, Schneck AN, Ziech DA, Ba M, Facemyer KC, Halayko AJ, *et al.* Direct evidence for functional smooth muscle myosin II in the 10S self-inhibited monomeric conformation in airway smooth muscle cells. *Proceedings of the National Academy of Sciences of the United States of America*. 2011;108(4):1421–1426.

[53] Yang S, Lee KH, Woodhead JL, Sato O, Ikebe M, Craig R. The central role of the tail in switching off 10S myosin II activity. *Journal of General Physiology*. 2019;151(9):1081–1093.

9

Heart and Blood: Co-evolved Intimate Partners

Introduction

The emergence of oxygen in the Earth's atmosphere 2.5 billion years ago made aerobic metabolism possible. Organisms responded by evolving larger and more complex bodies. Hearts were then needed to pump oxygen to increasingly distant tissues and a medium in which it could be transported. A variety of solutions have evolved for this purpose in animals; in humans and other vertebrates, blood provides this function. At first glance, this liquid organ may be thought of as a relatively simple solution in which oxygen and other substances are carried. However, this belies its complex structure, which evolved in ways that not only facilitate the transport of large amounts of oxygen to meet the needs of tissue with high energy needs, but also to exquisitely match the structure and function of the heart and circulation. Indeed blood can rightly be thought of as an intrinsic part of the circulatory system. Its structure and physiology have evolved to match those of the blood vessels in which it circulates and the heart which pumps it. The shape and size of erythrocytes (red blood cells) were central to how the circulation evolved in vertebrates [1]. Thus blood is an intrinsic element of the circulation and although not usually included in books primarily concerned with the heart, I believe a discussion of how blood evolved and how it functions within the circulation is essential in order to understand its crucial role in the physiology of the heart and circulation.

Like the word heart, blood is also widely used as a metaphor related to emotion and feelings. Unlike the heart, however, which we are aware of because we can feel it pumping within us, we cannot sense blood. Blood and its passage through arteries and veins are not palpable. When we feel a pulse, it is the action of the heart that comes to mind. Rather, it is the visual impact of blood which has such a powerful effect on us. For humans, red is perceived as a signal of aggression and dominance [2] and has strong associations with emotions such as anger and danger [3]. There has long been interest in the extent to which these perceptions are learned through culture and language or whether they have a biological origin. Given the widely shared associations which the colour red has among different cultures, it seems very likely that some of these are learned. However, recent evidence also points to biological origins. Thus, pre-language infants can already categorise colours as red, blue, yellow and purple [4]. This is significant since it cannot have been learned and because colour is in reality a continuous spectrum and its perception as distinct colours is a function of human biology. Infants also show an attention bias for red [5], suggesting a possible innate predisposition that could contribute to establishing such cultural links. In addition, our ancestors used ochre, an iron-based pigment, to create the colour red in their cave drawings. Hence the bright red colour of blood when exposed through injury is both dramatic and alarming. We associate its colour red with danger and this may have contributed to its international recognition, for example, as a "stop" sign. It is likely therefore that the shared powerful emotions we associate with the colour of blood underlie our harnessing it as a word to convey abstract meanings. And so, we use "warm-blooded" to describe a kind, generous person and "cold-blooded" for a cruel or uncaring one. We also use blood for family connections as in "blood relations" and "blood is thicker than water". The use of "blood and thunder" refers to violence and is likely to derive from the spilling of blood in battle and its dramatic visual impact.

Composition of Blood

The constituents of blood vary between species in ways that match the physiological needs of each. The composition of human blood is presented here and reference will be made to that of other species in the discussion which follows where appropriate. Humans on average are composed of around 60% water, varying from around 30% of bones to 83% of lungs. However, most of this water is contained within cells and the cells within organs that are rigid as in bones or semirigid structures. In contrast, blood can be thought of as a liquid organ, of which an average adult has about 5 L, accounting for 7% of an average adult of 70 kg.

Blood is composed of several cell types suspended in plasma.

(i) Plasma is a yellowish-straw coloured fluid comprising 55% of blood or 2.7–3.0 L in an average human. It is 92% water, the remaining 8% consisting of a wide range of proteins and other substances.

(ii) More than 99% of cells contained in blood are erythrocytes, also called red blood cells. They have the structure of biconcave (i.e., thinner at the centre) disks around 7 μm in diameter. Erythrocytes are the most numerous cells in humans; adults have 20–30 trillion of them and rough estimates suggest they amount to 83% of all cells. However due to their small size and volume this accounts for just 3 kg or 2% of a 70 kg adult by weight [6]. Erythrocytes are suspended in the plasma and constitute about 45% of blood by volume. They circulate in the blood for 100–120 days and are then engulfed by macrophages, broken down and recycled. New cells are produced in bone marrow at a rate of over two million each second. Erythrocytes are packed with haemoglobin which binds and carries oxygen in the blood. They are therefore a critical partner with the heart in delivering this vital gas which enables energy production in all parts of the body. This explains why it was necessary for haemoglobin and erythrocytes to evolve in a way that matches the structure and function of the circulation.

(iii) When blood is centrifuged in a tube, the cell components separate, erythrocytes falling to the bottom leaving plasma above. Between the two, a pale thin layer forms which is composed of leukocytes (from the Greek, λευκός meaning white and *KÚT-Oς* meaning cell), also called white blood cells. Leukocytes are larger than erythrocytes and adults have 4,000–11,000 leukocytes per microlitre of blood. They form part of the immune system and are important for the removal and destruction of invading bacteria and viruses. They also have an important role in tissue repair, such as following myocardial infarction (see Figures 6.4–6.6 and related text in Chapter 6). Leukocytes are composed of different cell types depending on their structure and function. Neutrophils and eosinophils attack invading parasites. Basophils produce histamine as part of inflammatory responses. Lymphocytes are of several types and are present in blood and in the lymphatic system. Some lymphocytes, known as memory cells, may live for many years and carry valuable immunological information about past infections. Lymphocytes attack viruses, contribute to antibody production and stabilise the immune system. Monocytes can migrate between the endothelial cells that line blood vessels into the surrounding tissues where they transform into macrophages which remove debris of broken cells and pathogens. They also have an important role in monitoring the health of the circulation and alerting the immune system in the case of injury or infection.

(iv) Platelets are also known as thrombocytes (from the Greek word *θρόμβος* for clot and *ΚΎΤ-Ος* for cell). Platelets do not have a nucleus and have the shape of biconvex disks (i.e., thicker at the centre). Platelets are 2–3 μm in diameter and are the smallest cells in blood. Blood contains between 150,000 and 400,000 platelets per microlitre of blood and this level is regulated by the clearance and replacement of around 10^{11} platelets per day. They are formed in bone marrow as fragments of much larger cells called megakaryocytes, each of which can produce between 2,000 and 5,000 platelets. Platelets together with various proteins in blood and tissues provide the coagulation system which maintains the integrity of the circulation should injury occur. Platelets rapidly change to an activated state if they come in contact with damaged areas of the endothelium lining of blood vessels. As a result, they clump together in bundles and interact with coagulant proteins to form a clot, which acts as a seal preventing blood loss. As discussed in Chapter 6, abnormal activation of platelets can occur due to contact with atheromatous plaques and is an important cause of arterial thrombosis resulting in heart attacks and stroke (see Figures 6.2 and 6.6 and related text in Chapter 6).

Transport Medium

Blood is responsible for transporting many substances around the body. Oxygen (O_2) is the most prominent of these and gives blood its dramatic red colour by its binding to haemoglobin. Blood devotes a huge effort to transporting O_2; the erythrocytes which contain its carrier, haemoglobin, make up 45% of blood and the heart and circulation are regulated to react quickly to changes in tissue needs for O_2 to ensure its efficient delivery (discussed in Chapters 3 and 5). Oxygen extracted from air in the lungs binds to haemoglobin in blood and is responsible for its bright red colour as it passes through arteries and capillaries. On reaching the tissues, haemoglobin gives up some of its O_2 and changes colour to a darker red as it travels in veins back to the heart and lungs. The availability of O_2 facilitates aerobic metabolism, which is a highly efficient method for generating energy from food and was a key step in the evolution of large-sized animals. The process of aerobic metabolism involves conversion of O_2 to carbon dioxide (CO_2) which is then transported back to the lungs where it diffuses into air and is exhaled. Oxygen is insoluble in plasma and therefore needs a carrier such as haemoglobin. Carbon dioxide is more soluble in plasma, but only about 7% of CO_2 is carried dissolved in plasma. Haemoglobin also binds CO_2 forming carbaminohaemoglobin and around 23% of CO_2 in blood is carried this way. CO_2 binds less avidly than O_2 and at a different site on the haemoglobin molecule. The remaining 70% of CO_2 is transported as bicarbonate ions. This is made possible by the action of the enzyme, carbonic anhydrase within erythrocytes, which combines CO_2 and water, to form carbonic acid (H_2CO_3). H_2CO_3 then rapidly dissociates into carbonate and hydrogen ions:

$$\begin{array}{c} \text{Carbonic anhydrase} \\ CO_2 + H_2O \leftrightarrow H_2CO_3 \leftrightarrow H^+ + HCO_3^- \\ \leftarrow \text{lungs} \qquad\qquad\qquad \text{tissues} \rightarrow \end{array}$$

This reaction is readily reversible depending on the concentration of CO_2. This means that when the concentration of CO_2 is higher (as in the tissues where it is produced), the reaction favours a rightward shift, but in the lungs where CO_2 is exhaled, reducing its concentration, the reaction favours a leftward shift, releasing more CO_2 for exhalation. In addition, the dissociation of H_2CO_3 to H^+ and HCO_3 releases H^+ ions which have an important role in buffering the pH of blood as well as on haemoglobin itself, known as the Bohr effect (discussed below). In addition to O_2 and CO_2, blood also transports many other substances. These include:

(i) nutrients derived from food such as glucose, amino acids and fatty acids;
(ii) waste products such as CO_2 and lactic acid;
(iii) hormones produced by endocrine glands;
(iv) chemical signals used by the immune system and cells and products of the immune system (such as antibodies) produced in response to infections or other challenges;
(v) cells and substances which regulate the coagulation system, converting blood from a liquid to a semisolid gel in response to vascular injury; and
(vi) redistributing blood to or from peripheral tissues to regulate body temperature.

Oxygen Consumption

Oxygen is the key determinant of energy production and therefore the efficiency of its delivery to tissues is a primary determinant of the maximum level of aerobic activity animals including humans are capable of. For example, the performance of a marathon runner primarily depends on the ability to deliver oxygen to the muscles involved and this is the case for all forms of physical or mental work. The converse also holds; the amount of oxygen consumed is the best measure of energy production. Oxygen consumption (VO_2) by tissues is reflected in the difference between the O_2 content of the arterial blood supplying them and that of the venous blood draining them and VO_2 by the body as a whole is measured as the difference between the oxygen content of arterial and venous blood. VO_2 is usually expressed per minute, and this is measured by multiplying the arterial–venous oxygen difference by the cardiac output per minute. In simple terms then, energy production is determined by the ability to deliver oxygen to tissues and the ability of tissues to extract it from blood. The delivery of oxygen is determined by the partnership of the heart and blood, in essence, the cardiac output and the level of haemoglobin in blood.

The amount of oxygen consumed in a resting human or animal is used as a measure of basal metabolic rate (BMR) and across a wide range of species it is related to body size; the larger the body, the more oxygen it needs (Figure 9.1(a)). During exercise or any form of activity, energy needs increase and so does oxygen consumption. To meet this demand, cardiac output increases with heart rate and oxygen uptake also increases with respiratory rate, so that oxygen delivery to tissues also greatly increases. The VO_2 measured at maximal activity (VO_2 max) is used as a measure of highest level of activity of which an animal or a human is capable. This also shows a significant relation to body size (Figure 9.1(a)). Thus, blood and the heart are closely coupled partners in the delivery of oxygen to tissues and therefore in determining the maximal level of activity humans and animals are capable of. This becomes even clearer when looking at the aerobic range of species.

Maximal oxygen consumption gives us a measure of energy use per minute at peak activity. Aerobic range is a measure of maximal energy utilisation with reference to the basal metabolic rate (BMR), i.e., by how many times can an animal or a human increase its/his BMR. This is measured as VO_2 max divided by BMR. As would be expected, aerobic range can vary within and between species. In humans, for example, trained footballers may have an aerobic range up to 20 times their BMR and marathon runners up to 22 times [7] compared to 12 in sedentary persons [8]. Thus by training, athletes can greatly increase their oxygen utilisation and therefore aerobic range. In much the same way, the lifestyles of animal species may affect their energy needs and their ability to meet them. This was examined in an interesting study by Charles Bishop in 1999 [9] by taking published data for aerobic range measured in several species of mammals and birds and relating it to body weight. The results of that study are illustrated in Figure 9.1(b), showing that the data are widely scattered with no significant relationship with body weight. This is not unexpected since their oxygen consumptions would more likely be related to their lifestyles than size. For example, a small hummingbird would have much higher energy need for its weight than most large mammals. By implication however, this interspecies variability in aerobic range must reflect a similarly variable ability to deliver oxygen to the tissues. When the data for aerobic range shown in Figure 9.1(b)

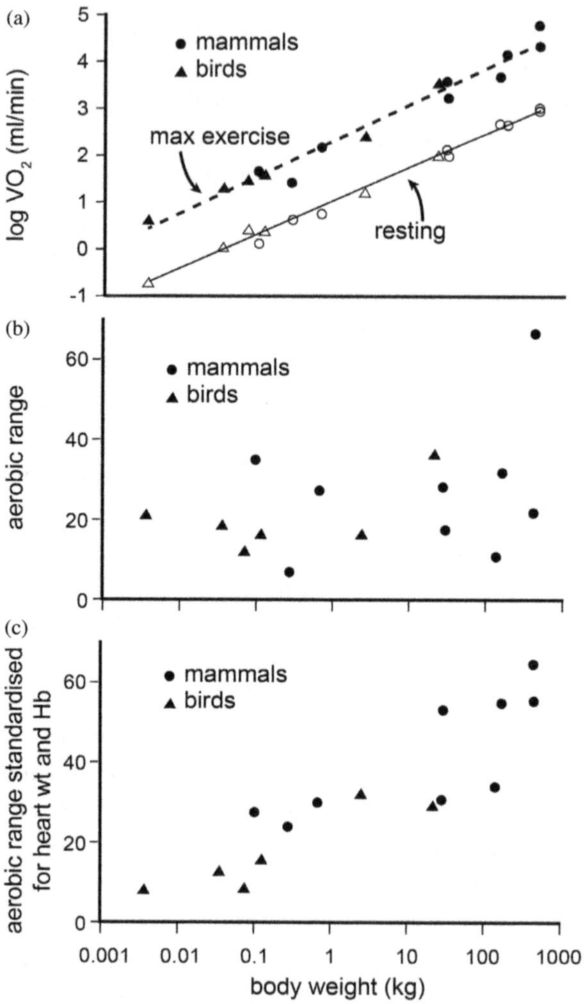

Figure 9.1: Aerobic capacity related to body weight and heart weight. (a) Oxygen consumption (VO_2) at rest (open symbols) and at maximal exercise (filled symbols) in mammals (circles) and birds (triangles) related to body weight. VO_2 is closely related to body weight both at rest and during maximal exercise. (b) Aerobic range (VO_2 at maximal exercise/resting VO_2) in relation to body weight. Aerobic range shows no apparent relationship to body weight among several species. (c) Although aerobic range is not related to body weight in birds and mammals, when standardised for heart weight and haemoglobin level, as shown here, it does show a significant relation to body weight. This indicates that most of the interspecies variation in aerobic range with body weight seen in (b) is attributable to heart weight and haemoglobin [9].

were standardised for heart weight and haemoglobin level (Figure 9.1(c)), a significant relation with body weight emerges [9]. It transpired that 85% of the interspecies variation in aerobic range shown in Figure 9.1(b) was due to differences in heart weight and 15% to differences in haemoglobin level. That study was of course limited to birds and terrestrial mammals. All of these animals have high metabolic requirements related to countering the effects of gravity and being homoeothermic. They also tend to have high haemoglobin levels with only modest variation between them. Aquatic and amphibian animals have much lower energy requirements because the effects of gravity are greatly reduced due to the buoyancy of water. In these, haemoglobin levels vary much more widely, Table 9.1. For example, hagfish, a primitive jawless vertebrate, has a haemoglobin level of 4.2 g/dL compared to 15 g/dL in humans and intermediate levels are found in many species. Thus, haemoglobin levels have also evolved widely to reflect the lifestyles and energy demands of species. In some, this includes a dynamic

Table 9.1: Haemoglobin levels in selected aquatic and terrestrial animals.

Species	**Hb g/dL**			**Reference**
Hagfish	4.10			[10]
Shark (sicklefin)	4.20	6.00#		[11]
Shark (black tip)	4.30	4.60#		[11]
Snake (cobra)	6.50			[12]
Alligators	7.70			[13]
Frog	8.80			[14]
Cow	11.00			[15]
Sheep	11.00			[15]
Elephant	11.70			[16]
Horse	13.00	15.47#		[17]
Killer whale	14.50	17.50*		[18]
Birds	15.00			[9]
Humans	15.00			[9]
Weddell	15.10	22.40a	25.40b	[19]

Notes: #Excited animals; *Offshore animals; aAfter short dive; bAfter long dive.

component as exemplified by sharks and horses which can increase the haemoglobin concentration in circulating blood when excited, due to contraction of the spleen which acts as a store of oxygenated haemoglobin. Similarly, in Weddell seals, which are capable of diving for 45 min and to depths of 600 m, haemoglobin can increase up to 25 g/dL during deep dives from a predive level of 15 g/dL. Killer whales which live in the deep ocean also have higher haemoglobin levels than coastal species, presumably also reflecting the energy needs of deep diving.

Thus, getting oxygen to tissues is the main determinant of an animal's capacity for activity and different species have evolved specific strategies to meet the energy needs of their lifestyle. Heart size is the largest contributor to this in birds and terrestrial mammals and among these, small animals tend to have much larger hearts relative to their size than large animals, for example, heart size in hummingbirds as a percent of body weight is more than twice that of humans [20]. Among aquatic animals, amphibians and reptiles, the evolution of higher haemoglobin concentrations has also been a major contributor to meeting increased energy needs by increasing oxygen carrying capacity. In addition, the ability to store and release oxygenated haemoglobin when required has allowed some species to dramatically increase their aerobic range for feeding and defence. In each case, the increasing availability of oxygen in the Earth's atmosphere was a key driver for the evolution of more efficient means to transport oxygen. As organisms became larger and more complex, the development of more efficient pumps (as discussed in Chapter 3) and improved oxygen carrying capacity were partners in this.

Oxidation of the Earth's Atmosphere

Oxygen is the most immediately pressing nutrient for almost all organisms including humans in that we become disabled quickly in its absence. It is for this reason that the oxidation of the Earth's atmosphere was a critical factor in the evolution of large and complex organisms and why blood pumped by increasingly powerful hearts was also needed to provide more efficient oxygen delivery to tissues as organisms developed more complex forms and greater energy needs (discussed below). When the Earth was formed around 4.5 Ga (giga-annum = billion years ago), the atmosphere contained almost no oxygen (Figure 9.2). The oxygen present in the Earth's atmosphere today is the result of oxygenic (oxygen producing) photosynthesis (production of carbohydrates from light, CO_2 and H_2O) which began around 2.5 Ga and involves splitting water into H^+ and O^{2-} ions ($2H_2O \rightarrow 4H^+ + 2O^{2-}$) by plants, algae and cyanobacteria.

Photosynthesis

Photosynthesis is a chemical reaction in which light is used as a source of energy by plants and other organisms to make substances such as carbohydrates which store chemical energy that can be used as fuel. Light energy, in the form of photons, is absorbed by the pigment chlorophyll, which causes it to lose an electron. That electron is then used as a source of energy by transforming other chemicals in a way that allows them to store the energy contained in the electron. The energy stored in this way can then be used as fuel to support the needs of the organism. A key step in photosynthesis is the need to replace the electron lost by the light sensitive pigment during photo-stimulation in order to allow the cycle to be repeated. The vast majority of this takes place in the world today through the splitting of water into H^+ and O^{2-} ions by an enzyme called photosystem II. This reaction frees an electron which is captured and used to replace that lost by photo-stimulation.

Prior to the evolution of oxygenic photosynthesis, single-celled organisms had appeared which were capable of non-oxygenic photosynthesis around 3.3 Ga. These early cells used alternative substances as electron sources. These included hydrogen sulphide which was split into sulphur and water ($CO_2 + 2H_2S \rightarrow CH_2O + 2S + H_2O$) and several organic compounds such as oxalate. In each case, the available substance was stripped of an electron which was then used to restore the light sensitive pigment. The transition from non-oxygenic to oxygenic photosynthesis had a profound effect on the evolution of life on Earth. This is because the oxygen released by oxygenic photosynthesis was responsible for increasing the level of oxygen in the atmosphere to the level it is today (~20%). This in turn made aerobic metabolism (chemical reactions which use oxygen) possible. This was important because aerobic metabolism produces about 18 times more energy per molecule of carbohydrate than anaerobic metabolism and as a result greatly increased evolutionary possibilities.

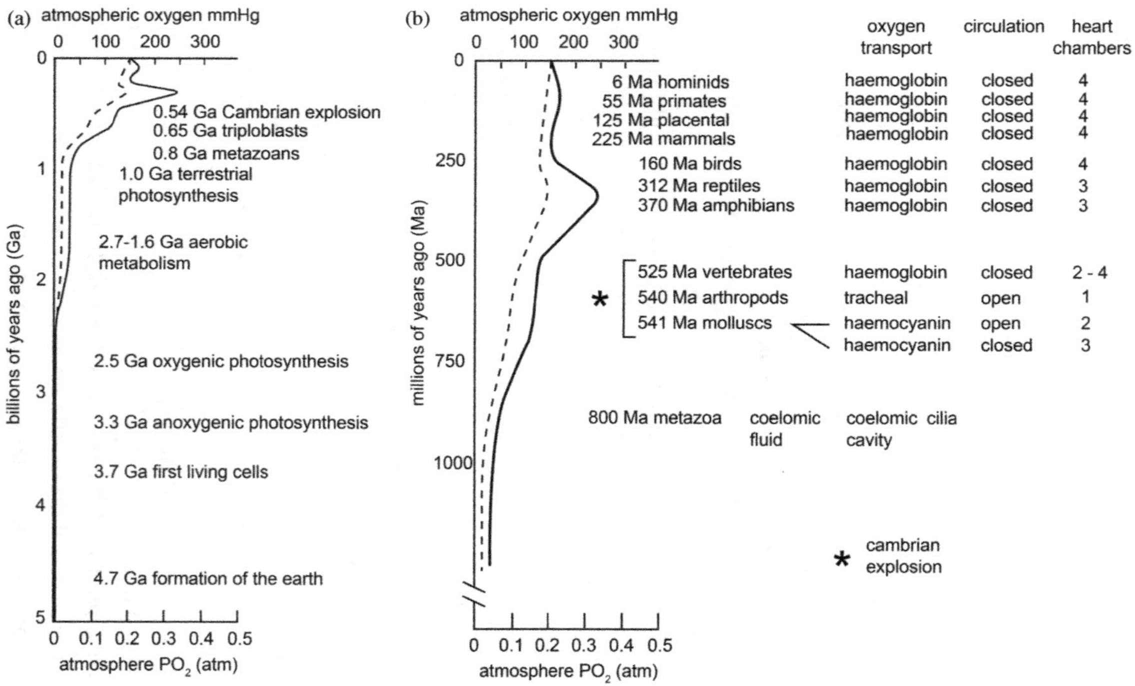

Figure 9.2: Oxygenation of the atmosphere and evolution of complex life forms. (a) Changes in atmospheric oxygen expressed as partial pressure (upper axis) and mmHg (bottom axis) over time (vertical axis) since the Earth was formed. Oxygen levels are shown for upper (solid line) and lower (dashed line) estimates [71]. After the Earth was formed 4.7 Ga, the atmosphere contained almost no oxygen. The first living cells appeared around 3.7 Ga and the ability for photosynthesis evolved at 3.3 Ga. Initially this was non-oxygenic, based on reductants such as hydrogen sulphide. Cells evolved oxygenic photosynthesis around 2.5 Ga, and for a period of 50–80 million years the oxygen produced by this reacted with chemical dissolved in the oceans, particularly iron, forming insoluble oxides which were deposited on the ocean floor. Once these oxygen sinks were consumed, oxygen began to increase in the atmosphere rising to a partial pressure of around

Oxygenic photosynthesis became possible by an evolutionary change in the enzyme photosystem II, a key protein involved in photosynthesis, which enabled it to catalyse the splitting of a water molecule (H_2O) into molecular oxygen (O_2) and hydrogen ions ($2H^+$) with release of electrons ($2e^-$). The electrons were then available to replace those lost during light stimulation of chlorophyll. This enzyme contains an inorganic core consisting of manganese, calcium, oxygen and carbonates. The transition to splitting water as a source of electrons is likely to have resulted from an evolutionary modification of this core, possibly by the introduction of calcium [21]. The emergence of a method for splitting water into electrons and protons meant that photosynthesis was no longer constrained by the availability of chemicals which could be used to replace pigment electrons; instead, water provided an unlimited source and resulted in a vastly increased rate of oxygen production. It also led to an immense increase in the formation of organic carbon and an explosion of plant life, literally changing the face of the Earth. This evolution took place in cyanobacteria around 2.5 Ga and resulted in photosynthesis which converted carbon dioxide and water into carbohydrates and oxygen ($CO_2 + 2H_2O \rightarrow CH_2O + O_2$). This process began in marine environments and for 50–80 million years, the oxygen released was captured by combining with dissolved substances in the oceans (known as oxygen sinks), most notably iron, forming iron oxides, which are insoluble and therefore precipitated and fell to the ocean floor. Over time, these hardened into characteristic rock

Figure 9.2: (*see figure on facing page*) 15 mmHg by 2.0 Ga and may have remained relatively stable for around a billion years. Atmospheric oxygen then increased more rapidly between 1.0 and 0.5 Ga from 15 mmHg to the present level of 150 mmHg as photosynthesising plants extended from the oceans on to land. A sharp increase in atmospheric oxygen occurred from about 380 Mya to 240 Mya, reaching a peak partial pressure of 250 mmHg at 300 Mya and then declining to the current level of 150 mmHg. (b) Expanded graph showing atmospheric oxygen and time related evolutionary events for the last 1300 million years. Species are listed with dates of their first appearance in the fossil record, together with their mechanism for oxygen transport, type of circulation and its form of propulsion. The evolution of complex life correlates with the rise in atmospheric oxygen with multicellular organisms appearing around 800 Mya and then a dramatic expansion of life forms during the Cambrian explosion after 500 Mya.

formations, which make up around 60% of iron deposits in the world today. Once these oxygen sinks were depleted, oxygen began to accumulate in the atmosphere (Figure 9.2) from almost nil to a partial pressure of around 15 mmHg by 2.0 Ga [22]. Although this was small compared to present levels, it had a dramatic effect, making aerobic respiration possible from then onwards and is known as the "great oxidation event" in the history of the Earth. It led to the first multicellular organisms which became widespread over the next billion years. During this period, animal life consisted of relatively simple organisms containing a limited number of cell types. Atmospheric oxygen was then increased more rapidly between 1.0 Ga and 0.5 Ga from 15 mmHg to the present level of 150 mmHg.

Atmospheric oxygen and the evolution of complex life forms

As the level of atmospheric oxygen increased, organisms evolved larger and more complex body plans. While this depended on many environmental factors, several lines of evidence indicate that the increase in atmospheric oxygen was most important.

(i) The development of increased numbers of different cell types is a feature of greater complexity in animals and roughly correlates with the rise in atmospheric oxygen. Thus, organisms with two cell types appeared around 2.5 Ga. More complex forms with around 10 cell types had evolved by 1.5 Ga and with 50 cell types by 1 Ga [23].

(ii) Polar gigantism refers to the tendency for animals to be larger at higher latitudes. Studies of animals over a wide range of species that live at various Earth latitudes show a strong correlation between body length and the local oxygen content of water [24]. Water holds more oxygen at lower temperatures and is therefore present at higher concentrations at higher latitudes.

(iii) Further evidence comes from what is known as insect gigantism. This refers to a period around 300 million years ago (Mya), during which the fossil record shows a dramatic increase in the size of insects followed by a later reduction. For example, dragonflies and mayflies evolved with wing spans of up to 71 cm and 20 cm, respectively. The giant Carboniferous dragonfly, *Meganeura monyi*, had a chest diameter of around 2.8 cm compared with the largest one today of 1 cm [25]. This corresponded with a transient increase in atmospheric oxygen [26], which began about 380 Mya and lasted until 240 Mya, reaching a peak partial pressure of around 250 mmHg at 300 Mya (Figure 9.2). This was due to the expansion of terrestrial plants which greatly increased the rate of oxygen production from photosynthesis. Not long afterwards, microbes such as fungi evolved which were capable of rapidly decomposing dead plants. In the process, they consumed oxygen which ended the period of atmospheric hyperoxia.

(iv) Changes in the oxygen content of transmembrane proteins were a key step in the evolution of eukaryote cells from prokaryotes and a major transition in the early evolution of life. Unlike prokaryotes such as bacteria, which have no separate intracellular compartmental structures, eukaryote cells, which are present in all animals and plants evolved compartments such as nuclei enclosing genetic material and mitochondria around 2.1–1.8 Ga [27]. All of these are surrounded by membranes which are impermeable to large molecules. While this afforded advantages in facilitating the building of specialised compartments within the cell, it also required the development of systems for transmembrane communication and transport to and from the cytoplasm to the interior of these compartments. These functions are undertaken by proteins that span the membranes of these compartments with one tail ending on each side so that signals acting on the outside of the membrane can be transmitted to the interior and vice versa. The development of such proteins was a crucial step in the evolution of eukaryotes and this appears to be related to the increase in atmospheric oxygen concentration which occurred from 2.5 Ga.

The background to this is that the function of proteins is determined by their structure and for proteins in the aqueous (and therefore the hydrophilic) environment of cells, the distribution of hydrophobic and hydrophilic amino acids is an important determinant of how they fold and their overall structure. Oxygen-rich amino acids tend to be hydrophilic and are important in this regard. In a study of transmembrane proteins, Acquisti *et al.* demonstrated that their oxygen content changed during the evolution of eukaryotes. In particular, the presence of oxygen-rich (and therefore hydrophilic) amino acids in the outside tails of transmembrane proteins increased with the appearance of eukaryotes [28]. This change correlated with the increase in atmospheric oxygen. During earlier anoxic periods, the incorporation of oxygen rich amino acids is likely to have been limited because the reducing effect of such an atmosphere would have made these proteins unstable. The removal of this constraint by a rise in atmospheric oxygen greatly expanded the potential folding possibilities of proteins and thereby the repertoire of functions they could undertake including the evolution of transmembrane proteins and subcellular compartments.

Complex Organisms Needed More Efficient Oxygen Delivery

Thus, increasing atmospheric oxygen was important in the evolution of multicellular organisms from around 2.2 Ga (Figure 9.2). The increasing complexity and size of organisms which followed, required more efficient systems for delivering oxygen and other nutrients. Initially, this involved cilia driven convective fluid circulation in coelomates and later as organisms became more complex and larger, open and closed circulations driven by pumps of increasing efficiency. Thus, hearts consisting of one, two, three and four chambers evolved as animals became more complex with higher energy needs, Figure 9.2 (see also Figures 7.1–7.3 and related text in Chapter 7). At the same time, circulatory transport media also evolved with improved ability to carry nutrients (Figure 9.2). Invertebrates evolved an open circulation and use haemolymph as a transport medium. This is well suited to their needs for transporting nutrient foods and signalling chemicals. In some invertebrates such as molluscs, haemolymph contains haemocyanin, a copper containing protein, as an O_2 carrier. Arthropods developed a tracheal system (extensions of the respiratory apparatus) for delivering oxygen and have only small amounts of haemocyanin. Vertebrates have higher energy needs and therefore greater demands for gas transport. For this they evolved haemoglobin which can transport greater amounts of oxygen and more powerful hearts to pump it. In order to achieve this, haemoglobin and the erythrocytes which produce it coevolved in a way that exquisitely matches the heart and circulation.

Evolution of Oxygen Use

The availability of oxygen in the atmosphere allowed animals to generate energy much more efficiently and this is observed in the emergence of animals with greater agility, particularly in locomotion. Diffusion of oxygen from the environment was sufficient for primitive small animals [29] (see Figure 7.1 and related text in Chapter 7). Medusae (jellyfish) which emerged before the Cambrian explosion (541 Mya) rely on direct oxygen diffusion from sea water. This allowed them to evolve movement based on propulsion and to achieve great size, for example, the giant lion's mane (*Cyanea capillata*) can grow up to 3.6 m. Scallops, cephalopods and squids evolved more powerful and faster jet propulsion mechanisms but requiring greater oxygen consumption for which they developed oxygen carriers and pumped circulations (discussed in Chapters 2 and 7).

Oxygen Transport

The importance of oxygen transport as a critical factor in this is exemplified by the emergence of three different oxygen carriers to meet the varied lifestyles and needs of these rapidly diverging evolutionary pathways. Depending on their oxygen binding characteristics, oxygen carriers are categorised in three groups: (a) haemocyanins, (b) haemerythrins and (c) haem proteins.

(a) Haemocyanins bind oxygen at an active site which contains two copper ions, each held within three histidine rings. They are present in arthropods and molluscs and might be thought to have evolved earlier than haemoglobin present in more advanced vertebrates. However, research has shown that haemoglobins present in primitive invertebrates very likely were the first oxygen carriers (discussed below).
(b) Haemerythrins are oxygen carriers present in primitive marine worms. They are protein polymers containing two Fe^{2+} ions which bind oxygen. However, they are distinct from haemoglobin and do not contain a haem group.
(c) Haem proteins contain iron within a porphyrin ring structure known as a haem group, which binds oxygen reversibly. The best known is of

course haemoglobin but they also function as oxygen binders very widely, including in plants, bacteria and animals.

Haemocyanins

Haemocyanins evolved from an oxidising enzyme called tyrosinase independently in molluscs and arthropods around 740 Mya and 600 Mya, respectively. Mollusc haemocyanin evolved further with the emergence of cephalopods around 520 Mya and again with more agile modern octopi, squids and cuttlefish around 420 Mya [30]. Tyrosinase is an oxidising enzyme found in plants, animals and fungi. In humans, one of its functions is in oxidising the amino acid tyrosine to make the pigment melanin. The active site of tyrosinase contains two copper ions that are also held within a structure of histidine rings. These bind oxygen in a highly reactive form which then oxidises substrates. Haemocyanin is believed to have evolved from tyrosinase by modifying its binding characteristics for oxygen in a way that made it reversible and therefore a useful oxygen transporter. The emergence of haemocyanin for oxygen transport was a crucial factor in the successful evolution of all of these animals and their increasing agility and adaptability depended on subsequent modifications of this oxygen carrier.

The structure of haemocyanins

Haemocyanins are very large protein polymers composed of several identical subunits. Despite the similarity in their functions, arthropod and mollusc haemocyanin have quite different amino acid sequences and overall structure.

Mollusc haemocyanins are formed of ten similar protein subunits (decamers) bound together forming cylindrical-shaped particles, 35 nm in diameter and 18 nm long and are readily visible under electron microscopy [31]. The subunits are proteins of around 420 amino acids and 45–50 kDa molecular weight (MW), which usually have eight oxygen binding sites, octopi and nautilus being unusual having seven. Thus, a complete decamer with 10 subunits each having eight binding sites has the capacity to bind 80 oxygen molecules. In addition, decamers are frequently joined as di-decamers and even multi-decamers having a tube-like structure.

Arthropod haemocyanins are composed of six subunits (hexamers). Having evolved independently, the amino acid sequence of subunits has no similarity to those of molluscs. Arthropod subunits have a MW of 75 kDa [32] and consist of three functional units, the second of which contains an oxygen binding site. Hexamers are present as singles and multiples from bi-hexamers in some spiders up to eight hexamers in the horseshoe crab.

Haemocyanins are effective oxygen transporters and have successfully evolved to allow species to populate the Earth very widely including, for example, the extreme cold conditions of the Antarctic. Low temperatures increase haemocyanin affinity for oxygen, which impairs the release of oxygen limiting its availability for tissues. Octopi which inhabit arctic waters have evolved a modification of haemocyanin to overcome this by modifying the amino acid sequence in an area close to the active binding site in a way that reduces affinity for oxygen and thereby facilitates it release [30].

Thus, haemocyanins have been crucial partners in the evolution of complex life for 750 million years by providing a mechanism for efficient transport of oxygen alongside the development of pumped circulations. As competition for habitats has increased, haemocyanins have continued to evolve supporting more successful defensive and predative lifestyles.

Haemerythrins

Haemerythrins are oxygen carriers present in primitive unsegmented (sipunculids and priapulids) and segmented (annelids) marine worms and in brachiopod shellfish, all of which are present in the fossil record of the earlier Cambrian period from around 540 Mya. Despite the name, they do not contain the oxygen binding haem group present in haemoglobin. The name comes from *haem*, the Greek word for blood and *erythrin*, red. Haemerythrins are polymers of protein subunits each of 113–118 amino acids and MWs of 13.5–13.9 kDa. Each subunit contains a single oxygen binding site consisting of two Fe^{2+} ions bound between five histidine rings and the carboxyl groups of nearby glutamate and aspartate amino acids [33]. Haemerythrins are most frequently present *in vivo* as octamers but dimers, trimers and tetramers have been reported in species of sipunculids [34]. Haemerythrins are colourless when deoxygenated and have a violet-pink colour when oxygenated. They are not extracellular carriers like haemocyanins, but are always present within nucleated cells called haemerythrocytes, sometimes also referred to as pink cells, in coelomic and vascular fluids.

Haemerythrin affinity for oxygen is 25% greater than haemoglobin, suggesting that it may function as a storage mechanism for oxygen as well as a transporter. This would also be consistent with lifestyles (living in burrows) that involve wide fluctuations in the ambient environmental partial pressure (pO_2) of oxygen. Thus at a pO_2 of 125 mmHg, oxygen concentration in coelomic fluid was 69% of the total carrying capacity and at 30 mmHg it was only 10% [35]. In these circumstances, the ability to store oxygen and release it slowly may be more important for survival than immediate transport. The species in which haemerythrins are present have diversified widely and secondary functions for haemerythrins may have contributed to this, such as iron storage, metal detoxification and immunity [36]. Thus the functional plasticity of haemerythrins may have been particularly well suited to the highly varied lifestyles of these early organisms which encountered periodic low oxygen availability.

Haemoglobin

Haemoglobins are ancient proteins found in all kingdoms of organisms (animals, plants, fungi, protists and bacteria). In humans, virtually all vertebrates and in some invertebrates, they have evolved to serve as oxygen carriers. The icefish (*Channichthyidae*) is an exception to this, being a vertebrate that lives in cold waters of the Antarctic, where it can survive on the high concentration of dissolved oxygen without any haemoglobin. In plants and bacteria, haemoglobins have also evolved as binding molecules that serve as oxygen stores, oxygen sensors or oxygen scavengers. In mammals, haemoglobin is present exclusively in erythrocytes, but in many invertebrates it is extracellular.

Structure of haemoglobins

The complete structure of haemoglobin varies widely between species. However, they all contain a haem group with an iron ion at its centre surrounded by a protein called globin. Despite some differences, all globins share characteristic structural features [37] (Figure 9.3(a)). The haem group consists of a porphyrin ring which holds a Fe^{2+} ion at its centre by four coordinate covalent bonds (chemical bonds in which two electrons are provided by just one of the participants, in this case nitrogen atoms of the porphyrin ring) (Figure 9.3(b)). Iron is capable of six of these coordinate bonds in all. A fifth is provided by a specific amino acid in the globin chain. In vertebrates, the sixth bond is provided by oxygen or by a loose association with water in deoxygenated haemoglobin. Oxygen binding to haemoglobin is reversible and it is this reversible association which gives haemoglobin its ability to function as an oxygen carrier. In some plant haemoglobins, a second amino acid in the globin chain fills the sixth coordinate position but does so loosely and reversibly and in a way that allows oxygen to bind to the site when present.

Globins are proteins of 17 kDa MW. All globins have either six or eight amino acid segments forming α-helices (labelled A to H) joined by short loops, which fold in the form of a tiered sandwich, known as the "globin fold" (Figure 9.3(a)). This is a feature of all known haemoglobins irrespective of where they occur in nature. The outward facing amino acids of globins are hydrophilic which makes haemoglobin soluble in water. In contrast, the inward facing amino acids are largely hydrophobic and form a pocket in which the haem group is tightly bound. Despite the wide distribution of haemoglobins in nature and the range of uses they undertake, the amino acid sequences of all globins have common features. Importantly too, a histidine residue at F8 (the eighth amino acid in the F helix) is invariant. This matters because the imidazole ring of this amino acid binds to the Fe ion in the haem group at its fifth coordinate position (Figure 9.3(b)). Thus the structure of globins is one of the most conserved of all proteins known and points to a shared ancient ancestor, most likely a bacterium that developed around 3.5 billion years ago [38].

In vertebrates, globin is present as a single subunit in myoglobin and as tetramers (four subunits) in haemoglobin. The former is present in skeletal muscle and acts as an oxygen store while the latter is the oxygen carrier in blood. These will be the focus here. However to put them in an evolutionary context, it is worth noting that invertebrate haemoglobins are present in a huge variety of polymeric sizes and subunit assemblies. Thus, 17 kDa globin subunit monomers can be present as intracellular or extracellular haemoglobins and form the largest group in some invertebrates

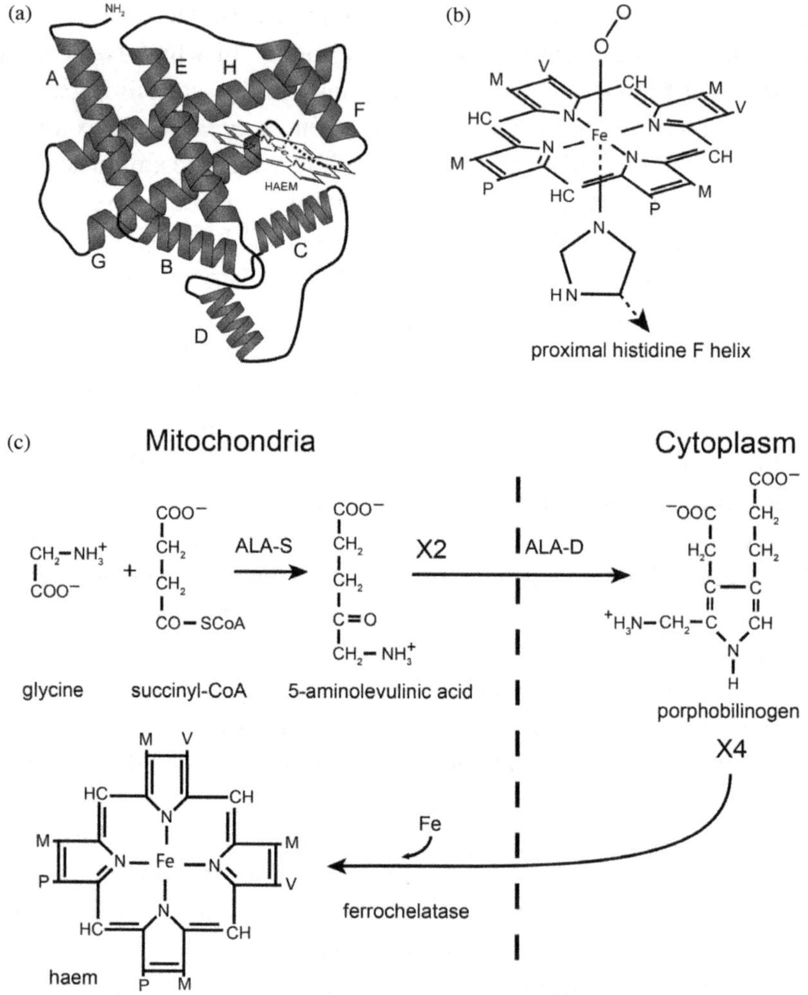

Figure 9.3: (a) Illustration of myoglobin structure. Myoglobin consists of a single globin molecule which is very similar in structure to each of the four globin subunits in haemoglobin. It consists of eight α-helixes (A to H) folded around the haem moiety which contains an Fe ion that binds oxygen. A highly conserved histidine amino acid at position 8 in the F helix provides a coordinate covalent bond which binds the Fe ion and haem group to the globin chain. (b) Illustration of oxygenated haem. Haem is shown as a planar structure with Fe bound to four N atoms in the centre of the porphyrin ring. Two additional bonds at a right angle to the plane are also shown; the fifth binds Fe to the N atom of an imidazole ring in a histidine amino acid in the F helix of the globin chain (also known as the proximal histidine) below the plane and a sixth binds oxygen above the plane. Each of these bonds is a coordinate covalent bond, meaning the Fe shares two electrons with each of the

including molluscs, annelids and nematodes. Dimers (two subunits) and tetramers are present within the erythrocytes of *Scapharca* (arc clams) [39]. Multiple polymeric assemblies are also widely present in many arrangements as, for example, in *Lumbricus terrestris* (earth worm) which has a giant 3,600 kDa haemoglobin consisting of an assembly of 12 dodecamers. That is, polymers consisting of 12 globin subunits which are then assembled in groups of 12, arranged in two layers of six; in all 144 globin subunits each with an oxygen binding site [40]. Increasing oxygen binding cooperativity (the property whereby haemoglobin oxygen affinity is increased by the initial binding of oxygen) seems to be an advantage of these large molecular assemblies [40].

Haemoglobin is also present in almost all plants. It was first discovered in the root nodules of leguminous (nitrogen fixing) plants in 1939 [41] on the basis of its similar spectroscopy (the frequencies at which it absorbs light) and its affinities for oxygen and other substances known to bind to haemoglobin, such as carbon dioxide, carbon monoxide and cyanide. These plant species are able to capture nitrogen from the atmosphere and incorporate it into organic chemicals needed to support plant growth. This nitrogen fixation is done by rhizobium bacteria which live symbiotically in the root nodules of leguminous plants. These haemoglobins, known as leghaemoglobin, are essential for this process for two reasons. First, as oxygen carriers, they supply oxygen for respiration in these bacteria. Second, the nitrogen fixing nitrogenase enzymes in rhizobium bacteria are inhibited by high levels of oxygen and leghaemoglobin prevents this by binding oxygen, which maintains low levels of free oxygen.

Figure 9.3: (*see figure on facing page*) substrates, but in each case both electrons are provided by the substrates. (c) Synthesis of haem. Synthesis of haem begins with the formation of 5-aminolevulinic acid (ALA) from glycine and succinyl coenzyme A by the enzyme aminolevulinic acid synthase in mitochondria. ALA is then transported into the cell cytoplasm where two molecules are joined to make porphobilinogen (PPBgen) by the enzyme aminolevulinic acid dehydrase. Four molecules of PPBgen are then combined to make the haem precursor coproporphyrinogen in a four step enzymatic process (not shown here for simplicity). This is then transported back into mitochondria where Fe is inserted to form haem.

Notes: Fe = Iron; N = Nitrogen; M = Methyl; P = Propionate; V = Vinyl; ALA-S = Aminolevulinic acid synthase; ALA-D = Aminolevulinic acid dehydrase.

In 1983, haemoglobin was also discovered in non-leguminous plant species and in 1988 in non-nodulating plant species. These haemoglobins are present in plant stems and leaves as well as roots and have now been reported to be present in virtually all plants. Plant haemoglobins have been categorised as symbiotic (those which occur in association with symbiotic bacteria) and non-symbiotic. More recently they have been categorised based on their functional properties into three classes.

1. Class I haemoglobins, which includes the non-symbiotic haemoglobins, are marked by their very high affinity for oxygen and are therefore unsuitable as oxygen carriers (high affinity limits their ability to release oxygen). Their primary role is now believed to be in binding nitric oxide (NO). Nitric oxide is known to regulate many physiological processes in plants and by scavenging NO, haemoglobin modulates these [42].
2. Class II haemoglobins, which include most of the symbiotic haemoglobins, have low oxygen affinity and function as oxygen carriers.
3. Class III haemoglobins have been described more recently and are noted for having a reduced globin structure. The pocket in which haem is fixed is made of a sandwich of just four helix loops rather than six or eight and gave rise to their name as truncated haemoglobins. They are evolutionarily related to bacterial haemoglobins. They have a very low oxygen affinity and are widely present in plants; however, their function remains to be clarified.

Synthesis of haemoglobin

Production of haemoglobin is described here as it occurs in humans and some references will be made to other organisms. Haemoglobin synthesis takes place in erythrocytes during their formation in bone marrow. Like all blood cells, erythrocytes develop from a shared primitive cell type and are released into the circulation after around two days. Haemoglobin production occurs throughout this process and is intense, bearing in mind that two million erythrocytes are produced every second. The synthesis of haemoglobin involves three separate but closely inter-related pathways:

(i) the delivery of Fe from plasma to developing erythrocytes, (ii) transcription of globin genes and (iii) synthesis of the haem molecule including the insertion of Fe^{2+} ions.

Each of these is carefully regulated and synchronised to avoid bottlenecks and accumulation of intermediate products. This is necessary due to the limited size of erythrocytes, but more importantly because some of the substances formed may be toxic. Fe is used in chemical reactions in almost all cells because its atomic structure makes it an effective electron donor and acceptor, meaning it is a highly reactive element. This is why it sits at the centre of the oxygen binding site of haemoglobin, but for the same reason it can be highly toxic to other cell components. Therefore Fe metabolism must be carefully managed and regulated. To achieve this, Fe is transported in plasma by a specific protein called transferrin and stored in another, ferritin. These chaperone Fe in a way that prevents its interaction with other substances in plasma and cells. In addition, this "protection" of Fe continues during the delivery of Fe into erythrocytes and right up to the point at which it is bound to the enzyme which finally inserts it into the haem precursor.

Haem synthesis

Haem synthesis begins within the mitochondria of erythrocytes with the formation of 5-aminolevulinic acid (ALA) from glycine (a widely distributed amino acid) and succinyl coenzyme A (a common product of carbohydrate metabolism) by an enzyme called ALA synthase. ALA is then exported from mitochondria into the cell cytoplasm where two molecules are combined to make porphobilinogen (PPBgen) by the enzyme aminolevulinate dehydrase (ALA-D) (Figure 9.3(c)). Four molecules of PPBgen are then combined in four enzymatic steps to form the haem precursor coproporphyrinogen. This involves the formation of the porphyrin ring structure of haem by combining the pyrrole rings of the four porphobilinogens. Coproporphyrinogen is then transported back into mitochondria where mature haem is produced. The final step is the insertion of Fe by the enzyme ferrochelatase. To manage this correctly, Fe is transported to the mitochondrial membrane by transferrin where it binds to a specific transferrin receptor. The combined transferrin-receptor complex is then surrounded by a membrane in which Fe is released by transferrin and delivered to the enzyme ferrochelatase by a mitochondrial transporter protein. In this way, the delicate machinery of the mitochondria is protected from harmful effects of free Fe.

Globin synthesis

Globin proteins are transcribed from genes arranged in two clusters. The α cluster is located on chromosome 16 and the β cluster on chromosome 11. The α cluster contains four genes, two copies of the ζ gene ($\zeta 1$ and $\zeta 2$) and two copies of the α gene ($\alpha 1$ and $\alpha 2$). The β cluster contains five globin genes, the ε gene, two copies of the γ gene ($\gamma 1$ and $\gamma 2$) and one each of the δ and β genes. In all, humans have the genetic possibility to synthesise nine different globin molecules. Synthesis takes place in the cytoplasm of erythrocytes and the haem molecule transported from mitochondria is then incorporated into the globin fold structure described above. Initially, the globin fold is in an open state with its hydrophobic pocket sufficiently exposed so that haem can gain access to it. The Fe^{2+} ion of haem then binds to the histidine residue at F8 and this induces a conformational change in globin which closes the fold resulting in binding haem firmly [43].

Human haemoglobin consists of a tetramer of four globin chains, each with its haem group. In 97–99% of adults, the four globins consist of two α and two β chains ($\alpha 2\beta 2$). The remainder have two α and two δ chains ($\alpha 2\delta 2$). The situation is quite different during embryonic life (Figure 9.4(a)). Synthesis of haemoglobin begins around two weeks gestation and initially takes place in the yolk sac (see Figure 2.3, Chapter 2). This embryonic haemoglobin consists of two ζ-globins and two ε-globins ($\zeta 2\varepsilon 2$). At five weeks of gestation, haemoglobin synthesis in the liver begins to replace the yolk sac as the source of haemoglobin and this corresponds with a change in the haemoglobin genes transcribed; expression of the ζ and ε genes declines and ceases and is replaced by transcription of the α and γ genes forming $\alpha 2\gamma 2$ haemoglobin. This is known as foetal haemoglobin and is the dominant form of haemoglobin

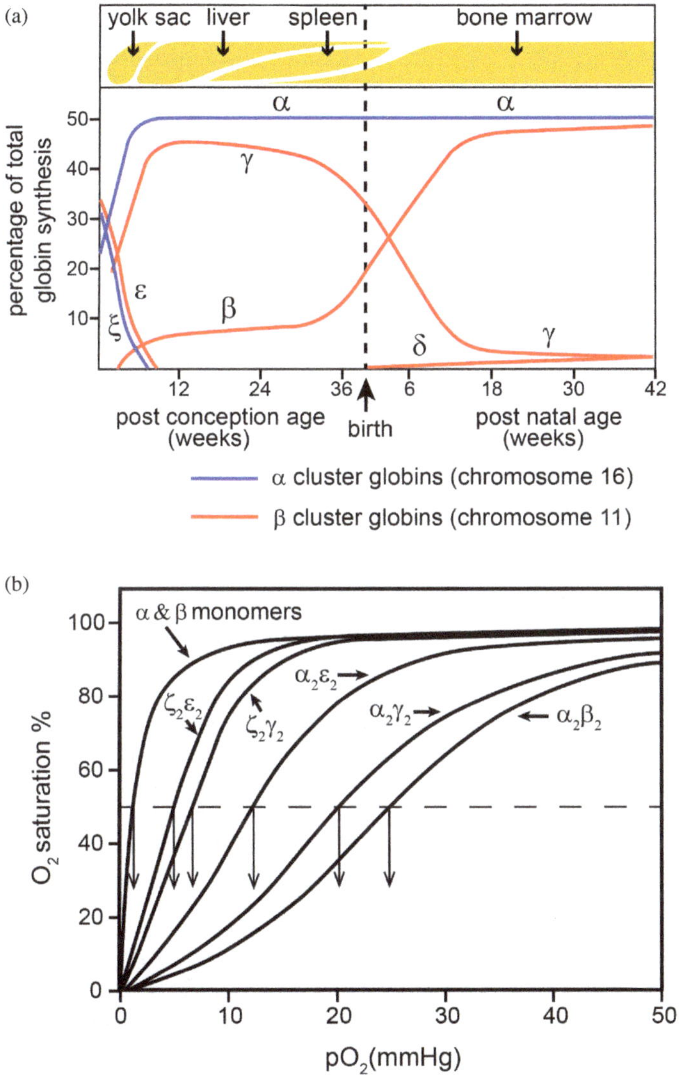

Figure 9.4: (a) Expression of globin subtypes during embryonic, foetal and adult life in humans. Curves represent the percentage that each globin subtype contributes to the total haemoglobin during gestation and early post-natal life. The site of haemoglobin synthesis is indicated by the yellow bars at the top. At all stages, haemoglobin consists of globin tetramers, composed of two heterodimers. In adults, 97–99% of haemoglobin contains two $\alpha\beta$ globin heterodimers making a tetramer $\alpha 2\beta 2$. The remainder is $\alpha 2\delta 2$. During gestation, haemoglobin synthesis begins in the yolk sac at two weeks and consists of $\zeta 2\varepsilon 2$ and $\zeta 2\gamma 2$ (known as embryonic haemoglobin). Then at around five weeks, production

from about eight weeks and for the rest of gestation [44]. At 11–12 weeks gestation, bone marrow is sufficiently mature to commence haemopoiesis and this becomes the major site during the last trimester. A further gradual change in haemoglobin synthesis takes place from around 32 weeks gestation until 12 weeks after birth. This involves a decline in γ-gene expression and its replacement by β-globin. This results in the adult form of haemoglobin, $\alpha 2\beta 2$. The production of embryonic and foetal forms of haemoglobin rather than adult haemoglobin confers important advantages to the growing foetus. This is due to their greater oxygen affinity which is better suited to the low oxygen concentrations present in the early embryo and in extracting oxygen from maternal blood in the placenta. The mechanisms underlying these properties are discussed below.

Figure 9.4: (*see figure on previous page*) switches to the liver and this corresponds to a decline in ζ and ε synthesis and their replacement with α and γ globins forming $\alpha 2\gamma 2$ haemoglobin (known as foetal haemoglobin), which becomes dominant from about eight weeks and persists for most of the rest of gestation. From 11 to 12 weeks, haemoglobin production begins in bone marrow, which is the main source during the third trimester. Finally from 32 weeks gestation, expression of γ-globin declines and is replaced by β-globin which increases rapidly until 12 weeks after birth and forms the dominant adult haemoglobin, $\alpha 2\beta 2$ [44]. (b) Illustration of haemoglobin oxygen dissociation curves for adult, foetal and embryonic haemoglobins. Each curve shows haemoglobin binding of oxygen as a % of 100% saturation (on the *Y*-axis) and the partial pressure of oxygen (pO_2) at which this occurs (on the *X*-axis). The dashed horizontal line crosses each curve to indicate 50% saturation and the downward arrows point to the pO_2 at which this occurs. Adult haemoglobin (labelled $\alpha 2\beta 2$, furthest to the right) is maximally saturated at 50 mmHg pO_2 and 50% saturated at 26 mmHg (vertical downward arrow). The curve for foetal haemoglobin (labelled $\alpha 2\gamma 2$) is shown to the left of the adult curve and the embryonic haemoglobins (labelled $\alpha 2\varepsilon 2$, $\zeta 2\gamma 2$ and $\zeta 2\varepsilon 2$) are shown progressively leftward. The left most curve shows dissociation curve for α and β globin monomers. Note that for all haemoglobins, maximal saturation occurs at a pO_2 of 50 mmHg. However, 50% saturation occurs at lower pO_2 values as the curves move leftward, indicating greater binding affinity for oxygen. Note that the adult haemoglobin curve shows a sigmoidal shape (meaning that the slope of the curve increases as haemoglobin oxygen binding increases) indicating that initial haemoglobin binding of oxygen increases its affinity. This becomes less as the curves shift leftward and is virtually absent in embryonic haemoglobin and globin monomers [49].

The regulation of globin gene expression has been of interest for several decades because persistent foetal haemoglobin expression in adults can be very helpful for patients with genetic abnormalities in synthesising adult globins, particularly sickle cell disease and thalassaemia. In 2008, Uda *et al.* identified a gene called BCL11A which suppresses foetal haemoglobin synthesis [45] and this is currently being explored as a possible therapeutic target to restore foetal haemoglobin synthesis in these patients.

Regulation of haemoglobin production

Transcription of globin genes depends on the presence of haem. Like all genes, those which code for globin are regulated by transcription factors (specific proteins which bind to the DNA sequence and control its transcription). Haem binds to globin transcription factors and in doing so modulates the synthesis of globins. Thus haem deficiency inhibits globin gene transcription and synthesis of protein. As discussed above, haemoglobin is composed of at least two different globins, i.e., $\alpha 2\beta 2$ in adults and it is important therefore that α-globin, β-globin and haem are produced proportionately to form haemoglobin. Haem does this by interacting with an enzyme called haem-regulated inhibitor kinase (HRI) [46]. A deficiency of haem causes increased expression of HRI and this in turn phosphorylates the protein eIF2α which is necessary to initiate DNA translation. In this way, haem deficiency serves as a feedback mechanism to inhibit all protein synthesis including both globins. In addition, the availability of Fe has a regulatory role in haem synthesis. Synthesis of 5-aminolevulinic acid (ALA) by the enzyme ALA synthase is a key step in haem synthesis and this enzyme is sensitive to the availability of Fe. RNA which carries the code for the enzyme ALA synthase requires the presence of Fe to initiate its translation to make ALA synthase protein [47]. In this way, the availability of Fe exerts a positive feedback control over haem synthesis. Thus, the three elements of haemoglobin production (availability of Fe, haem and globin synthesis) are very carefully regulated. The availability of iron controls haem synthesis and in turn the presence of haem is necessary to globin synthesis. In this way, synthesis of the components of haemoglobin are coordinated and balanced and potentially harmful accumulation of intermediate products is avoided.

Bearing in mind that haemoglobin is present in all forms of life, it follows that it has many synthetic pathways in nature. A detailed discussion of these is beyond the scope of this chapter, however it is worth noting that in leguminous plant root nodules, synthesis of haemoglobin involves a partnership not unlike that in human erythrocytes. The globin protein is synthesised by the plant cells, while the haem moiety is produced by bacteria present in the root nodule [48] and are essential for making leghaemoglobin as well as nitrogen fixation. These bacteria are aerobic bacteroids, meaning they depend on oxygen for energy. This is likely to be a stimulus for synthesising haem which is then transported through the bacterial wall into the plant cytoplasm where it is assembled into mature leghaemoglobin. This of course has similarities with the export of haem from mitochondria into the cytoplasm of erythrocytes where it is inserted into globin chains. It does not seem impossible that these share a common evolutionary history. The porphyrin ring, which holds Fe in haem, is an evolutionarily ancient chemical structure. It binds magnesium (rather than iron) to form the active site of chlorophyll on which photosynthesis depends. Haem itself also forms the active site of cytochrome enzymes which are ubiquitous in the metabolism of plants, animals and eukaryotic microorganisms. Since the haem molecule is the same in all haemoglobins, one may speculate that its synthesis originated in an ancient bacteria, descendants of which became symbiont with animals and plants, obligate in the case of mitochondria and facultative cyanobacteria in leguminous plant nodules.

In summary, haemoglobin is an ancient oxygen binding molecule which serves many functions in plants and animals and this is due to its ability to bind oxygen with a wide range of affinities. In humans, this plasticity in its binding characteristics allows it to serve relatively hypoxic tissues in embryonic life and oxygenated adult tissues. Thus, as the heart develops in the foetus, its increasing strength is accompanied by the oxygen carriers that perfectly match conditions of life *in utero* during the various stages of growth. This is made possible by changes at a molecular level in haemoglobin and although these are less obvious than the dramatic growth of the heart, they are nevertheless just as vital for survival.

Haemoglobin and Oxygen Transport

Haemoglobin oxygen binding characteristics are classically described in terms of its oxygen dissociation curve. When haemoglobin is exposed to increasing concentrations of oxygen, it binds more of it until fully saturated. Oxygen dissociation curves express this haemoglobin saturation, expressed as a percent (%) of the maximum in relation to the concentration of oxygen present. Oxygen saturation curves for embryonic, foetal and adult haemoglobin are illustrated in Figure 9.4(b). The % haemoglobin saturation is shown on the vertical axis and the concentration of oxygen on the horizontal axis. Oxygen concentration is usually given as the partial pressure (pO_2). Atmospheric pressure at sea level is nominally 760 mmHg and since oxygen makes up 20% of the atmosphere, pO_2 is 160 mmHg.

The curve furthest to the right in Figure 9.4(b) illustrates the oxygen dissociation curve for adult haemoglobin and shows an increase from zero to around 100% saturation at a pO_2 of 50 mmHg. The affinity of any given haemoglobin is expressed as the pO_2 at which 50% saturation occurs, which in adult haemoglobin is a pO_2 of ~26 mmHg (downward pointing arrow). The oxygen dissociation curves for foetal and embryonic haemoglobins are progressively to the left of adult haemoglobin, indicating that their oxygen affinity is much greater. Thus, foetal and embryonic haemoglobins achieve 50% saturation at much lower pO_2 values than adult haemoglobin, 20%, 12%, 6% and 4%. In other words, haemoglobin affinity for oxygen changes dramatically during embryonic and foetal life; this of course is ideally suited to the low oxygen availability during gestation.

Cooperative and allosteric binding

The changing slope of the oxygen dissociation curve for adult haemoglobin (furthest to the right in Figure 9.4(b)) is described as sigmoidal, meaning that as pO_2 rises and oxygen binding commences, the slope of the curve gradually increases, being steepest at around 50% saturation and then declines as oxygen saturation approaches the maximum. This is an example of cooperative binding and is due to the fact that initial binding of oxygen by haemoglobin increases its oxygen affinity and enhances further binding.

This is a very common feature of ligand binding in biology and is due to a change in the conformation (shape) of the haemoglobin molecule induced by its initial binding of oxygen, resulting in increased affinity for oxygen.

This sigmoidal feature of oxygen binding is progressively less noticeable with foetal and embryonic haemoglobins and is virtually absent in the earliest embryonic haemoglobins, for example, the dissociation curve for $\zeta 2, \varepsilon 2$ haemoglobin compared to $\alpha 2\beta 2$ haemoglobin in Figure 9.4(b). This association between higher oxygen affinity and loss of binding cooperativity led Manning *et al.* [49] to explore whether they are linked. In Figure 9.4(b), all of the oxygen dissociation curves shown with the exception of the leftmost illustrate binding of tetrameric haemoglobins. The leftmost curve shows binding of α or β globin monomers which clearly have far higher oxygen affinity than when combined into $\alpha 2\beta 2$ tetrameric haemoglobin. This strongly suggests that formation of globin polymers modifies the oxygen affinity of individual monomers. Formation of tetrameric haemoglobin initially involves combining monomers into heterodimers which then join to form tetramers. Thus for adult haemoglobin, two $\alpha\beta$ heterodimers are initially formed which then join as mature $\alpha 2\beta 2$ tetrameric haemoglobin. Manning *et al.* were able to show that oxygen affinity of mature tetrameric haemoglobins is inversely related to the strength of binding between its individual monomers and that this binding is weaker in foetal and embryonic haemoglobins. In other words, the strength of the bonding interfaces between the α and β globin subunits in adult haemoglobin is stronger than that between the corresponding ζ and ε subunits in embryonic haemoglobin. As a result, adult oxygen binding is much weaker than that of its monomers, whereas the weaker binding between embryonic globin subunits means that this foetal haemoglobin is little changed from that of its monomers. Thus, small changes in the amino acid sequence of the globin proteins which surround the oxygen binding site can alter its affinity in ways that match the prevailing oxygen levels.

In biology, the activity of many enzymes can be modulated by molecules that bind to them at sites other than the active site. This is known as allosteric (from the Greek *ἄλλος* meaning "other" and *στερεός* meaning "object" or in this case site) regulation and results from a change in the shape of the enzyme causing a change in its activity. Haemoglobin is not an enzyme, however the changes in oxygen binding induced by interaction with other globin subunits in tetrameric haemoglobin are an example of negative allosteric regulation. The cooperative binding of adult haemoglobin for oxygen (discussed above and illustrated by the sigmoidal curve of oxygen dissociation for $\alpha2\beta2$ haemoglobin) is positive in that initial binding of oxygen increases haemoglobin affinity. The mechanism here is that oxygen binding to one of the four globin units induces a change in its conformation which then modifies its binding to adjacent subunits of the tetramer causing an increase in their affinity for oxygen.

In Vivo Modulation of Haemoglobin Oxygen Affinity

The Bohr effect

Not surprisingly, haemoglobin binds many substances in blood as well as some important toxins. For example, cyanide and carbon monoxide bind to Fe at the haem site and inhibit oxygen transport and for this reason are potentially life-threatening toxins. Carbon dioxide is produced in tissues as a by-product of aerobic metabolism and is transported in blood to be exhaled by the lungs. This has two allosteric effects which modulate haemoglobin function, collectively known as the Bohr effect. (a) CO_2 binds to haemoglobin, forming carbaminohaemoglobin and about 23% of CO_2 in blood is transported in this way. CO_2 does not bind to haem; instead, a CO_2 molecule binds to the end of each of the four globin subunits so that each haemoglobin carries four CO_2 molecules. Binding of CO_2 to haemoglobin reduces haemoglobin affinity for oxygen and since this occurs in tissues where CO_2 is produced by aerobic metabolism, it also coincides with where oxygen is needed. Thus binding of CO_2 to haemoglobin facilitates the release of oxygen by lowering its affinity. (b) About 70% of CO_2 is transported in blood as bicarbonate ions due to the action of carbonic anhydrase. This enzyme converts $CO_2 + H_2O$ into carbonic acid H_2CO_3 which then dissociates into protons (H^+) and bicarbonate ions (HCO_3^-). Protons also bind to haemoglobin and cause tighter bonding between the α and β subunits. This in turn also reduces haemoglobin oxygen affinity.

Thus, the Bohr effect assists the release of oxygen from haemoglobin, due to its binding to CO_2 and by the action of carbonic acid produced from CO_2.

The Haldane effect

In the lung capillaries, the high oxygen concentration favours oxygen binding to haemoglobin. This results in oxygenation of haemoglobin which reduces its affinity for CO_2 and therefore enhances its release from haemoglobin. In addition, increased oxygen binding to haemoglobin reduces its affinity for H^+ ions, reversing the protonation that occurred at the periphery. The release of H^+ ions increases their concentration locally and this favours the reconversion of bicarbonate to carbonic acid and then to CO_2 and H_2O, pushing the reaction further to the left. In this way, production of CO_2 is increased which facilitates its exhalation.

$$\begin{array}{c} \text{carbonic anhydrase} \\ CO_2 + H_2O \leftrightarrow H_2CO_3 \leftrightarrow H^+ + HCO_3^- \\ \leftarrow \text{lungs} \qquad\qquad\qquad \text{tissues} \rightarrow \end{array}$$

This is known as the Haldane effect and in reality is the reverse of the Bohr effect. Together, they provide an oscillating mechanism which increases and decreases oxygen affinity in the lungs and at the peripheral tissues, respectively, and therefore optimises the physiological uptake and release of both gases.

2,3-Bisphosphoglyceric acid

In addition to these allosteric effects of CO_2 binding and pH, haemoglobin is sensitive to several organic phosphate molecules in erythrocytes. In humans and some other mammals, the most important of these is 2,3-bisphosphoglyceric acid (2,3 DPG), which is present in much higher concentrations in erythrocytes (~5 mmol/L) than in other tissues (micromolar amounts). This is because 2,3 DPG is a by-product of glycolytic glucose metabolism. Glycolysis is a form of non-aerobic metabolism which human erythrocytes rely on having lost their nuclei and with that, the ability to undertake oxidative metabolism. In other tissues, oxidative metabolism predominates and glycolysis plays minimal role. In species which have retained nucleated erythrocytes and with it their ability for oxidative metabolism, other organic phosphates are more abundant than 2,3 DPG in their erythrocytes. For example, adenosine triphosphate (ATP) is more abundant in birds, reptiles, amphibians and fish and has similar allosteric effects on haemoglobin to 2,3 DPG in humans [50].

The structure of 2,3 DPG is tiny compared with globin molecules; however, it is complementary to key elements of the β-globin chain and binds to specific amino acids within its pocket region, which also bind the haem group. This occurs preferentially when haemoglobin is deoxygenated and has the effect of pulling the E and F helixes closer together [51]. This is turn reduces haemoglobin affinity for oxygen and enhances its release to the tissues. On returning to the lungs where the high pO_2 oxygenates haemoglobin, 2,3 DPG is released, and its allosteric effect is reversed resulting in an increase in oxygen affinity. Thus 2,3 DPG provides yet another elegant mechanism which reversibly modulates haemoglobin affinity for oxygen, enhancing its uptake in the lungs and its release to the tissues which need it.

In summary, haemoglobin has an ancient heritage extending back to the oxidation of the Earth's atmosphere. During its long evolutionary history, it has continued to evolve to meet the needs of animals with many differing lifestyles. The plasticity of its oxygen binding allows it to serve humans from two weeks of gestation during great variations in oxygen availability and even in mature adults its affinity for oxygen is elegantly modulated minute by minute to bind and release oxygen as it is pumped by the heart through the lungs and around the body.

Evolution of Haemoglobin and Erythrocytes: Matching the Heart and Circulation

Human erythrocytes have a constant shape and size and are packed with haemoglobin. We might be tempted to think this is the same for blood in all animals, but nothing could be further from the truth. First, annelids

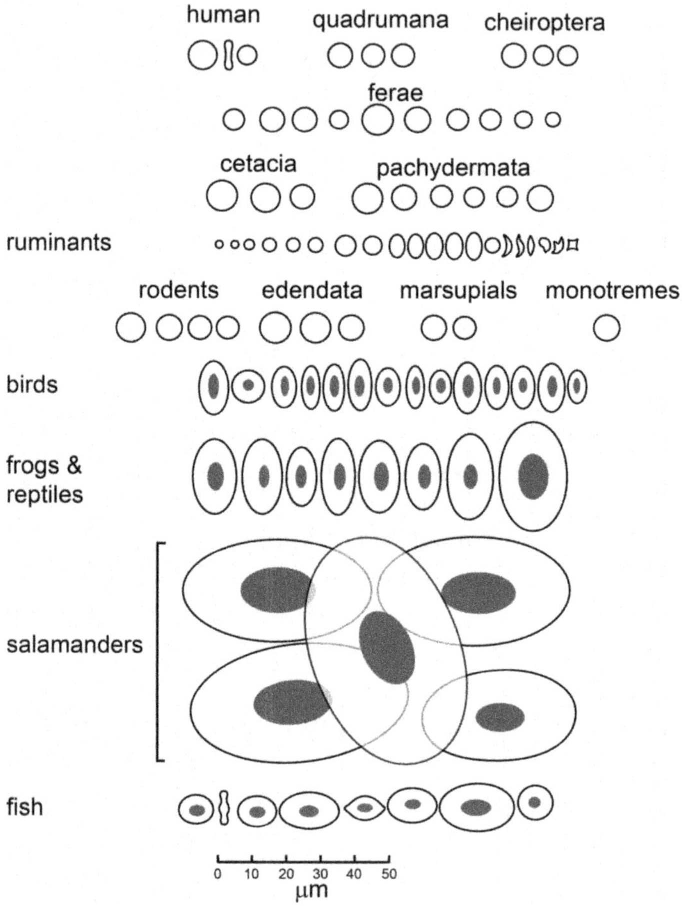

Figure 9.5: Erythrocytes in aquatic and terrestrial vertebrates, redrawn from Gulliver 1875 [52]. The size of erythrocytes varies among species, ranging from just 2 μm in mice compared to *amphiuma* salamander which can be up to 68 μm in length. Most animals have either spherical or ovoidal erythrocytes. In contrast, human erythrocytes have the shape of circular disks described as biconcave, meaning that the disks are narrower at the centre than the rim. Mammals extrude the nuclei of erythrocytes in the bone marrow prior to release into the circulation and are therefore enucleate when mature. In contrast, birds, reptiles, amphibians, salamanders and fish have nucleated erythrocytes. The large size of salamander erythrocytes is due to the size of their genome (and therefore nucleus), which is ten times larger than in humans. The size of erythrocytes is of great significance for the heart and circulation because they have to pass through the smallest blood vessels, which together form the largest component of the resistance to blood flow and therefore the work the heart has to do.

including segmented worms such as the common earth worm and *vestimentiferans* and *Siboglinidae* (marine tube worms) do not have erythrocytes, instead they have extracellular haemoglobins that are evolutionarily linked to ours. Second, erythrocytes vary greatly in size and shape among vertebrates as revealed by George Gulliver's well known study in 1875 (Figure 9.5) [52]. Thus, erythrocytes are most often spherical or ovoid disks, although many shapes are possible, including biconcave disks in humans. They vary in size from just 2 μm in mice up to 69 μm long in salamanders. In addition, although mammalian erythrocytes circulate as enucleated cells, having lost their nuclei during development in bone marrow, bird, reptile, salamander and fish erythrocytes retain theirs.

These features of blood have great importance for the circulation, not least because erythrocytes must be able to pass through the smallest capillaries and therefore must match them in size. Erythrocyte deformability also contributes to their ability to traverse narrow vessels. Both size and deformability of erythrocytes are affected by the presence of a nucleus because nucleated cells are (i) necessarily larger and (ii) less deformable. A brief examination of Figure 9.5 shows that nucleated erythrocytes are almost invariably larger than enucleated ones. Hence erythrocytes of larger size and those which retain nuclei are less deformable and offer greater resistance to blood flow [53]. As discussed in Chapter 7, vascular resistance is a major determinant of how hard the heart must work to pump blood through the

Figure 9.5: (*see figure on facing page*)

Notes: *Quadrumana*: Non-human primates

Cheiroptera: Bats

Ferae: Cats, hyenas, dogs, bears, seals

Cetacea: Whales, dolphins, porpoises

Pachydermata: (Now obsolete) includes elephants, horses, rhinoceroses, pigs, hippopotamuses

Ruminants: Cattle, goats, sheep, dear, antelope

Rodents: Mice, rats, squirrels, chipmunks, prairie dogs

Edentata: Anteaters, tree sloths, armadillos, aardvarks, pangolins

Marsupials: Kangaroos, koalas, wombats

Monotremes: Platypus, echidnas

circulation and also of blood pressure. Since the greatest portion of vascular resistance is imposed by the smallest blood vessels, it follows that erythrocyte size and structure and how well they match these vessels are critical factors for the normal function of the heart and circulation.

This is apparent from the close correlation between capillary diameter and erythrocyte size among species (Figure 9.6(a)). Thus, average capillary size in salamanders can be greater than 30 μm, compared with around 5 μm in mammals, while amphibians (e.g., frogs) and squamates (snakes and lizards) are intermediate, each closely matching the size of their erythrocytes. As a result, the microvascular structure is quite different in these species. Although capillaries are much smaller in mammals, they are also more numerous than in other species (Figure 9.6(b)). This also means that the distance between capillaries is much less in mammals than in other species and therefore that oxygen diffusion distance (between capillaries and tissues) is less (Figure 9.6(c)). The energy demands of amphibians are substantially less than warm blooded terrestrial mammals. It seems likely therefore that the emergence of smaller erythrocytes and a finer microvasculature were needed to improve oxygen delivery to tissues.

Of course, blood flow is the other crucial factor in delivering oxygen to the tissues and that is a function of vascular resistance and blood pressure. Clearly, smaller vessels offer higher resistance to flow but a greater number of vessels in parallel will tend to offset that. In practice, mammals have a much higher vascular resistance than amphibians (Figure 9.7(a)). This is because resistance to flow in a tube is related to the fourth power of its radius, meaning that small changes in diameter have a big effect on resistance. Consequently, mammals have needed to evolve hearts capable of pumping at higher blood pressures to perfuse tissues, as discussed in Chapters 2 and 7. It seems likely therefore that the need for higher rates of aerobic metabolism to meet the energy needs of terrestrial life was achieved by the co-evolution of smaller erythrocytes and a reduction in capillary size together with more powerful hearts to provide higher perfusion pressures.

The reduction in the size of erythrocytes and capillaries discussed above relates to amphibian and terrestrial animals. However, fish which evolved before amphibians have smaller erythrocytes, suggesting that large erythrocyte size is not simply a feature of primitive animals and that

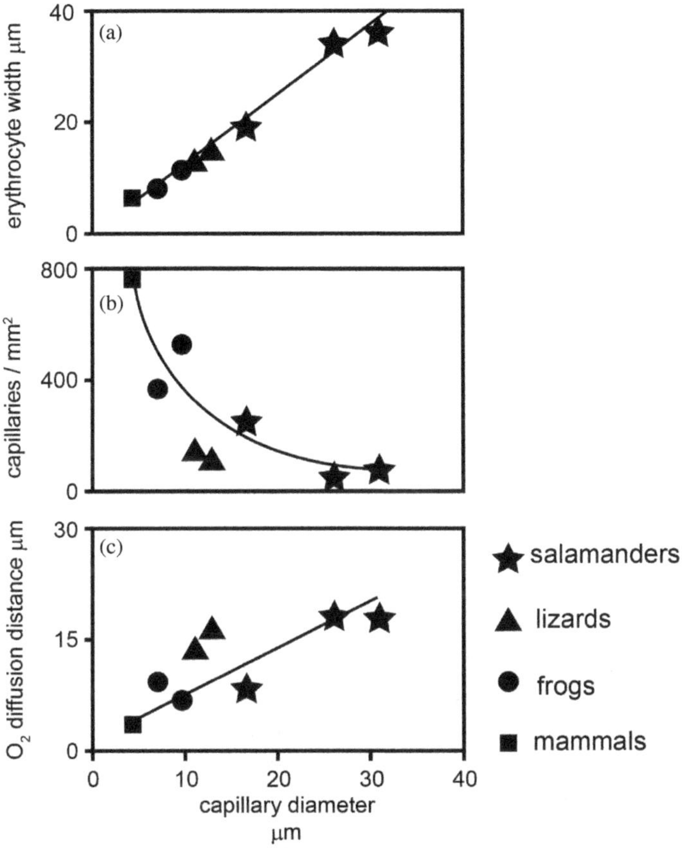

Figure 9.6: Erythrocyte size, capillary density and oxygen diffusion distance in salamanders, lizards, frogs and mammals. (a) Mammals including humans have smaller erythrocytes than amphibians such as frogs, lizards and salamanders. Human erythrocytes are about 7 μm in diameter compared to salamanders which can be as large as 36 μm at their narrowest point and 68 μm in length. Erythrocyte size in each species is closely matched to the diameter of capillaries, which varies from 5–10 μm in humans to 33 μm in salamanders. (b) Capillary density expressed as the number of capillaries per unit volume of tissue is calculated from the number observed in an area under microscopic examination. The density of capillaries is inversely related to their diameter, the smaller capillaries in mammals being much more numerous than the larger ones in amphibians. (c) The distance oxygen must diffuse from capillaries to the tissues which use it is directly related to the size of capillaries, being less with smaller more numerous capillaries [1].

the increased erythrocyte size in amphibians may have occurred as a specific evolutionary response. The most likely hypothesis is that the emergence of large erythrocytes and capillary diameters coincided with and facilitated the onset of air breathing [1]. Thus, jawless fish (such as hagfish and lampreys), elasmobranchs (such as sharks) and bony fish (such as swordfish and tuna) have smaller erythrocytes than amphibians which evolved later (Figure 9.7(b)). Fish rely exclusively on gills to obtain oxygen from water. They have a two-chambered heart which perfuses the gills and systemic circulation via a single circulation (see Figure 2.8 and related text in Chapter 2 and Figure 7.2 and related text in Chapter 7). As aquatic animals, their energy needs are relatively low and they can function with low blood pressures compared to terrestrials. Nevertheless, the more evolutionarily advanced fish developed higher pressures and resistances than earlier hagfish (Figure 9.7(b)).

This trend was reversed with the emergence of salamanders, which have lower vascular resistance than fish. Unlike mammals and birds which have four-chambered hearts that perfuse the pulmonary and systemic circulations separately at low and high pressures, respectively, early amphibians relied on a two-chambered heart and single circulation as fish do. This presents a problem because air-breathing requires that oxygen must be able to diffuse between air filled spaces (called alveoli) and blood across extremely thin-walled capillaries. Such capillaries are easily damaged by high pressures, which is why lungs are always perfused at lower pressure than the systemic circulation (as discussed in Chapter 7). Amphibians which have a single ventricle perfuse the systemic and pulmonary circulation at the same pressure and that has to be low in order to protect the pulmonary capillaries. To allow this, salamanders evolved a vasculature with a lower vascular resistance than sharks and bony fish (Figure 9.7(b)). This was made possible by having large capillaries (much larger than mammals' or fish), Figure 9.7(a), and erythrocytes of similar size. This had two advantages. First, large erythrocytes are more deformable than smaller ones and this would have been helpful in keeping resistance low; second, larger erythrocytes can fold in a way that brings their cell membranes close to the walls of comparably-sized capillaries, which reduces the oxygen diffusion distance. In this way, early air-breathers could perfuse their systemic and pulmonary circulations at low pressure by a single ventricle.

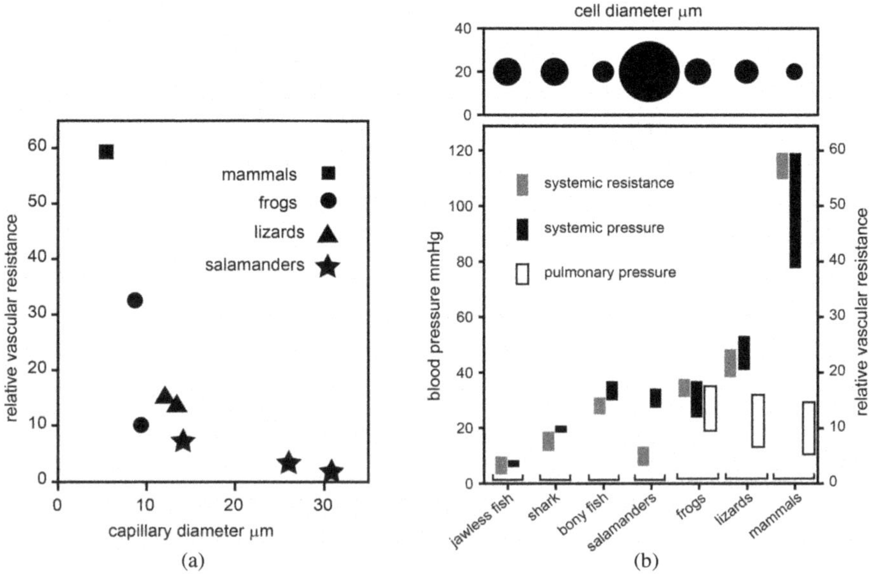

Figure 9.7: (a) Vascular resistance and capillary diameter in mammals, frogs, lizards and salamanders (vascular resistances are expressed relative to salamanders, which are arbitrarily set to 1). Resistance to blood flow is predominantly contained within the microvasculature. This is illustrated by the inverse relationship between capillary diameter and vascular resistance in mammals, frogs, lizards and salamanders. Vascular resistance is much higher in mammals which have small capillary size compared with species with larger capillaries. (b) Erythrocyte size, blood pressures and vascular resistance among aquatic, amphibian and terrestrial animals. Primitive fish such as hagfish have low vascular resistance and blood pressure which is sufficient for their energy needs. More advanced fishes like sharks and bony fish, e.g., tuna, evolved progressively higher pressures as their energy demands increased. As early air-breathers, salamanders still had the single circulation present in fish, which perfused both the lungs and systemic circulation and therefore had to maintain low pressure in each. This was possible by having large diameter capillaries to keep vascular resistance low. The emergence of animals with greater energy needs required increased oxygen delivery to tissues. The evolutionary solution to this was to increase blood flow, by increasing systemic pressures. This was finally achieved by the evolution of four-chambered hearts allowing separation of the pulmonary and systemic circulations. This process began with amphibians such as frogs which have three-chambered hearts that can direct deoxygenated blood to the lungs and oxygenated blood to the body, while maintaining similar pressures in each. Varanid lizards also have three-chambered hearts in which contraction effectively divides the ventricle into functionally separate left and right chambers. This allows the systemic circulation to be pumped at higher pressure than the pulmonary. Mammals evolved complete anatomical separation of the pulmonary and systemic circuits with four-chambered hearts allowing the perfusion of each with markedly different pressures. In each species, the diameter of erythrocytes (top) matched that of capillaries. The large diameter of salamander erythrocytes appears to have been a transient feature which evolved as a means to match the large diameter of capillaries in this species [1].

While this arrangement was sufficient to allow the development of air breathing lungs, low systemic pressure also limited oxygen and therefore energy delivery to the rest of the body. The evolutionary solution to increasing oxygen delivery and thereby greater energy production was achieved by (i) increasing the haemoglobin concentration in blood and (ii) increasing the rate of blood flow. The latter required increased systemic pressures, which was possible with the development of separate pulmonary and systemic circulations and four-chambered hearts. Thus, mammals have high systemic vascular resistance and pressure, but low pulmonary pressure (Figure 9.7(b)). Amphibians and squamates occupy intermediate positions between fish and mammals, summarised here and discussed in greater detail in Chapter 7. Anurans (such as frogs) have a three-chambered heart (separate atria and a single ventricle). However, they also have a spiral baffle in the outflow tract which directs flow preferentially to the pulmonary and systemic circulations (see Figure 7.2, Chapter 7 and Figure 2.9, Chapter 2). This is a significant improvement in directing deoxygenated blood to the lungs and oxygenated blood to the systemic circulation. However, pressure in both circuits is the same and within safe limits for the pulmonary alveoli (Figure 9.7(b)). Crocodiles have four-chambered hearts but have retained the ability to regulate outflow to the pulmonary and systemic circulations when diving, via apertures between the aorta and pulmonary arteries (see Figure 2.10 and related text in Chapter 2). Varanid lizards are exceptions in being terrestrial squamates with high energy needs. They also have a single ventricle, which in this case is functionally separated during systole. To achieve this, they have a septum between the left and right side of the chamber with a residual communication, which is sealed during contraction. As a result, contraction of the single ventricle generates low pressure in the pulmonary circulation reflecting its low vascular resistance, but high pressure in the systemic circulation where resistance is higher [54]. In mammals and birds, complete separation of the pulmonary and systemic circuits was made possible by the evolution of a four-chambered heart. In this way, they were released from the need to maintain low pressures throughout the circulation to protect the lungs and subsequently evolved high systemic pressures which allowed greater oxygen delivery and energy production. This change was also accompanied by reduction in the size of capillaries and erythrocytes (Figure 9.7(b)). Thus mammals adopted the strategy of increased blood flow to tissues in order to deliver more oxygen and allow greater energy production. At the tissue level, this was achieved by increasing the number of capillaries while reducing their size and those of erythrocytes.

Why Are Capillaries and Erythrocytes of Similar Size?

It is no accident that capillaries and erythrocytes are of similar size, erythrocytes usually being a little smaller than the minimal capillary diameter. Their matching has important advantages; one relates to the diffusion of oxygen. Erythrocytes are readily deformable which assists their motion through capillaries which are smaller in size [55]. Folding of disc or crescent-shaped cells also implies that the outer face of a folded cell can adopt a closer position in relation to the capillary wall (Figure 9.8(a)) and thus reduce the oxygen diffusion distance. This of course would apply to capillaries and erythrocytes of any size so long as they match, including those of the salamander. The alternative of a large number of smaller erythrocytes, having collectively the same cell volume, would result in (a) cells being widely dispersed within the capillary lumen, (b) a greater distance from the capillary wall and (c) a greater oxygen diffusion distance. Such an alternative arrangement (a larger number of cells having the same volume of a single large one) would increase the surface to volume ratio and therefore the area available for oxygen diffusion, but presumably not sufficient to outweigh the benefit of close apposition of erythrocyte membranes to the capillary wall. As mammals pursued a strategy of increased blood flow to deliver more oxygen to tissues and their capillaries became smaller and more numerous, a matching reduction in erythrocytes' size maintained the advantage of their close apposition. This is illustrated in Figure 9.8(a), which shows a markedly folded erythrocyte within a myocardial capillary resulting in close apposition of its cell membrane and the capillary wall. Reducing erythrocyte size improves oxygen diffusion further by reducing the oxygen diffusion distance between haemoglobin and the cell surface [56]. Additionally, erythrocyte surface to volume ratio increases with reduction in size creating a larger area for gas transfer [57]. Thus, the structure and size of erythrocytes are ideally suited to that of the microcirculation

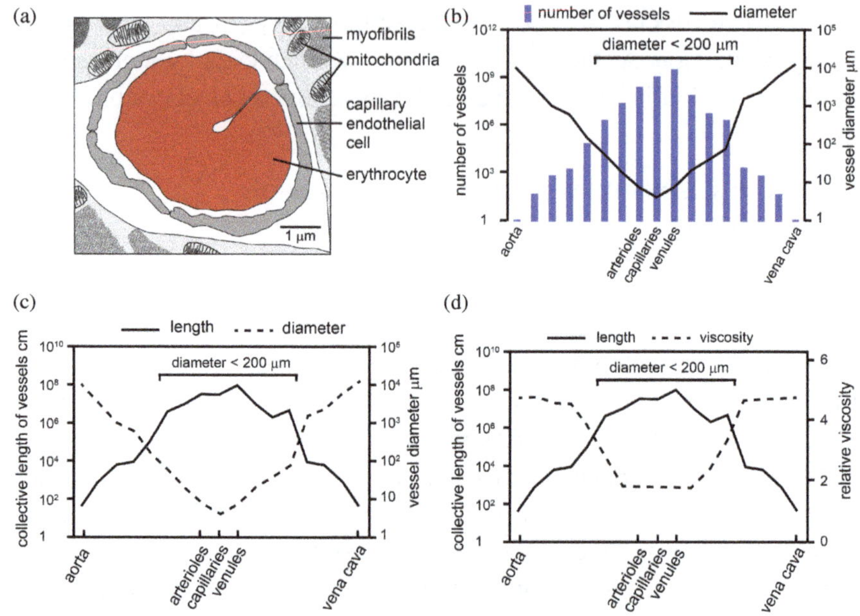

Figure 9.8: (a) Illustration of an erythrocyte within a myocardial capillary. The erythrocyte shows marked folding which facilitates transit through small vessels. This arrangement results in excellent apposition of the cell membrane with the endothelial cells of the capillary which facilitates rapid diffusion of oxygen and carbon dioxide between tissues and erythrocytes. The ability of erythrocytes to fold in this way also dramatically reduces the viscosity of blood within capillaries and resistance to flow. This in turn greatly reduces the work required by the heart to maintain blood flow [72]. (b) The number of vessels (blue bars, scale on the left axis) and their diameter (solid line, scale on the right axis) according to their size in the canine circulation. From just one aorta, the number of arteries increases dramatically as they divide into smaller branches until they reach capillary level. From there blood is collected in venules and then veins of increasing size and fewer number until finally returning to the heart in the vena cava. The horizontal bar indicates that the vast majority of vessels are small <200 μm. (c) The diameter (dashed line, right axis) and collective length of vessels (the average length x number of each size; solid line, left axis) according to their distribution in the circulation. Although smaller vessels are also shorter, their enormous number means that their collective length is much greater so that by far, the longest transit of blood in the circulation is through small (<200 μm) vessels. (d) The collective length of vessels (the average length x number of each size; solid line, left axis) and the viscosity of blood (dashed line, right axis) in vessels according to their distribution in the circulation. The viscosity of whole blood decreases sharply as it flows through narrow vessels (<200 μm). Known as the Fåhræus–Lindqvist effect, this greatly reduces the resistance to blood flow and the work performed by the heart [58].

in a way that facilitates efficient uptake and delivery of oxygen and carbon dioxide.

Erythrocyte structure, blood viscosity and the heart

Matching of erythrocyte and capillary size also provides a crucial energy sparing mechanism which greatly enhances the efficiency of the heart by modulating the viscosity of blood. Viscosity means the deformability of a fluid, more generally referred to as the stickiness or thickness of a liquid and the ease with which it flows, a property that derives from the degree of friction within a fluid. More viscous fluids are more resistant to deformation and require more effort to pump. Hence, viscosity of blood directly affects how much blood pressure the heart must generate to pump it. Consequently, it is advantageous for the heart to have a low viscosity of blood. Blood is a complex fluid containing dissolved and suspended particles including erythrocytes. These components increase its viscosity, which is higher than that of a simple liquid such as water. Relative viscosity is used to describe the viscosity of a solution as a ratio of its solvent. The viscosity of plasma is 1.8 times that of water and hence its relative viscosity is 1.8. The relative viscosity of whole blood depends on the haematocrit (the percentage of blood occupied by erythrocytes), being 4.0 at a normal haematocrit of 45%.

Factors affecting viscosity of blood

(i) Viscosity is altered by increases and decreases in haemoglobin level. Severe anaemia causing a 50% reduction of the haematocrit would lower the relative viscosity to about 2.5, while polycythaemia could increase it to 7. Such an increase significantly affects the circulation and increases the risk of clot formation.

(ii) Blood viscosity is also increased at lower temperatures and this can be a significant factor contributing to impaired blood flow in hypothermia, particularly in the hands and feet.

(iii) Interaction between the molecules and particles in a fluid which causes them to stick together tends to increase viscosity and this is especially important in blood. When blood is static or flowing slowly, these interactions are greater and erythrocytes can adhere to each other forming stacks of discs known as rouleau formation. It is for this reason that the viscosity of blood is affected by its flow rate, being higher when flowing slowly or static. It also explains why blood is described as a non-Newtonian fluid. For simple liquids such as water, the rate of flow in a tube is related to pressure driving it in a linear manner and this is described as Newtonian since it obeys Newton's equation. In the case of blood, however, its increased viscosity when static means that a minimum pressure is needed to initiate flow which then gradually increases with further pressure. In other words, the relationship between pressure and flow is non-linear, and hence blood is described as non-Newtonian.

The Fåhræus–Lindqvist effect

The viscosity of blood depends on its haematocrit and while this might be expected to be constant, in practice it varies within the circulation. This is because when flowing in small vessels such as arterioles and capillaries, erythrocytes tend to move towards the centre, leaving a cell free layer of plasma close to the vessel wall. This has the effect of lowering the haematocrit and reduces the viscosity of blood in these small vessels. This is known as the Fåhræus–Lindqvist effect and means that viscosity of blood decreases as the diameter of blood vessels decreases within a range of 10–300 μm. This is a very important property of blood for two reasons.

(i) Blood pressure in the aorta close to the heart is pulsatile and normally around 120/80 mmHg. As we move further away from the heart, the pressure becomes less pulsatile and lower, so that at capillary level, pressure is much less, typically around 10–20 mmHg. This also means that flow velocity is much less. This would be expected to increase viscosity for the reasons described above, however the Fåhræus–Lindqvist effects offset this and results in a net reduction in viscosity.

(ii) Blood viscosity directly affects the heart. An increase in blood viscosity creates a greater resistance to flow and therefore requires a higher blood pressure to maintain flow at a normal rate. This is potentially harmful because it increases the work demanded of the heart and also because higher pressures if prolonged are injurious to blood vessels. The size and structure of erythrocytes mitigate this through the Fåhræus–Lindqvist effect. The microscopic vessels which perfuse tissues are vastly more numerous than those we can see and feel (Figure 9.8(b)). For example, mammals are estimated to have over 1 million blood vessels of less than 10 μm in diameter for every artery or vein of visible size (>1 mm) [58]. An immediate consequence of this is that the total distance blood flows through small vessels is many times greater than through large arteries and veins (Figure 9.8(c)). This also explains why these vessels collectively offer the greatest resistance to flow. It is also why the factors that affect flow in these vessels, including erythrocyte size and structure, are so important. By reducing viscosity when flowing through such small vessels (Figure 9.8(d)), the Fåhræus–Lindqvist effect greatly reduces the level of blood pressure and cardiac work that would otherwise be needed. For example, Snyder [58] calculated that whereas a healthy dog can maintain normal blood flow at rest with a mean blood pressure of 103 mmHg, it would require 202 mmHg in the absence of the Fåhræus–Lindqvist effect.

This difference represents a dramatic reduction in the workload of the heart. It also underscores the importance of erythrocyte size and structure and the necessity that they match those of the small vessels.

Co-evolution of Erythrocytes and Haemoglobin

The ability of the heart to deliver oxygen to tissues depends on the rate of blood flow and the oxygen content of blood. In humans and other mammals, the latter depends on the haemoglobin concentration in blood. As discussed above, haemoglobin is widely used as an oxygen carrier in nature and is present at markedly different concentrations in animals (Table 9.1). In addition, different species have evolved haemoglobins having a wide range of quaternary structures, from globin monomers with a MW of 17 kDa to giant 3,600 kDa dodecamers [40]. In many invertebrates, haemoglobin is present as an extracellular oxygen carrier but in vertebrates it is always confined within erythrocytes. The reasons for this

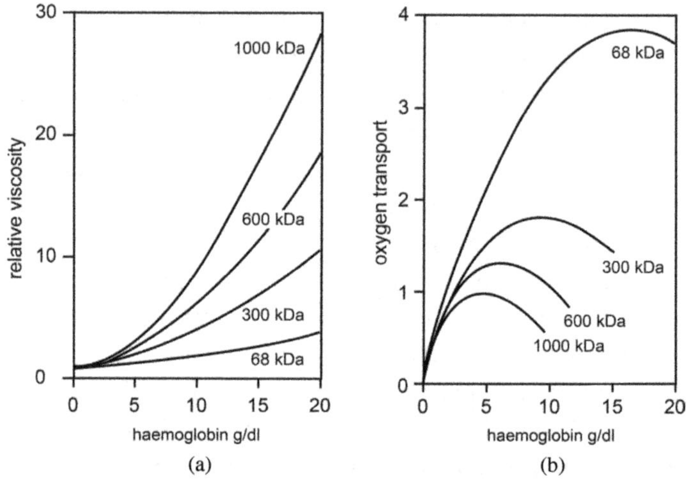

Figure 9.9: Viscosity and oxygen transport capacity of solutions of haemoglobin related to their MW and concentration. (a) Viscosity of solutions of haemoglobin increases with their MW so that at any given concentration of haemoglobin viscosity increases with MW. (b) The oxygen carrying capacity of solutions of haemoglobin increases as the concentration increases, irrespective of the MW of haemoglobin. However as the concentration increases, so too does the viscosity and this creates an increasing resistance to flow. Since oxygen carrying capacity also depends on the flow rate, it is limited by the effect of increasing viscosity, reaching a peak and then declining as the concentration increases further. The haemoglobin concentration at which oxygen transport is optimal is higher in haemoglobins of low MW. For this reason, animals with high energy and oxygen demands have evolved to have high concentrations of low MW haemoglobins [59].

relate to an interplay between the oxygen carrying capacity, viscosity and behaviour of haemoglobins of different MWs.

In solutions of haemoglobin, the viscosity increases with its concentration. However at any given concentration, viscosity also increases with the MW of haemoglobin. This means that whereas a solution of tetrameric haemoglobin (MW 68 kDa) of 15 g/dL has a viscosity of about 5, a solution of the same concentration of haemoglobin with a MW of 1,000 kDa would have a viscosity of 20 (Figure 9.9(a)). Such an increase in viscosity would offer a higher resistance to flow at any given pressure. This means that while oxygen transport increases with haemoglobin concentration, there is a trade off against increasing viscosity, so that oxygen carrying capacity actually declines beyond a certain concentration depending on the MW of haemoglobin (Figure 9.9(b)). Haemoglobins in solution therefore have an optimal oxygen carrying capacity and that optimal value is greater for low MW haemoglobins due to their lower viscosity. For example, the optimal oxygen carrying capacity for haemoglobin of 68 kDa occurs at a concentration of 15 g/dL, compared to 5 g/dL for haemoglobins of 1,000 kDa [59]. Mammals such as humans need a high concentration of haemoglobin to meet their energy needs (typically 15 g/dL), for which haemoglobin consisting of globin tetramers with a MW of 68 kDa is ideally suited. However, 68 kDa haemoglobins are small enough to pass through the kidneys' filters, which would result not only in their passing into urine and being lost, but also damage to the kidneys. It is for this reason that low MW haemoglobins are always located in erythrocytes and also why the extracellular haemoglobins present in invertebrates have large MWs. Thus, the high energy and oxygen demands of mammals require high concentrations of haemoglobin for which molecules of low MW are best suited and these in turn have to be isolated within cells to avoid leakage from the circulation.

For many years, it was assumed that haemoglobin located within erythrocytes also provided a lower viscosity than a similar concentration in solution. This was clearly shown not to be the case in experiments in which haemoglobin was released from haemolysed erythrocytes and caused a marked drop in viscosity [60]. However, unlike whole blood, the viscosity of haemoglobin solutions is constant, meaning that its contribution to resistance to flow would be the same in vessels of all sizes and the Fåhræus–Lindqvist effect would be absent. The effect of this is that while a container of haemoglobin in solution has a lower viscosity than a similar container of blood, the Fåhræus–Lindqvist effect gives whole blood a lower viscosity in the circulation. Thus, it seems likely that haemoglobin and erythrocytes co-evolved to optimise oxygen carrying capacity in animals with high energy needs and this process was also closely coordinated so that the size and structure of erythrocytes are exquisitely suited to that of the circulation.

Biconcave Disc Erythrocytes

Erythrocytes by and large have the shape of flattened discs. Human erythrocytes are biconcave discs, being narrower at the centre. This structure has a number of physiological benefits which protect the circulation:

(i) Unlike a sphere in which most of the mass (or weight) is at the centre, the shape of biconcave discs places most of the mass where they are widest, i.e., around the periphery. This changes the ease with which they can be made to rotate. In other words, more energy (known as the moment of inertia) is needed to rotate a biconcave disc than a spherical cell. For this reason, biconcave discs are less prone to rotation in the circulation [61]. This helps to minimise turbulent flow which protects the circulation from platelet activation and diseases such as atheroma.

(ii) The biconcave shape also facilitates erythrocyte folding, meaning that it requires less energy to bend than other shapes and the biconcave shape of human erythrocytes appears to be optimal in this regard [62]. This reduces resistance to flow in small vessels and thereby pressure needed to maintain the circulation.

(iii) Erythrocytes persist in the circulation for around 120 days. Elements of the immune system can recognise changes in the composition of erythrocyte cell membranes and initiate their removal. In addition, a filtering process located in the spleen can detect changes in the structure, shape or deformability of erythrocytes, which are features of aging cells, and select them for removal. Erythrocytes are required to pass through a narrow opening, called an inter-endothelial slit, and cells which fail to transit successfully are retained and removed [63].

(iv) Recent research has also shed new light on the functional importance of disc-shaped erythrocytes [64]. The normal haemoglobin concentration in whole blood in humans is 15 g/dL. However, since it is packed into erythrocytes which normally amount for just 45% of blood volume, the concentration of haemoglobin within erythrocytes, expressed as the mean corpuscular haemoglobin concentration (MCHC), is much higher, being 33–35 g/dL. This has important effects on the behaviour of the cytosol of erythrocytes. As discussed above, oxygen and carbon dioxide are transported in blood and both diffuse into and from erythrocytes during this process. In the case of carbon dioxide, only 7% is dissolved in plasma and the remainder is bound to haemoglobin (23%) or as carbonate ions (70%). Until recently it was thought that diffusion of gasses within erythrocytes was rapid as it is in water. However, Richardson and Swietach [64] tracked the speed of carbon dioxide diffusion within erythrocytes and found it to be just 5% of that in water. The main cause of this restricted diffusion appears to be the high concentration of haemoglobin within erythrocytes, diffusivity being halved for every 7.5 g/dL increase in haemoglobin concentration. Due to this restriction, the diffusion distance within erythrocytes is highly significant and is reduced in disc-shaped cells compared to spheres. It transpires that the thickness of erythrocytes across a wide range of species is inversely related to the MCHC (supplementary data in Ref. [64]), suggesting that the disc shape was an evolutionary mechanism to compensate for the high MCHC in animals with increasing energy needs.

In summary, the size and shape of erythrocytes are intimately matched to the circulation. The biconcave shape of human erythrocytes facilitates cell folding, which minimises resistance to flow in small vessels and thereby reduces the pressure needed to drive the circulation. It also reduces flow turbulence which protects the circulation. Specialised microvessels in the spleen have evolved which filter and remove ageing erythrocytes based on their deformability, size and shape and recent evidence shows that the disc shape of erythrocytes is necessary to allow rapid diffusion of gases to and from the lungs and peripheral tissues. These characteristics underscore the central role that erythrocytes play in the physiology of the circulation. Recent evidence also suggests that this matching of erythrocytes to the circulation has a long evolutionary history which predates the arrival of animals living today.

Erythrocyte Size, Metabolic Activity and Genome Size

It has been known since the early 1950s that the genome size within the nuclei of animal species varies greatly [65] and since then species in which genome size varies more than 200,000 fold have been reported. Yet this enormous variation appears to have no relationship with animal complexity, which has puzzled scientists for several decades. It is well established that the vast majority of the genome is not expressed, for example, over 98% of the human genome is non-coding DNA and its purpose is unclear. In contrast to the wide variations in genome size among different species, it is much more constrained within species, for example, varying less than twofold in birds and fourfold in mammals. Genome size also correlates positively with the size of erythrocytes in vertebrates over a wide range of values [66] and this is also true for mammalian erythrocytes which have ejected the nucleus during development [67] (Figure 9.10(a)). Of special interest is an association between genome size and erythrocyte size with the intensity of metabolic activity among species. Thus, amphibians and lungfish, which have low metabolic rates, also have the largest known genome and erythrocyte sizes, whereas birds and bats, which have high metabolic rates related to flight, have small genome and erythrocyte sizes. Although birds and mammals have small erythrocytes, mature erythrocytes in birds, despite retaining the nucleus, are generally smaller than those of mammals, which eject the nucleus during development (Figure 9.10(b)). Both appear to have pursued a strategy of erythrocyte diminution. In the case of birds, this is facilitated by a small genome and nuclear size. By ejecting the nucleus, mammalian erythrocytes can adopt a biconcave disc shape, which by enhancing folding allows them to navigate vessels smaller than their diameter. Why birds should have retained their nuclei is unknown; one possibility is that nucleated erythrocytes may have a greater capacity to respond to hypoxia associated with high altitude [68].

As discussed above, relative heart size (measured as a ratio of body weight) correlates closely with the aerobic range of species (Figure 9.1(c)) making it a good indicator of metabolic activity. To investigate this further, Vinogradov and Anatskaya [69] used relative heart weight as a surrogate marker for metabolic intensity and showed that genomic size in a large number of species of reptiles, mammals and amphibians correlates negatively with metabolic intensity (Figure 9.10(b)). In other words, animals appear to have evolved a smaller genome size as their metabolic intensity increased and this may have been a necessary change to facilitate

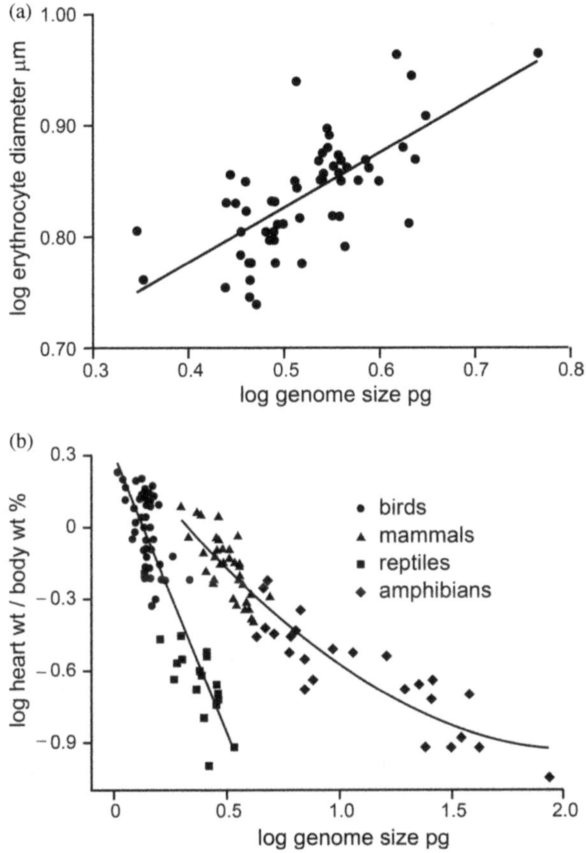

Figure 9.10: (a) Genome size and erythrocyte size among 67 species of mammals show a highly significant correlation [67]. (b) Genome size shows a striking negative correlation with heart weight to body weight ratio among birds, reptiles, mammals and amphibians. Birds and mammals which have a larger heart weight: body weight ratio also have smaller genome sizes. The correlation is steeper for birds and reptiles than mammals and amphibians, possibly related to their different evolutionary lineages [69].

a comparable reduction in erythrocyte size. Bearing in mind that erythrocyte size also correlates with minimal capillary size, this raises the intriguing possibility that reduction in genome size and erythrocyte size occurred alongside the evolution of the circulation to meet increased energy needs by facilitating a reduction in the size of capillaries and in turn to increase density of their distribution in tissues.

In order to study the evolutionary origin of changes in erythrocyte size, Huttenlocker and Farmer [70] used fossilised bones to study the microvasculature in extinct ancestors of mammals and dinosaurs.

Diminution of Capillaries and Erythrocytes in Mammals and Bird

Studies of the fossils of animals, known as palaeo-biology, can often reveal remarkable details about the biology of extinct animals. Scientists have learned a great deal about the size, shape, lifestyles, habitat and feeding habits of animals that lived millions of years before humans emerged. The study of fossilised bones has shed light on the coevolution of erythrocytes and the microcirculation. Bone, like other tissues, needs a blood supply. To allow this, bones have small passages known as canals through which even the smallest blood vessels pass. Studies of the microscopic structure of bone in animals living today show that the smallest of these canals corresponds closely with the size of capillaries and also of the erythrocytes which circulate in them [70]. With this information in mind, palaeobiologists have studied the canal structure and size in the fossil bones of several extinct forerunners of modern mammals and birds and used that information to deduce the size of their erythrocytes [70].

Mammals, birds and reptiles are derived from amniotes which divided into synapsids and sauropsids around 310–320 Mya. The synapsids would later give rise to the earliest mammals around 225 Mya and the sauropsids to modern warm blooded birds around 80 Mya. Studies of the bone structure of several extinct species that lived between 320 Mya and 100 Mya show that capillary and by inference erythrocyte size underwent a gradual diminution [70], indicating that this was a gradual and continuous process in the forerunners of mammals and birds. As discussed above, reduction in the size of capillaries allowed them to grow in denser networks which facilitated more rapid diffusion of oxygen and thereby more energetic lifestyles. Thus, it seems most likely that evolutionary pressure to increase aerobic capacity was a stimulus to develop more efficient oxygen delivery to tissues and this was achieved by the evolution of smaller and denser networks of capillaries which in turn needed smaller erythrocytes and genomes. Acting together with more powerful four-chambered hearts, these adaptations provided the energy needed for flight in birds and the emergence of high-energy-using mammals including *Homo sapiens*.

Conclusions

Increasing energy production has always been a strategic element in the evolution of animals and availability of oxygen has been a key aspect of this. Hearts evolved to distribute oxygen to tissues in body plans of increasing size. In doing so it coevolved with several partners which greatly improve its efficiency and protect it from harm. Coevolution of blood with the heart and circulation was central to this.

Haemoglobin is present in all life forms and in multiple molecular conformations suited to the needs of each. In humans, its tetrameric molecular structure is ideal for the high levels of oxygen transport needed as well as minimising the viscosity of blood. Erythrocytes provide the necessary packaging for the low MW haemoglobins present in mammals and prevents its loss from the circulation. Erythrocytes vary greatly in size and shape between species but are ideally suited to the circulations in each. In mammals, the nucleus of erythrocytes is ejected during development and in humans their biconcave shape facilitates folding which lowers the viscosity of blood in small vessels and resistance to blood flow, thereby greatly mitigating the work required of the heart. This exquisite functional matching of blood and the cardiovascular system is the result of millennia of coevolution.

References

[1] Snyder GK, Sheafor BA. Red blood cells: Centerpiece in the evolution of the vertebrate circulatory system. *American Zoologist*. 2015;39(2):189–198.

[2] Wiedemann D, Burt DM, Hill RA, Barton RA. Red clothing increases perceived dominance, aggression and anger. *Biology Letters*. 2015;11(5): 20150166.

[3] Montoya P, Campos JJ, Schandry R. See red? Turn pale? Unveiling emotions through cardiovascular and hemodynamic changes. *The Spanish Journal of Psychology*. 2005;8(1):79–85.

[4] Skelton AE, Catchpole G, Abbott JT, Bosten JM, Franklin A. Biological origins of color categorization. *Proceedings of the National Academy of Sciences*. 2017;114(21):5545.

[5] Franklin A, Pitchford N, Hart L, Davies I, Clausse S, Jennings S. Salience of primary and secondary colours in infancy. *British Journal of Developmental Psychology*. 2010;26:471–483.

[6] Sender R, Fuchs S, Milo R. Revised estimates for the number of human and bacteria cells in the body. *PLoS Biology*. 2016;14(8):e1002533.

[7] Saltin B, Astrand PO. Maximal oxygen uptake in athletes. *Journal of Applied Physiology*. 1967;23(3):353–358.

[8] De Lorenzo A, Tagliabue A, Andreoli A, Testolin G, Comelli M, Deurenberg P. Measured and predicted resting metabolic rate in Italian males and females, aged 18–59 y. *European Journal of Clinical Nutrition*. 2001;55(3):208–214.

[9] Bishop CM. The maximum oxygen consumption and aerobic scope of birds and mammals: Getting to the heart of the matter. *Proceedings. Biological Sciences*. 1999;266(1435):2275–2281.

[10] Fänge R. Hagfish blood cells and their formation. In: *The Biology of Hagfishes*. Dordrecht: Springer; 1998. p. 287–299.

[11] Schwieterman GD, Bouyoucos IA, Potgieter K, Simpfendorfer CA, Brill RW, Rummer JL. Analysing tropical elasmobranch blood samples in the field: Blood stability during storage and validation of the HemoCue® haemoglobin analyser. *Conservation Physiology*. 2019;7(1):coz081.

[12] Parida SP, Dutta SK, Pal A. Hematology and plasma biochemistry of wild-caught Indian cobra *Naja naja* (Linnaeus, 1758). *Journal of Venomous Animals and Toxins Including Tropical Diseases*. 2014;20:14.

[13] Jensen FB, Wang T, Jones DR, Brahm J. Carbon dioxide transport in alligator blood and its erythrocyte permeability to anions and water. *American Journal of Physiology*. 1998;274(3):R661–R671.

[14] Packard GC, Stiverson RK. Blood hemoglobin concentration in chorus frogs (*Pseudacris triseriata*): Relationship to body size and altitude. *The American Midland Naturalist*. 1976;96(2):482–487.

[15] Allcroft WM. Observations on the haemoglobin level of cows and sheep. *The Journal of Agricultural Science*. 1941;31(3):320–325.

[16] Silva ID, Kuruwita VY. Hematology, plasma, and serum biochemistry values in domesticated elephants (*Elephas maximus ceylonicus*) in Sri Lanka. *Journal of Zoo and Wildlife Medicine*. 1993;24(4):440–444.

[17] Fegler G. Haemoglobin concentration, haematocrit value, and sedimentation rate of horse blood. *Quarterly Journal of Experimental Physiology and Cognate Medical Sciences*. 1948;34:129–139.
[18] Cornell LH. Hematology and clinical chemistry values in the killer whale, *Orcinus orca* L. *Journal of Wildlife Diseases*. 1983;19(3):259–264.
[19] Qvist J, Hill RD, Schneider RC, Falke KJ, Liggins GC, Guppy M, *et al.* Hemoglobin concentrations and blood gas tensions of free-diving Weddell seals. *Journal of Applied Physiology (1985)*. 1986;61(4):1560–1569.
[20] Crile G, Quiring DP. A record of the body weight and certain organ and gland weights of 3,690 animals. *The Ohio Journal of Science*. 1940;40: 219–259.
[21] Dismukes GC, Klimov VV, Baranov SV, Kozlov YN, DasGupta J, Tyryshkin A. The origin of atmospheric oxygen on Earth: The innovation of oxygenic photosynthesis. *Proceedings of the National Academy of Sciences of the United States of America*. 2001;98(5):2170–2175.
[22] Koch LG, Britton SL. Aerobic metabolism underlies complexity and capacity. *Journal of Physiology*. 2008;586(1):83–95.
[23] Hedges SB, Blair JE, Venturi ML, Shoe JL. A molecular timescale of eukaryote evolution and the rise of complex multicellular life. *BMC Evolutionary Biology*. 2004;4:2.
[24] Chapelle G, Peck LS. Polar gigantism dictated by oxygen availability. *Nature*. 1999;399(6732):114–115.
[25] Graham JB, Aguilar NM, Dudley R, Gans C. Implications of the late Palaeozoic oxygen pulse for physiology and evolution. *Nature*. 1995; 375(6527):117–120.
[26] Berner RA, Petsch ST, Lake JA, Beerling DJ, Popp BN, Lane RS, *et al.* Isotope fractionation and atmospheric oxygen: Implications for phanerozoic O_2 evolution. *Science*. 2000;287(5458):1630.
[27] Han TM, Runnegar B. Megascopic eukaryotic algae from the 2.1-billion-year-old negaunee iron-formation, Michigan. *Science*. 1992;257(5067): 232–235.
[28] Acquisti C, Kleffe J, Collins S. Oxygen content of transmembrane proteins over macroevolutionary time scales. *Nature*. 2007;445(7123):47–52.
[29] Dor RK, Webber DM. Invertebrate athletes: Trade-offs between transport efficiency and power density in cephalopod evolution. *Journal of Experimental Biology*. 1991;160(1):93.

[30] Oellermann M, Strugnell JM, Lieb B, Mark FC. Positive selection in octopus haemocyanin indicates functional links to temperature adaptation. *BMC Evolutionary Biology*. 2015;15(1):133.

[31] Markl J. Evolution of molluscan hemocyanin structures. *Biochimica et Biophysica Acta*. 2013;1834(9):1840–1852.

[32] Beintema JJ, Stam WT, Hazes B, Smidt MP. Evolution of arthropod hemocyanins and insect storage proteins (hexamerins). *Molecular Biology and Evolution*. 1994;11(3):493–503.

[33] Mangum CP. Oxygen transport in invertebrates. *American Journal of Physiology*. 1985;248(5 Pt 2):R505–R514.

[34] Vanin S, Negrisolo E, Bailly X, Bubacco L, Beltramini M, Salvato B. Molecular evolution and phylogeny of sipunculan hemerythrins. *Journal of Molecular Evolution*. 2006;62(1):32–41.

[35] Mangum CP, Kondon M. The role of coelomic hemerythrin in the sipunculid worm Phascolopis gouldi. *Comparative Biochemistry and Physiology: A, Comparative Physiology*. 1975;50(4):777–785.

[36] Costa-Paiva EM, Whelan NV, Waits DS, Santos SR, Schrago CG, Halanych KM. Discovery and evolution of novel hemerythrin genes in annelid worms. *BMC Evolutionary Biology*. 2017;17(1):85.

[37] Weber RE, Vinogradov SN. Nonvertebrate hemoglobins: Functions and molecular adaptations. *Physiological Reviews*. 2001;81(2):569–628.

[38] Appleby CA, Bogusz D, Dennis ES, Peacock WJ. A role for haemoglobin in all plant roots? *Plant Cell & Environment*. 1988;11:359–367.

[39] Chiancone E, Vecchini P, Verzili D, Ascoli F, Antonini E. Dimeric and tetrameric hemoglobins from the mollusc *Scapharca inaequivalvis*. Structural and functional properties. *Journal of Molecular Biology*. 1981;152(3):577–592.

[40] Vinogradov SN, Sharma PK, Qabar AN, Wall JS, Westrick JA, Simmons JH, *et al.* A dodecamer of globin chains is the principal functional subunit of the extracellular hemoglobin of *Lumbricus terrestris*. *Journal of Biological Chemistry*. 1991;266(20):13091–13096.

[41] Kubo H. Uber das Hamoprotein aus den Wurzelknollchen von Leguminosen. *Acta Phytochimica Tokyo*. 1939;11(1):195–200.

[42] Gupta KJ, Hebelstrup KH, Mur LA, Igamberdiev AU. Plant hemoglobins: Important players at the crossroads between oxygen and nitric oxide. *FEBS Letters*. 2011;585(24):3843–3849.

[43] Rose MY, Olson JS. The kinetic mechanism of heme binding to human apo-hemoglobin. *Journal of Biological Chemistry*. 1983;258(7):4298–4303.
[44] Wood WG. Haemoglobin synthesis during human fetal development. *British Medical Bulletin*. 1976;32(3):282–287.
[45] Uda M, Galanello R, Sanna S, Lettre G, Sankaran VG, Chen W, *et al.* Genome-wide association study shows BCL11A associated with persistent fetal hemoglobin and amelioration of the phenotype of beta-thalassemia. *Proceedings of the National Academy of Sciences of the United States of America*. 2008;105(5):1620–1625.
[46] Chen J-J. Regulation of protein synthesis by the heme-regulated eIF2 kinase: Relevance to anemias. *Blood*. 2006;109(7):2693–2699.
[47] Melefors O, Goossen B, Johansson HE, Stripecke R, Gray NK, Hentze MW. Translational control of 5-aminolevulinate synthase mRNA by iron-responsive elements in erythroid cells. *Journal of Biological Chemistry*. 1993;268(8):5974–5978.
[48] O'Brian MR, Kirshbom PM, Maier RJ. Bacterial heme synthesis is required for expression of the leghemoglobin holoprotein but not the apoprotein in soybean root nodules. *Proceedings of the National Academy of Sciences of the United States of America*. 1987;84(23):8390–8393.
[49] Manning LR, Popowicz AM, Padovan JC, Chait BT, Manning JM. Gel filtration of dilute human embryonic hemoglobins reveals basis for their increased oxygen binding. *Analytical Biochemistry*. 2017;519:38–41.
[50] Bunn H. Regulation of hemoglobin function in mammals. *American Zoologist*. 1980;20:199–211.
[51] Arnone A. X-ray diffraction study of binding of 2,3-diphosphoglycerate to human deoxyhaemoglobin. *Nature*. 1972;237(5351):146–149.
[52] Gulliver G. Observations on the sizes and shapes of the red corpuscles of the blood of vertebrates, with drawings of them to a uniform scale, and extended and revised tables of measurements. *Proceedings of the Zoological Society of London*. 1875;31:474–495.
[53] Chien S, Usami S, Dellenback RJ, Bryant CA. Comparative hemorheology — Hematological implications of species differences in blood viscosity. *Biorheology*. 1971;8:35–57.
[54] Burggren WW. Form and function in reptilian circulations. *American Zoologist*. 2015;27(1):5–19.

[55] Noguchi H, Gompper G. Shape transitions of fluid vesicles and red blood cells in capillary flows. *Proceedings of the National Academy of Sciences of the United States of America*. 2005;102(40):14159–14164.

[56] Holland RA, Forster RE. The effect of size of red cells on the kinetics of their oxygen uptake. *Journal of General Physiology*. 1966;49(4):727–742.

[57] Vandegriff KD, Olson JS. Morphological and physiological factors affecting oxygen uptake and release by red blood cells. *Journal of Biological Chemistry*. 1984;259(20):12619–12627.

[58] Snyder GK. Erythrocyte evolution: The significance of the Fåhraeus–Lindqvist phenomenon. *Respiration Physiology*. 1973;19(3):271–278.

[59] Snyder GK. Blood corpuscles and blood hemoglobins: A possible example of coevolution. *Science*. 1977;195(4276):412–413.

[60] Schmidt-Nielsen K, Taylor CR. Red blood cells: Why or why not? *Science*. 1968;162(3850):274–275.

[61] Uzoigwe C. The human erythrocyte has developed the biconcave disc shape to optimise the flow properties of the blood in the large vessels. *Medical Hypotheses*. 2006;67(5):1159–1163. Proc. R. Soc. 2016; B283: 20162201.

[62] Canham PB. The minimum energy of bending as a possible explanation of the biconcave shape of the human red blood cell. *Journal of Theoretical Biology*. 1970;26(1):61–81.

[63] Pivkin IV, Peng Z, Karniadakis GE, Buffet PA, Dao M, Suresh S. Biomechanics of red blood cells in human spleen and consequences for physiology and disease. *Proceedings of the National Academy of Sciences*. 2016;113(28):7804.

[64] Richardson SL, Swietach P. Red blood cell thickness is evolutionarily constrained by slow, hemoglobin-restricted diffusion in cytoplasm. *Scientific Reports*. 2016;6:36018.

[65] Mirsky AE, Ris H. The desoxyribonucleic acid content of animal cells and its evolutionary significance. *Journal of General Physiology*. 1951;34(4):451–462.

[66] Olmo E. Nucleotype and cell size in vertebrates: A review. *Basic and Applied Histochemistry*. 1983;27(4):227–256.

[67] Gregory TR. Nucleotypic effects without nuclei: Genome size and erythrocyte size in mammals. *Genome*. 2000;43(5):895–901.

[68] Barve S, Dhondt AA, Mathur VB, Cheviron ZA. Life-history characteristics influence physiological strategies to cope with hypoxia in Himalayan birds. *Proc. R. Soc.* 2016;B283:20162201.

[69] Vinogradov A, Anatskaya O. Genome size and metabolic intensity in tetrapods: A tale of two lines. *Proceedings Biological sciences/The Royal Society*. 2006;273:27–32.

[70] Huttenlocker A, Farmer CG. Bone microvasculature tracks red blood cell size diminution in Triassic mammal and dinosaur forerunners. *Current Biology*. 2016;27:48–54.

[71] Holland HD. The oxygenation of the atmosphere and oceans. *Philosophical Transactions of the Royal Society of London: Series B, Biological Sciences*. 2006;361(1470):903–915.

[72] Reinhart WH, Huang C, Vayo M, Norwich G, Chien S, Skalak R. Folding of red blood cells in capillaries and narrow pores. *Biorheology*. 1991;28:537–549.

10

Preserving the Heart

Introduction

Undertaking the research for a book on the heart, it is impossible not to be deeply moved by the elegance of its physiology in meeting the demands placed upon it as a continuously working pump, while at the same time constantly renewing its contractile machinery. The key to its success is that it does all this without our conscious involvement; we do not control how it functions. What then if anything can we do to protect it from injury and harm? In terms of its minute-by-minute function, there is little we can do to assist it. However, taking a longer-term view there is much we can do; this relates to, on the one hand, how the heart has evolved to function optimally and, on the other, the way we live our lives and whether the demands this places on the heart are compatible with how the heart has evolved to work most efficiently.

The heart has a long evolutionary history. Primitive hearts appeared soon after the Cambrian explosion around 540 Mya (million years ago) and over the next 250 Mya, four-chambered hearts pumping closed circulations in mammals evolved (Figure 9.2). Hominids which include primitive apes appeared around 55 Mya and hominins including humans and chimpanzees (our closest relatives) diverged from them 6 Mya. Fossils of early humans such as homo habilis indicate their presence around 1.8 Mya. Homo sapiens evolved in Africa about 200,000 YA (years ago), from where they began to disperse around 180,000 YA, reaching Eurasia 60,000 YA, Europe 50,000 YA and the Americas 30,000 and 12,000 YA. Modern humans have been the most successful species in occupying the planet and in the process, displaced many other species to the point of extinction. This is due to their superior ability for abstract thought, social development and manual dexterity, features which have often led to the idea of human exceptionalism, the idea that humans are fundamentally different from other species. From a genetic perspective however, this is quite wrong. In fact, the human genome differs remarkably little compared to other hominins. Among the many genes we share with our closest relatives, chimpanzees, their DNA sequences are almost 99% identical to ours and if deletions and insertions are considered they are still 96% the same [1]. In other words, only small genetic differences have occurred between humans and chimpanzees since we both diverged from hominids 6 million years ago. This also illustrates the slow rate of evolution, when considered in the context of our experience of time. By the same token, it means that genetic changes have been negligible since the arrival of modern anatomical humans around 200,000 YA.

In contrast to the slow rate of genetic evolution, the physiological and metabolic requirements of humans living in Western countries today have changed dramatically compared to our hunter gatherer ancestors. This change began with the introduction of agriculture 10,000 YA but accelerated rapidly during the past two centuries since the industrial revolution. Given the slow rate of evolution, the genetic changes that would have occurred during that time are negligible. From a genetic perspective therefore, humans evolved to meet the challenges faced by our stone age ancestors, or as we have been described, "genetically, we are Palaeolithic preagricultural hunter gatherers" [2]. This divergence between the lifestyle for which we have evolved and that of humans living now underlies many of our modern diseases.

Of course, many aspects of industrialisation have benefited human health. The wealth created by industry has funded better housing, sanitation and health care which have greatly reduced mortality due to infections and trauma. As a result, average life expectancy today is about double that of our preagricultural ancestors. On the other hand, the lifestyles of current humans also differ from those of our forebears in ways that are detrimental to health, especially in relation to our diets and activity. We are no longer dependent on nourishment available in our immediate environments but can access foods from almost anywhere in the world and have much greater control over what we consume. In addition, we have nothing like the need for the amount of exercise undertaken by our hunter gatherer ancestors. Moreover, humans living today can easily access harmful products such as tobacco and alcohol. Collectively, these have led to an increasing prevalence of ill health due to obesity, diabetes, cancers, cardiovascular and degenerative diseases. All of these can be considered the result of a discordance between the lifestyles of humans living today and the biological behaviour for which our genome evolved.

Diets Then and Now

Estimates of the diets of palaeolithic humans have been made based on analysis of the composition of foods that would have been available to them at the time. These consisted mainly of wild game and uncultivated vegetables [2]. As hunter gatherers and in the absence of organised agriculture, consumption of animal protein must have been high and since wild animals are very lean, fat intake was likely to be low. Studies of hunter gatherers living in the last century indicate that their diets consisted of about 35% meat and 65% vegetable foods. On that basis, their energy intake from protein and fat was 34% and 21%, respectively. In contrast, Western peoples now derive 12% of their energy from protein and 42% from fat [3]. Carbohydrate intake among hunter gatherers would have varied seasonally with availability but on average was similar to modern humans. Intake of dietary fibre has important effects on health, particularly bowel cancers, and would have been 5–7 times greater by hunter gatherers than is consumed today. Apart from its content in other foods, sodium was unknown to many hunter gatherers [4] and amounted to about a quarter of that consumed by Western peoples today. Thus, our ancestors consumed more protein and fibre, but far less fat and sodium.

Physical Activity Then and Now

A hunter gatherer lifestyle necessarily involves prolonged periods of physical activity in foraging and hunting and such peoples would be expected to have high levels of fitness. This is supported by studies of the bone structure of preagricultural humans which had prominent muscular insertion sites and thick cortical bone shafts indicating large muscular frames [5]. Evidences of strength and stamina are also frequent characteristics of hunter gatherers living today. Although the lifestyles of industrialised peoples are often largely sedentary, humans are exceptionally well evolved for long distance running. For example, hunters in many cultures are known to have caught their prey by running them down over prolonged periods, often over several days to the point of exhaustion [6]. Thus, although most game animals are faster sprinters than man, few have better endurance capacity. There are several reasons for this. (i) Heat dissipation is crucial for distance running. The ability of humans to sweat in addition to being hairless offers much more efficient cooling than panting in which heat loss is limited to the respiratory tract. (ii) Respiration and oxygen consumption increase rapidly with running and upright posture is advantageous in this respect. The forelimbs of quadruped animals strike the ground with each stride and the energy of this is in part absorbed by the thorax, which undergoes some compression and expansion. As a result, running quadrupeds invariably limit respiration to one breath per stride. Bipedal animals including humans are not limited by fore limb impact when running and can adopt higher ratios of breaths per stride, from 1:1 to 4:1 [7]. This allows greater flexibility to adjust lung ventilation with running speed and explains why humans can run well over a wider range of speeds with little change in ventilatory efficiency than quadrupeds which tend to have a narrower range of running speeds [6]. Thus, our ancestors were well adapted for hunting and would have high levels of fitness as a result and this is also the case for hunter gathers living today who have greater ventilatory capacity than people living in industrialised countries [2, 8].

Tobacco and Alcohol Then and Now

Consumption of alcohol and tobacco are important factors in human health. It is quite possible that our preagricultural forebears may also have experienced the effects of alcohol as a result of natural fermentation of products such as honey or fruits; however, it is unlikely to have been widely or regularly available. Among preliterate societies studied in the last century, only about half were able to manufacture alcohol [9] and consumption of alcohol tended to be regulated and restricted to specific social events [10]. Addictive solitary drinking appears to be a feature of contact with modern industrialised societies.

Some palaeolithic peoples appear to have used tobacco although the extent of this is unknown. The spread of tobacco to Europe and beyond followed contact with the Americas where it had been in use, possibly for 5,000 years. Australian aborigines are also known to have chewed wild tobacco; however, its use among many hunter societies living now was unknown prior to contact with industrialised cultures. Originally smoked in pipes and cigars, the introduction of cigarettes led to its widespread use throughout the world in the early 20th century and also made it acceptable among women. Chewing tobacco has significant adverse effects on oral health, however inhalation of smoke which became the norm with the use of cigarettes increased the risks of tobacco manyfold. Thus, human lifestyles in rich Western nations differ greatly from those of our preagricultural ancestors. We are far more sedentary; our diets differ substantially, and we consume large quantities of harmful products. These differences account for the high rates of cardiovascular disease in humans today.

Cardiovascular Disease Then and Now

Given the shorter life span of our preagricultural ancestors, it is reasonable to ask did they live long enough to develop cardiovascular diseases. Cardiovascular disease is known to have occurred in ancient times, for example, it has been observed in Egyptian mummies. Symptoms indicative of diabetes, a major risk factor for developing it, were also described by the Egyptian physician Hesy-Ra around 1500 BCE and by the Greek physician Aretaeus in 150 CE. It is unlikely that many pre-agriculturalists would have lived long enough to sustain a high prevalence of cardiovascular disease. However even among young persons in current preliterate groups, cardiovascular disease is rarely seen [2]; in contrast, evidence of early coronary atherosclerosis has been found in 12% of Western adolescents (aged 16–20 years) and in 28% of young adults (aged 21–25 years) [11].

Diabetes Then and Now

Diabetes, one of the most important risk factors for developing cardiovascular disease, was known to have occurred in earlier times, however there is striking evidence that its prevalence has increased dramatically in recent times. Thus, between 1866 and 1923, deaths related to diabetes among those aged 45 and over increased tenfold in New York City and this trend has continued globally on a massive scale. The prevalence of diabetes around the world is projected to increase to 592 million by 2035 from 382 million in 2013 [12]. Evidence that this is related to industrialisation is indicated by the fact that increases are greatest among populations which have rapidly adopted Western lifestyles either through economic advancement or migration [2]. The prevalence of diabetes among palaeolithic humans cannot be ascertained, however among preliterate peoples studied during the 20th century diabetes was rare or uncommon, varying from none in nomadic pastoralists to 1.9% among Alaskan Eskimos. This contrasts with a global average of 4.0% in 1995, which is projected to increase to 5.4% by 2025 [13].

Obesity Then and Now

Obesity is also a major risk factor for many illnesses including diabetes, cardiovascular disease, many cancers and arthritis. We cannot accurately know its prevalence among stone age peoples, however studies of non-industrialised societies in the 20th century have found significantly lower body fat and body weight than in industrialised nations [2, 14]. Skin fold thickness (the thickness of a gently pinched skin fold) is a useful measure of adiposity and ranged from 4 to 6.3 mm among 15 groups of hunter gatherers and simple farmers compared with 9–11.2 mm in industrialised Westerners. Indeed, obesity is one of the most pressing health problems in the world today. Globally, obesity has more than doubled since 1975. In 2016, more than 1.9 billion adults were overweight and over 650 million were obese [15]. Industrialised lifestyles appear to be a primary reason for this growing epidemic. Evidence for this comes from the fact that the greatest increases have occurred in countries where wealth and income have increased most rapidly, The Middle East, Pacific Islands, Southeast Asia and China.

Studies of the lifestyle of hunter gathers living now not only confirm they have lower body weights than those in industrialised countries, but also shed some light on the reasons for this. For example, the Hazda, a group of hunter gathers living in northern Tanzania, have lower body weights, body mass index and body fat compared with Western populations [14]. Possible explanations for these differences include greater energy intake and lower energy expenditure by Westerners. Although hunter gatherers such as the Hazda are more physically active than Western peoples, nevertheless their total energy expenditure is similar [14]. This is important because it suggests that dietary habits and total energy intake in particular are most important. Western diets are richer in sugar and have greater energy per volume than traditional ones based on analysis of foods consumed [3]. Comparison of people consuming diets with high or low energy density over a five-day period showed that eating times to reach a feeling of fullness were longer with diets low in energy density. In addition, the total energy intake among people who consumed diets low in energy density was about half that observed in those on high energy diets [16]. As a result, after eating to a feeling of fullness, hunter gathers consume fewer calories. Furthermore, pre-agriculturalists drank water when thirsty, whereas most drinks consumed now contain significant amounts of energy. For example, a can of soda may contain 139 calories or 7% of an adult's daily requirement and some high energy cups of coffee as much as 400 calories or 20% of a day's needs. Thus, dietary habits in the industrial world are markedly different from those of our ancestors. This is largely due to the higher energy content of manufactured foods and is a major factor in the growing epidemic of obesity.

Blood Pressure Then and Now

Hypertension is now recognised as the greatest preventable risk for cardiovascular disease and deaths from any cause in the world. However, this was poorly understood until the middle of the 20th century. In 2010, 31.1% of the world's adult population had hypertension (systolic pressure of ≥140 mmHg and or diastolic blood of ≥90 mmHg) [17, 18] and in 2015 an estimated 7.8 million deaths worldwide were associated with hypertension [19]. The prevalence of hypertension is also increasing globally; however, the increase is greatest in lower- and middle-income countries, reflecting increased exposure to Western lifestyles coupled with greater awareness and treatment in high-income countries [20]. In addition, hypertension generally increases with age which contributes to the rise in prevalence as populations live longer.

Lifestyle habits are also important contributing factors to the prevalence of hypertension, especially exposure to high salt diets and low levels of physical activity. In contrast to the industrialised world, both hypertension and cardiovascular disease are rare or uncommon in hunter gatherer societies [21]. In a study of 1,489 primitive horticulturists living in the western highlands of New Guinea, just 3% of males were hypertensive [22] compared to around 30% globally. It should be noted that in this early study (published in 1976), the threshold for hypertension (>160 mmHg) was higher than in more recent studies; nevertheless, the prevalence is much lower than in Western countries or globally. Thus, industrialised nations appear especially vulnerable to risk factors associated with cardiovascular disease compared to preliterate societies. Hunter gatherers and simple farmers are more physically active, fitter, leaner, and have lower body weights than people living in high-income countries today. They are also much less likely to have hypertension or diabetes. Not surprisingly, this lower prevalence of risk factors is associated with lower rate of cardiovascular disease [2].

Replication of the lives of our pre-industrialised forebears is not an attractive or realistic proposition for Western societies today. The advantages that have accrued due to increased wealth in terms of health and longevity are substantial and not to be discarded. However, access to higher incomes has also been associated with some harmful effects on health. In particular, patterns of food consumption which initially might have seemed beneficial have turned out to be unhealthy. Until the 1980s, the global nutritional challenge was almost exclusively the lack of adequate food in many parts of the world, particularly in middle- and low-income countries. However, during the past three decades, obesity has emerged as a major challenge and is now an equal or greater threat to health in many regions; in 2004 the World Health Organisation estimated that 29% and 9% of children in the world had stunted growth and wasting, respectively, while 42% and 12% were overweight and obese, respectively [23]. Undernutrition and overnutrition are both important risk factors for ill health including death. In 2004, there were 2.2 million deaths attributable to being underweight compared to 2.8 million attributable to being overweight [23]. Thus, obesity is now recognised as a major risk factor for cardiovascular disease as well as the most important cause of diabetes, which itself carries a high risk of cardiovascular disease.

Impact of Lifestyle Changes on Longevity

Humans living today can expect to live about twice as long as our preagricultural forebears. This is largely due to lower infant mortality rates, better control of infections and reduced trauma. As discussed in Chapter 6, the causes of mortality and morbidity of modern humans have changed markedly during the past half century. Between 1990 and 2010, deaths due to all infections [24] declined globally by 32%, except for HIV which increased by 50% during this period, falling later by 35% between 2010 and 2017 as treatments were developed and distributed [25]. During the same period, the number of deaths from cardiac and circulatory diseases increased from 11.9 million to 15.6 million (31%) and this includes a 35% increase in deaths from coronary artery disease [24].

Thus, on a global scale cardiovascular disease and cancers have increased substantially in recent times. Geographical variations in the distribution of mortality causes support the concept that this is due to changes in lifestyle. While cardiovascular diseases rank as the first or second most frequent cause of death in most high-income countries, this falls to 16th or 17th in central, eastern and western sub-Saharan Africa [25]. Cancers accounted for about 15% of all deaths globally in 2010 but also showed a marked geographical variation, being more frequent in high-income countries. This was especially marked for cancers of the colon and rectum, which ranked 4th–10th in high-income countries compared with 56th–63rd in central, eastern and western sub-Saharan Africa.

Evidence that lifestyles have important effects on health also comes from changes in the rates of cardiovascular disease associated with migration between cultures. Risk factors associated with cardiovascular disease are known to change following national [26, 27] and international [28] migration. One of the best examples of this is Japanese migration to the USA. Japan has one of the lowest rates of cardiovascular deaths in the world and the USA one of the highest, but Japanese migrants to the USA experience an increase in cardiovascular mortality [28] within the span of just one generation. Thus, some aspects of modern industrialised lifestyles are poorly suited to preserving cardiovascular health and this appears to reflect a divergence from the physiological and metabolic demands for which we are optimally evolved. Cardiovascular disease is now the leading cause of death in the world and in countries where the prevalence is high, almost one third could be prevented by adopting healthier lifestyles [30]. The most important lifestyle factors that affect the risk of cardiovascular disease include dietary habits, levels of physical activity, tobacco use, body weight and blood pressure.

Learning from the Past

Although we would not wish to revert to being hunter gatherers, the absence among them of the diseases which now cause most ill health in the world is a good reason to consider how our lifestyles differ from theirs and whether we can learn from this. Cardiovascular disease carries a high burden in the world today and yet an estimated 70% of these deaths can be attributed to risk factors that we could change [29]. Some of these risks are biological, while others relate to our behaviour and lifestyles. Of the former, high blood pressure, high levels of cholesterol and diabetes account for around 35% of cardiovascular disease globally, while tobacco use, poor diet and physical inactivity account for about 16%. These numbers refer to deaths within populations and the size of the effect depends on the impact of the risk on individuals and its prevalence in the population. So, a risk with a lower individual impact can have a bigger effect on the population if its prevalence is high. High blood pressure carries the greatest effect on cardiovascular disease for the world's population. However, tobacco has an even greater adverse effect on individuals who smoke but because high blood pressure is more frequent its impact on the population is greater. It should also be stressed that some of these biological risks are affected by our behaviours. For example, hypertension, obesity and diabetes are all affected by what we eat, our levels of physical activity and alcohol consumption. Therefore, the lifestyles we chose are important for the health of our hearts. The genome we inherited and share with our ancestors evolved to meet the challenges they faced rather than those of us living in the industrialised 21st century. Can we learn from them how to live in ways that would better match it and protect our hearts? In the following, aspects of our behaviours that have important health benefits are discussed.

What and How We Eat

For our hunter gatherer ancestors, access to adequate food was the primary preoccupation of their daily lives and this remains true for many living in low-income countries today. Much of the work undertaken by men and women was directed to foraging and hunting for food or preparing their equipment for that purpose. Food was consumed soon after it was obtained or as part of family or community events. The idea of food as a commodity that is constantly available would have been impossible. In contrast, in Western countries and increasingly in middle- and lower-income countries food is always accessible. The composition of food was also different from that of most Western diets today. The crucial question then is should humans living in the 21st century attempt to match the diets of our stone age ancestors; would modification of our diets and eating habits to match theirs more closely be helpful? What changes could we make and is there evidence to indicate whether or not they would be beneficial?

Interest in all aspects of food has increased to an extraordinary extent in recent decades. Numerous books and magazines are devoted to cooking and recipes. We have daily radio and TV programs devoted to it and baking and cooking competitions are highly popular. Specialist diets based on the inclusion or exclusion of specific dietary elements receive a lot of attention. It is beyond the scope of this chapter to discuss these in any detail. It is entirely possible to have a healthy diet while avoiding meat and or dairy foods although adherents to them are well advised to ensure that nutrients they contain are adequately replaced using alternatives. Here, I propose to focus on the standard components of human food, their relative contribution to our diets and the ways in which food is prepared. In addition, for the purpose of this discussion, I will assume that humans are omnivores as this was how our ancestors evolved.

Dietary Modification to Reduce Cardiovascular Disease

Excess energy intake, meaning consumption of more calories than are expended, over a prolonged period inevitably leads to overweight and obesity. Restricting calorie consumption has undoubted benefits in terms of glucose metabolism, weight reduction and obesity [31]. However, achieving weight loss through dieting has proven to be very difficult for most people and many who are initially successful have difficulty in sustaining it. It is for this reason that recommended dietary modifications have moved away from simple calorie restriction in favour of adjusting the composition and quality of diets. The best evidence available comes from observational studies. These are studies in which the health outcomes of groups who adhere to one dietary regime are compared to others who do not.

Mediterranean diet

Diets consumed in parts of southern Europe and particularly around Crete and Greece tend to be rich in fruits, vegetables, whole grains and olive oil, with a regular consumption of seafood. In a study of seven countries that included North and South Europe, the United States and Japan, populations who live in those areas whose diets are of the Mediterranean type tended to have lower mortality rates from cardiovascular disease [32]. This led to great interest in the composition of these diets and whether they could help prevent cardiovascular disease in other regions. The main features of the Mediterranean diet are:

- Daily consumption of fruits, vegetables, grains, nuts and seeds, which make up the majority of food;
- Fat is mostly provided by olive oil, which may account for up to 40% of calories per day;
- Daily portions of fish, poultry or eggs and small portions of cheese or yogurt;
- Occasional consumption of red meat; and
- Modest amounts of wine are taken with meals.

Based on these early studies and research related to the metabolic effects of food components, a number of dietary recommendations have been proposed and investigated for possible effects on cardiovascular disease. These include:

- Healthy Eating Index (HEI): The Healthy Eating Index is a measure of diet quality based on a summation of 10 components of a diet and how much they contribute to daily intake [33]. Originally published in 1995, it was intended as a measure of diet quality, providing a score which could be used by researchers investigating the dietary habits of population groups. Details of HEI are summarised in Table 10.1.

- Alternate Healthy Eating Index (AHEI): The Alternate Healthy Eating Index (AHEI) is a modification of HEI developed in 2002 with the view to extending dietary benefits to include a range of chronic diseases as well as cardiovascular diseases [34]. AHEI includes foods believed to affect the risk of developing chronic diseases and is based on total intake of food components rather than their relative contribution. A summary of AHEI is shown in Table 10.2.
- Dietary Approaches to Stop Hypertension (DASH): DASH is a pattern of diet aimed at reducing levels of hypertension and reducing the risks of cardiovascular disease. DASH is based on an increased

Table 10.1: Healthy Eating Index.

Food group	**Scoring**	**Perfect score (10)**	**Minimum (0)**
1. Grains	1–10	6–11 servings*	0 servings
2. Vegetables	1–10	3–5 servings*	0 servings
3. Fruits	1–10	2–4 servings*	0 servings
4. Milk	1–10	2–3 servings*	0 servings
5. Meat*	1–10	2–3 servings*	0 servings
Diet guidelines			
6. Total fat	1–10	Fat energy <30%	Fat energy >45%
7. Saturated fat	1–10	Saturated fat energy <10%	Saturated fat energy >15%
8. Cholesterol	1–10	<300 mg	>450 mg
9. Sodium	1–10	<2,400 mg	>4,800 mg
10. Variety	1–10	16 different food items in 3 days	6 or less different food items in 3 days

Notes: *Depending on the recommended energy intake.
**For HEI, meat servings included legumes, and soya products.
The Healthy Eating Index provides a scoring system for researchers to measure diet quality. For each dietary element, the score ranges from 1 to 10. The number of servings of each food type to achieve the maximum and minimum scores is shown in the two right hand columns, i.e., 6–11 servings of grain containing foods scores 10 while none consumed scores 0.

Table 10.2: Alternative Healthy Eating Index.

Diet component	Minimum score	Maximum score
Vegetables, servings per day	0	≥5
Fruit, servings per day	0	≥4
Whole grains, grams per day		
Women	0	75
Men	0	90
Sugar-sweetened beverages and fruit juice, servings per day	≥1	0
Nuts and legumes, servings per day	0	≥1
Red/processed meat, servings per day	≥1.5	0
Trans Fat, percent of energy	≥4	≤0.5
Long-chain (n – 3) fats, mg per day	0	250
PUFA, percent of energy	≤2	≥10
Sodium, mg per day	Top decile	Bottom decile
Alcohol, drinks per day		
Women	≥2.5	0.5–1.5
Men	≥3.5	0.5–2.0
Total	0	110

Note: The Alternative Healthy Eating Index provides a scoring system for researchers to measure diet quality. For each dietary element, the minimum score is 0 and the maximum 10. The number of servings of each food to achieve the minimum and maximum scores is shown in the two right hand columns, i.e., no vegetables eaten scores 0 and 5 or more scores 10.

consumption of fruits, vegetables, grains, nuts, seeds, fish and poultry and a reduction in sweets and snacks. Short term clinical trials have demonstrated a significant lowering of blood pressure among adherents to DASH [35], which appears to be greater in individuals with higher levels of blood pressure, older subjects and those with a higher intake of sodium [35, 36]. Table 10.3 shows a summary of DASH recommendations related to total daily calorie intake.

A striking feature of all these dietary patterns is that they all contain an increased consumption of fruits, vegetables, legumes, fish and poultry and less contribution of red meats and saturated fats.

Table 10.3: Dietary Approaches to Stop Hypertension (DASH).

Food groups	1,200 calories/day	1,400 calories/day	1,600 calories/day	1,800 calories/day	2,000 calories/day	2,600 calories/day	3,100 calories/day	Serving sizes
Grains	4–5	5–6	6	6	6–8	10–11	12–13	1 slice of bread 1 oz dry cereal ½ cup cooked rice, pasta or cereal
Vegetables	3–4	3–4	3–4	4–5	4–5	5–6	6	1 cup of raw leafy vegetable ½ cup cut up or raw vegetables ½ cup vegetable juice
Fruits	3–4	4	4	4–5	4–5	5–6	6	1 medium fruit ¼ cup dried fruit ½ cup fresh, frozen or canned fruit ½ cup fruit juice
Low fat dairy	2–3	2–3	2–3	2–3	2–3	3	3–4	1 cup milk or yogurt 1½ oz cheese
Lean meat, poultry, fish	3 or less	3–4 or less	3–4 or less	6 or less	6 or less	6 or less	6–9	1 oz cooked meats, poultry or fish 1 egg
Nuts, seeds, legumes	3 per week	3–4 per week	3–4 per week	4 per week	4–5 per week	1	1	cup or 1½ oz nuts 2 Tbsp peanut butter 2 Tbsp or ½ oz seeds ½ cup cooked legumes

(Continued)

Table 10.3: (*Continued*)

Food groups	**1,200 calo-ries/day**	**1,400 calo-ries/day**	**1,600 calo-ries/day**	**1,800 calo-ries/day**	**2,000 calo-ries/day**	**2,600 calo-ries/day**	**3,100 calo-ries/day**	**Serving sizes**
Fat and oil	1	1	2	2–3	2–3	3	4	1 tsp soft margarine 1 tsp vegetable oil 1 Tbsp mayonnaise 2 Tbsp salad dressing
Sweets and added sugar	3 or less per week	3 or less per week	3 or less per week	5 or less per week	5 or less per week	< 2	< 2	1 Tbsp sugar 1 Tbsp jelly or jam ½ cup sorbet, gelatin dessert 1 cup lemonade
Maximum sodium	2,300 mg/day	2,300 mg/day	2,300 mg/day	2,300 mg/day	2,300 mg/day	2,300 mg/day	2,300 mg/day	

Note: Dietary Approaches to Stop Hypertension (DASH) diet is intended to provide guidance related to the total calorie intake per day. The table shows recommended servings of each food type for recommended energy intakes from 1,200 to 3,100 calories per day. Table values show servings per day unless otherwise stated.

Individual and Population Dietary Challenges

The importance of dietary habits for health is most easily understood at a population level where deaths and disability can be measured and presented. In 2013, for example, over 12 million deaths globally were attributable to poor dietary habits (low intake of fruits, vegetables, whole grains, nuts and seeds and high intake of sodium) [37]. This was the third most common risk factor for death globally and the first or second in many high-income countries. Governments and bodies such as the World Health Organisation and the World Bank increasingly recognise improving diets as a major factor for promoting health [38, 39].

Eating patterns are learned in childhood and throughout our lives. As well as being essential for life, they form an important element of how families and societies function and consequently are influenced by many factors such as taste preferences, eating patterns with family and friends, concerns about body image, media advertising, the cost of foods, education level and income. Therefore, changing less healthy eating habits which have been learned over a lifetime is not simple, and the challenge is much greater for some than others. In general, more healthy foods cost more and take longer to access and prepare than some industrially produced "fast foods". Families working long hours on low incomes will find it harder to comply with dietary health guidelines. Education is also important because it tends to be associated with higher incomes and more opportunities to access knowledge about the importance of diet quality. For example, students, staff, alumni and friends of universities are more likely to have access to simple and effective information about healthy diets than low skilled workers on zero hours contracts [40]. Given the enormous burden of ill health and deaths due to suboptimal diets, it follows that the potential to improve health by improving foods consumed is also great. New scientific evidence about healthy eating is just one element of the many reasons for our food choices but at an individual level may have little impact [41]. For this reason, efforts to promote healthy eating have shifted towards communities, nations and populations in order to alter the environment in which we access and consume foods. Changing the eating habits of children in school and at home, creating conditions where healthy foods can be more easily accessed and encouraging the production and distribution of more healthy foods may offer a better prospect for improving dietary health. Methods for which some effectiveness has been demonstrated [38] include:

- education and media campaigns to promote consumption of healthy foods and/or reduce consumption of less healthy foods and drinks;
- requiring foods offered for sale to display nutritional facts to influence producers to formulate more healthy products;
- subsidies and tax incentives to encourage the production transport and marketing of more healthy foods over sustained periods;
- school interventions to encourage nutritional education as part of programs to improve physical activity and diet. Provision of free fruits and vegetables as part of normal school activity;
- increased workplace availability of healthy food and beverage options and high nutrition standards for foods served;
- improving access to local supermarkets;
- restricting marketing of less healthy foods to children on TV and near schools. Setting nutrition standards for all foods marketed to children;
- policies to reduce specific nutrients in foods. Examples could be salt and certain fats, especially trans fats (fats produced by treating vegetable oils to form solid products like margarine); and
- regulation of agriculture to improve production, storage and distribution of healthy foods of proven value, for example, fruits and vegetables.

Collectively these measures make it easier for everyone to access and consume more healthy foods. Ultimately, however, our health is affected by what we eat as individuals. One reason for writing this book is that by showing how exquisitely the heart functions, we might value it more and be motivated as individuals to protect its health.

As well as essential for life, food should be enjoyed. We all have taste preferences and the social circumstances in which we eat are also very important. So dietary changes will work best and be more sustainable if they are pleasurable as well as healthy. It may take a little time to explore what suits us best, but it is a worthwhile endeavour. For this reason, a flexible approach that allows choices which suit us is far better than focusing

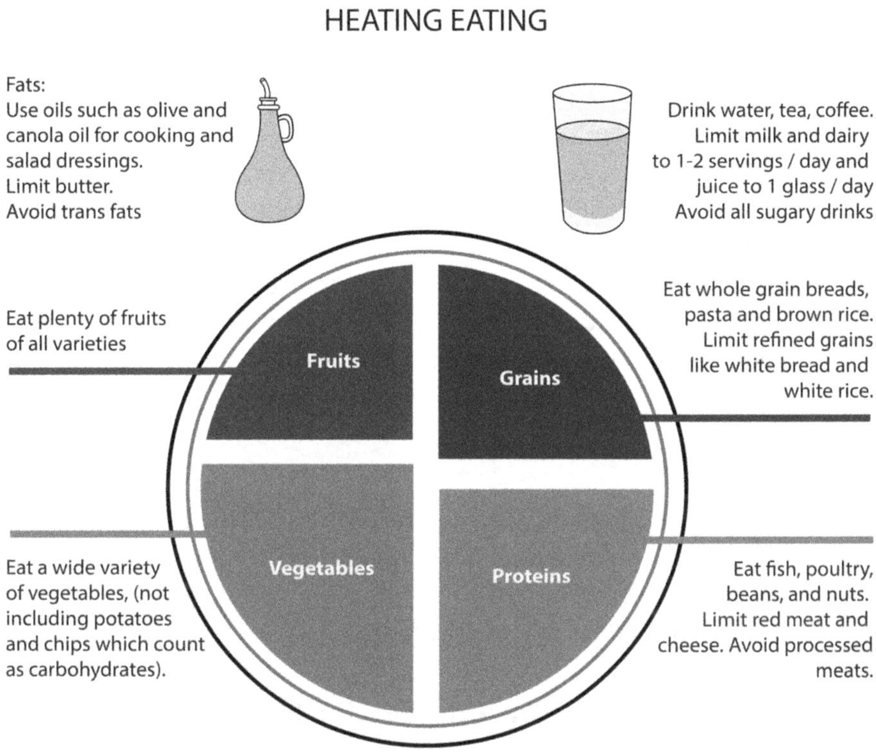

Figure 10.1: Illustration of food choices for healthy eating based on current guidelines. The focus here is to provide information to help prioritise healthy food choices in order to promote better eating habits that offer a wide variety of foods. Adopting a healthier pattern of eating that is also enjoyable will be more sustainable.

on specific nutrients. An approach based on dietary guidelines discussed above is illustrated in Figure 10.1 and can be summarised as follows:

- eat more fruits of all varieties;
- eat a wide variety of vegetables, excluding potatoes which are best counted among carbohydrates;
- eat whole grain bread, pasta and brown rice where possible rather than white bread and rice;
- increase protein intake from fish, poultry, beans and nuts and limit red meat and processed meats;

- drink water, tea or coffee. Limit dairy to 1–2 servings/day and juice to 1 glass/day. Avoid all sugary drinks;
- use healthy oils such as olive oil or canola oil for cooking and salad dressings. Limit butter and avoid trans fats;
- reduce salt intake. Unprocessed foods like fruits and vegetables are generally low in salt and most of our salt intake comes from industrially processed foods such as breads, cold cuts, cured meats, pizza, soups, sandwiches, cheese, egg dishes and packaged savoury snacks. Prepared sauces, salad dressings and gravies are also usually high in sodium. Salt intake can easily be limited by reducing the portion size and number of processed foods and relying more on home cooked meals.

The options here contain a myriad of possibilities that should allow a truly exciting dietary adventure. A healthy diet does not have to be boring or unenjoyable. Indeed, we may look back and think how limited and uninteresting our diet has been in the past.

Physical Activity and Cardiovascular Health

Interest in the effects of physical activity on heart health first arose in the 1950s when it was discovered that people whose work required physical activity had lower rates and less severe coronary heart disease than sedentary workers. A study of London transport workers in 1953 showed that the drivers of double decker buses suffered more heart attacks than their conductor colleagues and similarly, telephonists were more likely to develop coronary heart disease and die from heart attacks than postmen [42]. This prompted a much larger study of the presence of coronary heart disease among 3,800 middle aged men (45–70 years) who had died from non-coronary causes and had post-mortem examinations. Although heart disease was not the immediate cause of death, post-mortem evidence of heart damage due to coronary disease was present in 10% of them. However it was substantially less among those whose work involved heavy physical activity (6.8%), compared to those who did light physical work (13.4%) [43]. Thus, jobs involving higher levels of physical activity appear to offer protection from coronary heart disease.

Evidence that physical inactivity is associated with an increased risk of premature death is now beyond doubt and reflects increased occurrence of coronary heart disease, strokes, diabetes as well as osteoporosis and bone fractures [23, 44]. The risk of developing heart disease is estimated to be 50% higher in people who are physically inactive compared to those who exercise regularly and about 20% of coronary heart disease present in populations of high-income countries is attributable to physical inactivity [44]. There is also strong evidence that increasing physical activity helps to prevent the development of coronary heart disease and to slow its progression in those who have it [45]. Furthermore, deaths from all causes are lower in people who are more physically active [46]. Consequently, illness associated with low levels of physical activity represents a substantial burden on societies and this has prompted several nations and organisations to develop guidelines for being physically active. The most recent recommendations [47–50] represent a significant evolution since the 1980s as the result of new research findings.

Physical Activity Guidelines

Physical activity guidelines originated as training programs for rehabilitation of patients with coronary heart disease in the 1980s. Prior to that, patients who had heart attacks had been treated with bed rest in the hope that it would allow the heart to recover more fully. However studies of exercise training in such patients showed that heart muscle recovery was improved by earlier mobilisation and training [51]. Subsequently, guidelines aimed at preventing the development of heart disease were developed. These initially focused on continuous aerobic exercise but later evolved to include moderate physical activity such as walking as part of everyday living [52]. Research over the past two decades has led to a further refinement of current guidelines.

Any Is Good, the More the Better

In the 1980s, advice was based on 15–60 min periods of vigorous physical activity accumulating more than 60 min per week. Since then, evidence

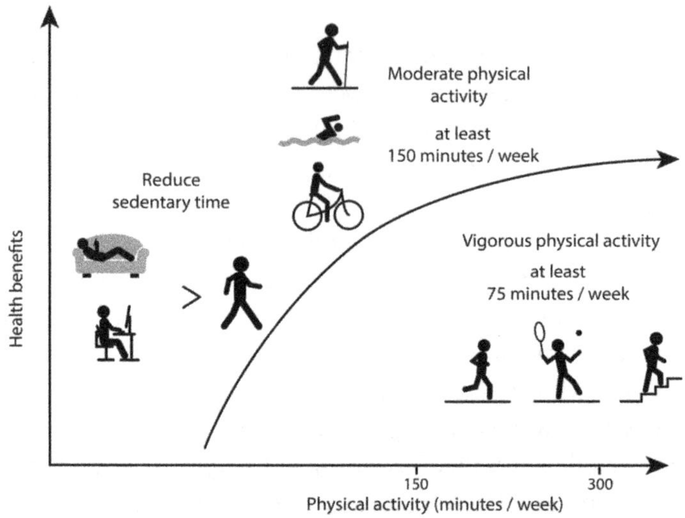

Figure 10.2: Illustration of the health benefits of being physically active. Research since the 1950s has clearly demonstrated a positive correlation between physical activity and health outcomes. However, studies during the past two decades have also shown that this relationship is curvilinear, meaning that the benefit is not the same over the whole range of activities. As illustrated here, the benefit is greatest (where the graph is steepest) for those who increase their activity from a largely sedentary lifestyle. This has led to a change in recent recommendations with an emphasis on reducing sedentary time and recognition that activity of any duration is beneficial; the more we do the better. Ideally adults should undertake at least 150 min of moderate physical activity or 75 min of vigorous physical activity per week or a combination of both.

emerged that the health benefits of exercise show a curvilinear relationship [44]. This means that the benefit of increasing physical activity depends on the baseline from where we start and it turns out that those who are least active derive the greatest benefit by increasing their physical activity (Figure 10.2). In addition, research showed that prolonged periods of sedentary behaviour are detrimental to health [53]. This has led to a change in current guidelines to emphasise that any physical activity is good, not just fixed periods of aerobic exercise. In the 1980s, the advice was to undertake vigorous physical activity (VPA) such as running or playing tennis for at least 60 min each week in bouts of 15 min or more and by 2000 the value of moderate physical activity (MPA), such as walking, was also recognised. Guidelines published in 2010 recommended either accumulating 150 min of MPA per week or 75 min of VPA or a mixture of the two [54]. The emphasis in current guidelines has shifted once again by adding a focus on reducing sedentary time and incorporating physical activity as part of everyday life. The new motto is "any is good and the more the better" instead of requiring fixed minimum periods of activity [47–49]. In practice, this means replacing sedentary time with light physical activity (LPA) such as housework or MPA in bouts of any length. For added benefit, adults should undertake at least 150 min of MPA or 75 min of VPA or an equivalent combination of the two. In addition, muscle strengthening and balancing exercises are recommended two days each week.

Physical Activity for Life

The guidelines discussed above refer to adults, however physical activity is beneficial for health throughout life and specific guidance for all age groups is widely available [47–49]. Children of all ages need to be physically active to grow normally and low levels of activity are an important link to obesity. For example, children under 5 years should be active for at least 180 min each day and 5–18-year-olds should spend at least 60 min exercising per day. Physical activity is also recommended for older people in order not just to protect the heart but also to maintain good balance and maintain muscle and bone strength. Here again the emphasis is on reducing sedentary time with the aim of building up to 150 min of MPA and for those already physically active including some vigorous physical activity. For older subjects, exercises which improve balance and muscle strength (such as dancing or playing bowls) are also important for reducing fragility and the risk of falls.

Tobacco and the Heart

Suspicions about the harmful effects of tobacco emerged in the late 1940s. The background to this was an increase in deaths from lung cancer during the first half of the 20th century and several studies suggested that this was linked to tobacco smoking. In the UK between 1922 and 1947, recorded deaths from lung cancer increased from 612 to 9,287, an astonishing fifteenfold increase. Prompted by this, Doll and Hill hypothesised that smoking, which had also increased in the preceding decades, was causally related to lung cancer and used modern statistical analysis to investigate it [55]. Their landmark study demonstrated a strong link between lung cancer, smoking and the amount of tobacco consumed. They also showed a strong temporal association between the increase in smoking from 1900 to 1950 and the rise in lung cancer deaths. It is now well established that smoking increases the risk of cancer of multiple organs [56].

Around this time too, there was increasing concern that tobacco smoking might be linked to heart disease. In 1964, the US Surgeon General's Report raised the possibility of a link between tobacco and heart disease [57]. A detailed statistical analysis of smoking habits and deaths among 34,440 UK doctors between 1951 and 1971 demonstrated an unequivocal association between deaths from coronary heart disease and cigarette smoking. There were around 60% more such deaths among smokers than non-smokers and the risk of death was positively associated with the amount of tobacco smoked [58]. This has since been confirmed in many studies [59] and a causal link is well established.

The Health Burden of Tobacco

As I write, 1 year after the outbreak of the COVID-19 pandemic, 2.5 million deaths have occurred worldwide. Yet tobacco kills more than 8 million people every year, around 7 million directly and 1.2 million non-smokers exposed to second-hand smoke. Currently about 1.3 billion of the world's population smoke and over 80% live in low-income and middle-income countries [60], where it diverts funds from other much needed areas including health care. Tobacco causes many diseases; in high-income countries (where the prevalence of smoking has declined in recent years), it still accounts on average for 20% of all deaths for men and 8% for women. Among those dying in middle age (35–69 years), tobacco smoke accounts for even more, 30% for men and 14% for women and each death results in about 22 years of life lost. In the case of lung cancer however, the cause can be attributed to tobacco in over 90% of cases in men and 70% of women. Vascular diseases, which includes coronary heart disease, are caused by many factors including tobacco smoke. The latter accounts for around 18% of vascular deaths in men and 5.3% in women of all ages but 30% of men and 9% of women dying aged 35–69 years [61]. This means that smoking is especially important as a cause of premature deaths and is the single most important reversible cause of vascular deaths. It should also be noted that six times more people die from vascular diseases than lung cancer.

The prevalence of smoking varies greatly throughout the world [62]. For example, in 2010 around 60% of Russian men and 54% of Chinese men smoked compared with around 20% in India and Brazil. In many high-income countries, smoking has declined. In the USA, the UK, Canada and France, smoking rates have halved to around 20% in recent decades but have risen sharply in many others, particularly in low- and middle-income countries. While cigarette consumption fell by 50% in the USA between the 1960s and 2010, it increased in Indonesia and China three and fourfold, respectively, during the same period. The impact of changes in smoking rates also translates into health effects. Thus, smoking related deaths have also fallen in countries where smoking prevalence has declined. In the UK, around 35% of all deaths were attributed to smoking in the late 1960s compared to just over 20% in 2000 [61] and a similar pattern occurred in the USA. It is beyond doubt that smoking tobacco is a major health hazard and is justifiably described as the greatest man-made epidemic.

Recollections of an Ex-smoker

It is not immediately obvious that small pieces of finely chopped leaves wrapped in paper could be so harmful, and this coupled with its powerful addictive properties makes tobacco exceptionally dangerous. As an ex-smoker, I would argue those who have never smoked cannot appreciate the strength of tobacco addiction. Being told that something is addictive gives little idea of what it is like to experience it. Consequently, young people are especially vulnerable to be drawn into this health catastrophe. And of course, the success of tobacco as a profitable "innovation" is due to its ability to "hook" users for life. In my own case, I started to smoke at the age of 14 and continued through school and my early years of medical school. I succeeded in stopping at the age of 20, by which time I was a seasoned user of 20 cigarettes per day. In my case, a number of factors coincided which made it possible for me to quit.

- The insight I gained from studying human physiology and pathology was important in that I learned how lung function depended on maintaining the delicate anatomy of the airways and alveoli. The idea of inhaling tar containing smoke into these exquisite structures became repellent.
- I was shocked by the effect of smoking on the lungs of long-term users at post-mortem examinations, reinforcing the adverse health effects.
- A sense that I was damaging something of great value, that had been gifted to me, became troubling.
- As a result, the idea of being free from all the necessary activity and "clean" of the habit became strongly appealing.

All of these feelings came together and developed into the strong motivation to quit. One evening after returning to my apartment, I realised I had no cigarettes and would have to go out to get some. Somehow this suddenly seemed to epitomise my dependency and added a sense of resentment. I also remember a sense of dread and alarm at what withdrawal would be like. But I really wanted to give it up. So, I went out and bought a pack of my usual brand and put it in a cupboard as a kind of emergency in case the craving really got too much. It may seem strange but knowing that eased my alarm. Nevertheless, I managed to keep my resolve in place and persisted in quitting. One further method that I found useful was running as fast and far as I could. By doing this, I found that my immediate craving to smoke was reduced and knowing that I could do this was helpful and reassuring during the immediate period after quitting.

I should also stress that giving up tobacco is not a quick fix. It needs to be a long-term project, a lifelong commitment. In the medium-term, the craving to smoke recedes but that too has its dangers. As the need to smoke eases, there is the risk of falling into the "I know I can do this" mind set which can undermine resolve and the entire project. And then there are the social occasions where smoking is common and the idea of "why not just for this evening" arises. These can be very difficult because as time passes and craving reduces but the need to maintain a defence against it remains. At this point, suddenly faced with a sudden unexpected full-on craving in a relaxing social occasion can be extremely challenging and for many ex-smokers is one of the most vulnerable times. In my case, this was the setting of a recurring nightmare for many years, each time I would wake with a terrible sense of loss at having smoked again. I found the best way to combat this was to regard each day free of tobacco as a significant achievement, something worthwhile and to be valued and not easily lost. This helped me to resist relapsing. I suspect that many if not most ex-smokers will recall similar experiences on quitting. After more than 50 years, my recollections remain intense and detailed, reflecting the power of tobacco addiction, something never to be underestimated.

Despite the difficulties, many people have succeeded in quitting tobacco particularly in high-income countries. In the USA in 1965, 20% of men reported to be ex-smokers and this had risen to 30% in 1980, comparable figures for women were 8% and 15% [59]. In 2008–2010, 26%, 24% and 25% of the adult populations of Canada, Australia and the UK also reported to be former smokers. The comparable figures for some low- and middle-income countries are much lower, for example, 5%, 6%, 3% and 11% in China, Bangladesh, India and the Philippines [62].

Working Towards a Tobacco-free World

Quitting smoking by individuals is an important element of freeing the world of tobacco. However, reducing recruitment of new smokers will also be crucial in ending this global catastrophe. In 2003, the World Health Organisation proposed the Framework Convention on Tobacco Control (WHO FCTC) which became a treaty adopted by the 56th World Health Assembly. The treaty recognised the devastating global health, social, environmental and economic consequences of tobacco and enacted a set of standards aimed at limiting all types of its use worldwide. It was ratified by 181 countries by 2005 and provides obligations to govern the production, sale, distribution, advertising and taxation of tobacco. The treaty includes measures to:

- use taxation and other measures to reduce the demand for tobacco and protect people from passive smoke;
- require plain packaging and health warnings on packs;
- increase public awareness of tobacco harms;
- ban tobacco advertising;
- restrict sales to children;
- provide programs to assist quitting;
- reduce and eliminate illegal trading of tobacco.

FTCT was a landmark success to initiating efforts towards reducing the harms of tobacco. However, progress has been limited and uneven. In the decade since it was ratified, tobacco has killed 50 million people, while profits made by the tobacco industry have continued to rise, fuelled by increased sales in many low- and medium-income countries, as smoking declined in high-income countries [63]. The challenge ahead is exemplified by countries such as China where one in three of the worlds smokers live and yet tobacco sales have increased despite anti-smoking policies [64]. Clearly more effective global action will be needed to free the world of tobacco [63].

Thus, tobacco continues to be a major health challenge for the world. However, there is ample evidence that quitting is possible, and recruitment of new smokers can be reduced. For those who read these chapters, some may still be using tobacco. My hope would be that understanding how the heart works and the exquisite mechanisms which it has evolved to adapt to the demands it faces may help to motivate a sense that it really is worth protecting. Tobacco use is the most important reversible risk for cardiovascular disease for any individual [29] and avoiding it or quitting saves many lives.

Alcohol and the Heart

There have been many studies of the effects of alcohol and the heart, the results of which have sometimes been controversial. However, there is general agreement among researchers that heavy drinking increases the risk of cardiovascular disease and deaths. For this purpose, heavy drinking would be defined as more than 14 units of alcohol per week on a regular basis. The definition of 1 unit depends on the alcoholic strength of drinks; for example, a small (125 mL) glass of wine would be 1.4 and 1.8 units for wines containing 11% and 14% alcohol, respectively; or a pint of lager would be 1.6 and 2.7 units for alcohol contents of 2.8% and 4.8%, respectively. For heavy drinkers, the risk is greater for those who drink in binges than those who spread their intake over several days. However, all forms of heavy drinking are best avoided.

For non-drinkers and moderate drinkers, the evidence is more complex. Many studies have reported a "J-shaped curve" relating alcohol consumption and risk, meaning that the risk is lowest for moderate drinkers, a little higher for non-drinkers and much higher for heavy drinkers. This has been challenged on the basis of a possible bias for non-drinkers. For example, non-drinkers could include former drinkers who stopped for health reasons which would mean they were already at higher risk. In addition, non-drinkers are more likely to be among lower socioeconomic groups, which also increases health risks. A recent study of almost 2 million adults aged over 30 years who had no cardiovascular disease at baseline examined this question by relating alcohol consumption and the later occurrence of cardiovascular disease. This study also looked separately at never drinkers and former drinkers [65]. The results confirmed the J-shaped curve with never and non-drinkers having slightly higher risks for cardiovascular disease than moderate drinkers. It also confirmed a substantially increased risk among heavy drinkers.

The risk identified in these studies was for all cardiovascular deaths combined as well as non-cardiovascular deaths. There are many forms of cardiovascular disease for which individually the risk may be different. One illness which has received particular attention is a rhythm disturbance known as atrial fibrillation (AF). This results from rapid activation of the atria and results in an irregular heart rhythm. Atrial fibrillation increases the risk of other problems such as heart failure and strokes and is important to recognise because it can be readily diagnosed and treated. The incidence of AF increases quite steeply after the age of 60, several studies reporting a prevalence of at least 10% in those aged 80 years. The risks for developing AF are similar to those for heart disease in general, so most cases of AF can be attributed to hypertension, diabetes, smoking or pre-existing coronary heart disease. Heavy drinking would be expected to increase the risk of AF since it also increases the risk of coronary disease and heart failure. Recent evidence suggests that the relationship between alcohol and AF differs from that of others forms of heart disease. In a study of 107,845 people followed for a median of 13.9 years, 5,854 developed AF. The results showed that the risk of developing AF increased progressively with the amount of alcohol consumed [66]. Thus, the association between alcohol consumption and AF showed no evidence of a J-shaped curve and the risk increased with drinking just one unit per day.

Can We Drink Safely?

There is clear evidence that heavy drinking is harmful for the heart and health. This means that it is advisable to avoid consuming 14 or more units per week on a regular basis. Moderate alcohol consumption appears not to be harmful generally except for the risk of developing AF. For those with other potential risks for developing AF, such as hypertension, diabetes or existing heart disease, special care should be exercised to restrict consumption to low levels and this may be more important for people over 60 years of age. There are no currently agreed guidelines for this group as yet but consuming less than 1 unit per day would appear appropriate.

Summary and Conclusions

Our heart works largely in the background of our lives. It provides for our needs without our conscious involvement, automatically endeavouring to meet whatever demands we place on it. We cannot interact directly with it to preserve it or protect it from harm. Nevertheless about 70% of cardiovascular disease globally is due to modifiable risk factors that we can change. High blood pressure, high levels of cholesterol and diabetes account for around 35% of cardiovascular disease globally. Behavioural risks such as tobacco use, poor diet and physical inactivity account for about 16% of global deaths but their risk to individuals is much greater. Tobacco, for example, kills about half of all long-term users. In addition, poor diets and sedentary lifestyles are causally related to high blood pressure, obesity and diabetes, so there is much that we can do that would protect the heart and prolong our lives.

The genes of humans living today are virtually identical to those of our stone age ancestors. This means that we have evolved to meet the challenges for survival which they faced and yet our lifestyles have diverged dramatically from theirs. This is especially the case in high-income countries, but increasingly so also in middle- and low-income nations. We are less physically active; our diets are richer in energy, and we consume large amounts of tobacco and alcohol. These differences account for a heavy burden of cardiovascular and other diseases in the world today. As a result, the World Health Organisation and many nations have developed guidelines for healthy living in an effort to reverse this. These can be summarised as:

- reduce sedentary time and become more physically active. Any activity is good; ideally spend at least 150 min per week in moderate physical activity such as walking or 75 min per week in vigorous physical activity like running or competitive games;
- increase consumption of fruits and vegetables. Rely more on poultry, fish, beans and nuts for proteins and less on red meat and cheese. Eat whole grain bread, pasta and brown rice and less white bread and rice. Use olive oil for cooking and salads instead of butter and avoid trans fats. Drink water, tea and coffee and avoid all sugary drinks. Avoid processed foods that contain large amounts of salt, sugar and preservatives;
- do not use tobacco in any form and avoid inhaling second-hand smoke. For those who smoke, quitting is difficult but possible and saves many lives. Electronic cigarettes are sometimes recommended to help quitting tobacco, but they too rely on nicotine addiction for continued sales and are best avoided;
- avoid heavy drinking (>14 units alcohol per week). Moderate drinking appears not to be risk for coronary heart disease; however, people over the age of 60 and who have other risks for developing atrial fibrillation should limit their alcohol intake to less than 1 unit per day.

These recommendations will be more challenging for some than others. People with addictions to alcohol and tobacco may need help and support to reduce or quit. Increasing physical activity and changing dietary habits are best tackled as lifestyle changes with the aim of finding solutions that are interesting and enjoyable as well as beneficial for health. Adopting these recommendations would align us more closely to the lifestyles of our ancestors as hunter gatherers and to how we are genetically evolved. They would also dramatically reduce the global burden of heart and other diseases. Can we do it? Yes, we just need to take the first step in that direction.

References

[1] Waterson RH, Lander ES, Wilson RK, The Chimpanzee Sequencing and Analysis Consortium. Initial sequence of the chimpanzee genome and comparison with the human genome. *Nature*. 2005;437(7055):69–87.

[2] Eaton SB, Konner M, Shostak M. Stone agers in the fast lane: Chronic degenerative diseases in evolutionary perspective. *American Journal of Medicine*. 1988;84(4):739–749.

[3] Eaton SB, Konner M. Paleolithic nutrition — A consideration of its nature and current implications. *New England Journal of Medicine*. 1985;312(5):283–289.

[4] Clastres P. The Guayaki. In: Bicchieri MG, editor. *Hunters and Gatherers Today: A Socioeconomic Study of Eleven Such Cultures in the Twentieth Century*. New York: Holt, Rinehart and Winston, 1972. p. 138–174.

[5] Smith P, Bloom RA, Berkowitz J. Diachronic trends in humeral cortical thickness of near eastern populations. *Journal of Human Evolution*. 1984;13(8):603–611.

[6] Carrier DR. The energetic paradox of human running and hominid evolution [and Comments and Reply]. *Current Anthropology*. 1984;25(4):483–495.

[7] Bramble DM, Carrier DR. Running and breathing in mammals. *Science*. 1983;219(4582):251.

[8] O'Keefe JH, Vogel R, Lavie CJ, Cordain L. Exercise like a hunter-gatherer: A prescription for organic physical fitness. *Progress in Cardiovascular Diseases*. 2011;53(6):471–479.

[9] Bacon MK, Barry H III, Child IL, Snyder CR. A cross-cultural study of drinking. V. Detailed definitions and data. *Quarterly Journal of Studies on Alcohol* 1965;26(Suppl 3):78–111.

[10] Horton D. The functions of alcohol in primitive societies: A cross cultural study. *Quarterly Journal of Studies on Alcohol*. 1943;4:199–320.

[11] Velican D, Velican C. Atherosclerotic involvement of the coronary arteries of adolescents and young adults. *Atherosclerosis*. 1980;36(4):449–460.

[12] Forouhi NG, Wareham NJ. Epidemiology of diabetes. *Medicine*. 2014; 42(12):698–702.

[13] King H, Aubert RE, Herman WH. Global Burden of Diabetes, 1995–2025: Prevalence, numerical estimates, and projections. *Diabetes Care*. 1998; 21(9):1414.

[14] Pontzer H, Raichlen DA, Wood BM, Mabulla AZ, Racette SB, Marlowe FW. Hunter-gatherer energetics and human obesity. *PLoS One*. 2012;7(7):e40503.

[15] World Health Organisation; Obesity and overweight: World Health Organisation; updated April 2020. Available from: https://www.who.int/en/news-room/fact-sheets/detail/obesity-and-overweight.

[16] Duncan KH, Bacon JA, Weinsier RL. The effects of high and low energy density diets on satiety, energy intake, and eating time of obese and non-obese subjects. *The American Journal of Clinical Nutrition*. 1983;37(5): 763–767.

[17] GBD 2017 Risk Factor Collaborators. Global, regional, and national comparative risk assessment of 84 behavioural, environmental and occupational, and metabolic risks or clusters of risks for 195 countries and territories, 1990–2017: A systematic analysis for the Global Burden of Disease Study 2017. *Lancet (London, England)*. 2018;392(10159):1923–1994.

[18] GBD 2017 Causes of Death Collaborators. Global, regional, and national age-sex-specific mortality for 282 causes of death in 195 countries and territories, 1980–2017: A systematic analysis for the Global Burden of Disease Study 2017. *Lancet (London, England)*. 2018;392(10159):1736–1788.

[19] Forouzanfar MH, Liu P, Roth GA, Ng M, Biryukov S, Marczak L, *et al.* Global burden of hypertension and systolic blood pressure of at least 110 to 115 mm Hg, 1990–2015. *JAMA*. 2017;317(2):165–182.

[20] Mills KT, Bundy JD, Kelly TN, Reed JE, Kearney PM, Reynolds K, *et al.* Global disparities of hypertension prevalence and control: A systematic analysis of population-based studies from 90 countries. *Circulation*. 2016; 134(6):441–450.

[21] Denton D. Can hypertension be prevented? *Journal of Human Hypertension*. 1997;11(9):563–569.

[22] Sinnett PF, Whyte HM. Epidemiological studies in a total highland population, Tukisenta, New Guinea. Cardiovascular disease and relevant clinical,

electrocardiographic, radiological and biochemical findings. *Journal of Chronic Diseases*. 1973;26(5):265–290.

[23] WHO. Global health risks; Mortality and burden of disease attributable to selected major risks. World Health Organisation, 2009. Available from: https://www.who.int/healthinfo/global_burden_disease/GlobalHealthRisks_report_full.pdf.

[24] GBD 2017 HIV collaborators. Global, regional, and national incidence, prevalence, and mortality of HIV, 1980–2017, and forecasts to 2030, for 195 countries and territories: A systematic analysis for the Global Burden of Diseases, Injuries, and Risk Factors Study 2017. *The Lancet HIV*. 2019;6(12):e831–e859.

[25] Lozano R, Naghavi M, Foreman K, Lim S, Shibuya K, Aboyans V, *et al.* Global and regional mortality from 235 causes of death for 20 age groups in 1990 and 2010: A systematic analysis for the Global Burden of Disease Study 2010. *Lancet (London, England)*. 2012;380(9859): 2095–2128.

[26] Hernández AV, Pasupuleti V, Deshpande A, Bernabé-Ortiz A, Miranda JJ. Effect of rural-to-urban within-country migration on cardiovascular risk factors in low- and middle-income countries: A systematic review. *Heart*. 2012;98(3):185–194.

[27] Unwin N, James P, McLarty D, Machybia H, Nkulila P, Tamin B, *et al.* Rural to urban migration and changes in cardiovascular risk factors in Tanzania: A prospective cohort study. *BMC Public Health*. 2010;10:272.

[28] Marmot M, Syme L. Acculturation and coronary heart disease in Japanese Americans. *American Journal of Epidemiology*. 1976;104:225–247.

[29] Yusuf S, Joseph P, Rangarajan S, Islam S, Mente A, Hystad P, *et al.* Modifiable risk factors, cardiovascular disease, and mortality in 155 722 individuals from 21 high-income, middle-income, and low-income countries (PURE): A prospective cohort study. *Lancet (London, England)*. 2020;395(10226):795–808.

[30] Yoon PW, Bastian B, Anderson RN, Collins JL, Jaffe HW, Centers for Disease C, *et al.* Potentially preventable deaths from the five leading causes of death — United States, 2008–2010. *MMWR*: *Morbidity and Mortality Weekly Report*. 2014;63(17):369–374.

[31] Soare A, Weiss EP, Pozzilli P. Benefits of caloric restriction for cardio-metabolic health, including type 2 diabetes mellitus risk. *Diabetes/Metabolism Research and Reviews*. 2014;30(S1):41–47.

[32] Keys A, Menotti A, Aravanis C, Blackburn H, Djordevic BS, Buzina R, *et al.* The seven countries study: 2,289 deaths in 15 years. *Preventive Medicine*. 1984;13(2):141–154.

[33] Kennedy ET, Ohls J, Carlson S, Fleming K. The healthy eating index: Design and applications. *Journal of the American Dietetic Association*. 1995;95(10):1103–1108.

[34] Chiuve SE, Fung TT, Rimm EB, Hu FB, McCullough ML, Wang M, *et al.* Alternative dietary indices both strongly predict risk of chronic disease. *The Journal of Nutrition*. 2012;142(6):1009–1018.

[35] Appel LJ, Moore TJ, Obarzanek E, Vollmer WM, Svetkey LP, Sacks FM, *et al.* A clinical trial of the effects of dietary patterns on blood pressure. DASH Collaborative Research Group. *New England Journal of Medicine*. 1997;336(16):1117–1124.

[36] Akita S, Sacks FM, Svetkey LP, Conlin PR, Kimura G. Effects of the Dietary Approaches to Stop Hypertension (DASH) diet on the pressure-natriuresis relationship. *Hypertension*. 2003;42(1):8–13.

[37] Collaborators GBDRF, Forouzanfar MH, Alexander L, Anderson HR, Bachman VF, Biryukov S, *et al.* Global, regional, and national comparative risk assessment of 79 behavioural, environmental and occupational, and metabolic risks or clusters of risks in 188 countries, 1990–2013: A systematic analysis for the Global Burden of Disease Study 2013. *Lancet (London, England)*. 2015;386(10010):2287–2323.

[38] Afshin A, Micha, R, Webb, M, Capewell, S, Whitsel, L, *et al.* Effectiveness of dietary policies to reduce noncommunicable diseases. In: D. Prabhakaran, S Anand, T Gaziano, J Mbanya, Y Wu, R Nugent, editors. *Disease Control Priorities* (third edition): Volume 5, *Cardiovascular, Respiratory, and Related Disorders*. 5. Washington, DC: World Bank, 2017.

[39] Global Health Risks. Mortality and burden of disease attributable to selected major risks: World Health Organization, 2009. Available from: https://www.who.int/healthinfo/global_burden_disease/global_health_risks/en/.

[40] Nutrition sources. Harvard School of Public Health. Available from: https://www.hsph.harvard.edu/nutritionsource/.

[41] Howard BV, Van Horn L, Hsia J, Manson JE, Stefanick ML, Wassertheil-Smoller S, *et al.* Low-fat dietary pattern and risk of cardiovascular disease: The Women's Health Initiative randomized controlled dietary modification trial. *JAMA*. 2006;295(6):655–666.

[42] Morris JN, Heady JA, Raffle PA, Roberts CG, Parks JW. Coronary heart-disease and physical activity of work. *Lancet (London, England)*. 1953;262(6796):1111–1120; concl.

[43] Morris JN, Crawford MD. Coronary heart disease and physical activity of work; evidence of a national necropsy survey. *British Medical Journal*. 1958;2(5111):1485–1496.

[44] Warburton DE, Charlesworth S, Ivey A, Nettlefold L, Bredin SS. A systematic review of the evidence for Canada's physical Activity guidelines for Adults. *International Journal of Behavioral Nutrition and Physical Activity*. 2010;7:39.

[45] Warburton DE, Katzmarzyk PT, Rhodes RE, Shephard RJ. Evidence-informed physical activity guidelines for Canadian adults. *Canadian Journal of Public Health*. 2007;98(Suppl 2):S16–S68.

[46] Paffenbarger RS, Hyde R, Wing AL, Hsieh C-C. Physical activity, all-cause mortality, and longevity of college alumni. *New England Journal of Medicine*. 1986;314(10):605–613.

[47] WHO. WHO 2020 guidelines on physical activity and sedentary behaviour 2020. Available from: https://apps.who.int/iris/bitstream/handle/10665/336656/9789240015128-eng.pdf.

[48] US Department of Health and Human Services. *Physical activity guidelines for Americans*, 2nd edition. Washington, DC: US Department of Health and Human Services, 2018.

[49] Department of Health and Social Care. *UK Chief Medical Officer's physical activity guidelines*. London: Department of Health and Social Care, 2019.

[50] Ross R, Chaput JP, Giangregorio LM, Janssen I, Saunders TJ, Kho ME, *et al.* Canadian 24-hour movement guidelines for adults aged 18–64 years and Adults aged 65 years or older: An integration of physical activity, sedentary behaviour, and sleep. *Applied Physiology, Nutrition, and Metabolism*. 2020;45(10 (Suppl 2)):S57–S102.

[51] Barletta GA, Fattirolli F, Bisi G, Bertini G, Malfanti PL, Gallini C, *et al.* Short-term responses to cardiac rehabilitation after acute myocardial

infarction. Cardiac function evaluation before and after physical training at rest and during stress test. *European Heart Journal*. 1983;4(11):761–772.

[52] Pate RR, Pratt M, Blair SN, Haskell WL, Macera CA, Bouchard C, *et al.* Physical activity and public health: A recommendation from the Centers for Disease Control and Prevention and the American College of Sports Medicine. *JAMA*. 1995;273(5):402–407.

[53] Saunders TJ, McIsaac T, Douillette K, Gaulton N, Hunter S, Rhodes RE, *et al.* Sedentary behaviour and health in adults: An overview of systematic reviews. *Applied Physiology, Nutrition, and Metabolism*. 2020;45(10 (Suppl 2)):S197–S217.

[54] WHO. Global recommendations on physical activity for health. Geneva: World Health Organization, 2010. Available from: https://www.who.int/dietphysicalactivity/global-PA-recs-2010.pdf.

[55] Doll R, Hill AB. Smoking and carcinoma of the lung; preliminary report. *British Medical Journal*. 1950;2(4682):739–748.

[56] Sasco AJ, Secretan MB, Straif K. Tobacco smoking and cancer: A brief review of recent epidemiological evidence. *Lung Cancer*. 2004;45(Suppl 2): S3–S9.

[57] Smoking and Health Report of the The advisory committee to the Surgeon General of the Public Health Service 1964. Available from: https://collections.nlm.nih.gov/ext/document/101584932X202/PDF/101584932X202.pdf.

[58] Doll R, Peto R. Mortality in relation to smoking: 20 years' observations on male British doctors. *British Medical Journal*. 1976;2(6051):1525–1536.

[59] The Health Consequences of Smoking; Cardiovascular Disease: A report of the Surgeon General, 1983. Available from: https://collections.nlm.nih.gov/ext/document/101584932X318/PDF/101584932X318.pdf. (Accessed on 3/3/2021).

[60] WHO Tobacco; Key Facts. Available from: https://www.who.int/news-room/fact-sheets/detail/tobacco. (Accessed on 4/3/2021).

[61] Richard Peto ADL, Boreham J, Thun M. Mortality from smoking in developed countries 1950–2000. Available from: https://www.ctsu.ox.ac.uk/research/mortality-from-smoking-in-developed-countries-1950-2005-or-later. (Accessed on 4/3/2021).

[62] Jha P. Avoidable deaths from smoking: A global perspective. *Public Health Reviews*. 2011;33(2):569–600.

[63] What will it take to create a tobacco-free world? *Lancet (London, England)*. 2015;385(9972):915.

[64] Luo H, Zhao Y. Increasing tobacco sales under the anti-smoking policy in China. *Lancet (London, England)*. 2020;395(10240):1834.

[65] Bell S, Daskalopoulou M, Rapsomaniki E, George J, Britton A, Bobak M, *et al.* Association between clinically recorded alcohol consumption and initial presentation of 12 cardiovascular diseases: Population based cohort study using linked health records. *BMJ*. 2017;356:j909.

[66] Csengeri D, Sprünker NA, Di Castelnuovo A, Niiranen T, Vishram-Nielsen JK, Costanzo S, *et al.* Alcohol consumption, cardiac biomarkers, and risk of atrial fibrillation and adverse outcomes. *European Heart Journal*. 2021;42(12):1170–1177.

Index

A
acetylcholine, 142, 189
acoelomates, 279
actin binding proteins, 352, 356
caldesmon, 353
calponin, 354
tropomyosin, 352
actin myosin interaction, 101, 111, 126–129, 335–336, 340, 353–355
actin polymerisation, 357
action potential, 104, 112, 118, 127, 136–138, 141–143, 148–153, 156–157, 159, 162–165, 167–169, 171, 173, 248–249
active changes in venous capacity, 315
acute coronary syndromes, 242–243
adrenergic signalling, 125
aerobic metabolism, 367, 371, 377, 399, 401, 404
aerobic range, 373–376, 418
Akkadian, 16, 19–20, 22, 45–46
Alcmaeon of Croton, 24
alcohol, 432, 438, 442, 456–459
amphibian, 63, 68–69, 71–72, 219, 283, 285, 287, 375–376, 401–407, 418–419
anaerobic metabolism, 192, 287
Analects of Confucius, 13–14
anaphylactic shock, 183
anatomy of the circulation, 290
ancient Egypt, 22–23, 26–28, 47
angina, 18, 190, 208–209, 211, 239–240, 242–243
Anrep effect, 97, 119, 120–121, 128, 130
aortic valve, 79, 81, 83, 96, 205, 207, 211, 294
arterial stiffness, 307
arteries, 58, 67, 69–70, 86, 186–187, 189–190, 192, 208, 234–235, 239, 241–243, 277, 280, 282, 285–287, 290–291, 295, 300–308, 312–313, 315–318, 324, 328–330, 347, 349–351, 357, 408, 410
distensibility, 297
elastic modulus, 295
eutrophic, 307
hypertrophic, 307
invertebrate arteries, 298
muscular, 302
non-linear elasticity, 295
resistance, 301
structure of arteries, 297
Aššurbānipal, 16
atherosclerosis, 238–242, 303, 433
atrial action potentials, 152, 162
atrial fibrillation, 154, 164, 457, 459
atrio ventricular node, 151–153

Atum of Heliopolis, 27
autonomic nervous system, 31, 96, 123, 165, 167, 181, 183, 231, 303–304, 318–319, 324, 327, 339, 341–342, 347
autophagy, 212, 215–217

B

balloon angioplasty, 86
baroreflex, 303, 305
basal metabolic rate, 373
Basque, 5–7, 42
bio-communication, 135
bi-polar and side-polar myosin, 337
birds, 71, 283, 288–290, 373–376, 401–402, 406, 408, 418–421
Blalock, Alfred, 85
blood-based vascular systems, 279
blood brain barrier, 309
blood pressure regulation, 289, 302
blood viscosity, 411
Book of Odes, 13
brachial hearts, 282
broken heart syndrome, 229–233, 235
bundle branches, 154
bundle of His, 154

C

Ca^{2+} signalling, 143, 145
calcium transient, 118, 120
calmodulin, 122, 216, 339–340, 344–345, 350, 353
calpain, 216
capillary filtration and transport, 308, 310
  paracellular transport, 310
  passive diffusion, 310
  receptor mediated transport, 310
cardiac output, 65, 67, 72, 97–98, 167, 181–185, 192, 195, 200, 211, 245–247, 266, 288–289, 303, 311, 313–315, 317–318, 372–373
cardiac rupture, 246–247
cardiac tamponade, 245–246
cardiogenic shock, 244, 246–247
cell-to-cell conduction, 155, 248
cephalopods, 65–66, 282, 284, 383–384
chemokines, 256
Chinese, 7–9, 11, 13, 453
Chinese medicine, 10–12
circulation, 3, 22, 26, 57–59, 65, 67, 69, 71
circulation's reservoir, 312
circulation time, 136
circulatory responses to exercises, 183
closed circulatory systems, 281
coelomates, 278, 382
coelomic circulation, 278
coeur, 2
conduction velocity, 157, 171
conductive arteries, 305
confucius, 13, 15
connexin, 155–157, 163–164, 195
connexon, 156
conus arteriosus, 68, 286
conus cordis, 55, 57, 78
corazón, 2
coronary artery disease, 86, 207, 209–210, 230, 234, 244, 437
coronary heart disease, 18–19, 208, 220, 233–235, 238, 448–449, 452–453, 457, 459
coronary thrombosis, 173, 235, 240, 242

crocodile, 71–74, 287–288, 408
cuneiform, 8, 16, 19, 23
cuore, 2
cytokines, 255

D

danger associated molecular pattern (DAMP), 253
demethylation, 52
dense bodies, 332
depolarisation, 112, 146–147, 152–153, 157, 159, 161, 163, 167, 169, 209, 237, 249, 346, 350
diastolic depolarisation, 165–168
diet, 430–431, 433, 439–442, 445–448, 458
  alternate healthy eating index, 441
  dietary approaches to stop hypertension, 441
  guidelines, 447, 449
  healthy eating index, 440
  Western, 439
dietary, 435, 438
dinosaurs, 289
  four-chambered heart, 289
distribution of blood flow during exercise, 183
dominant pacemaker, 168
Drosophila, 57–63, 74, 280–281
dynamic cytoskeleton, 333

E

early repolarisation, 152, 159, 162, 164–165
Edwin Smith Papyrus, 26
Egypt, 8, 22–23
Egyptian, 22–23
Egyptian culture, 27, 29, 47
Egyptian medicine, 23
elastic fibre assembly, 299
elastic modulus, 295–296, 298
elastin fibres, 298–301
electrical anatomy of the heart, 151–152, 155
electrical gradient, 136, 138, 237
electrical triggering, 117
electrochemical gradients, 140
embryogenesis, 49, 52, 55
emotional meanings, 3
emotion and feeling, 1, 6, 33, 368
epiblast, 53–54, 56
epigenetic, 50–53
epigenetically, 220
eruthrocyte
  biconcave disc, 416
  capillaries, 409
  diminution, 420
  evolution, 404, 413
  size, 401–403
erythrocytes, 367, 369, 371, 386, 389, 401–406, 408–411, 413, 415–418, 420–421
  evolution, 401
eucoelomates, 279
evolution, 49, 57, 64, 74, 89–90, 106, 131, 140–141, 146–147, 149, 277–278, 287–290, 299, 301, 371, 377–379, 381, 383, 385, 408, 421, 430
evolutionary, 70, 72, 98, 137, 285, 407, 417, 419–420
evolution of muscle, 351
exercise hypertrophy in older hearts, 202
exercise-induced hypertrophy, 199, 201

F

f-actin, 357
Fåhræus–Lindqvist effect, 412
Fallot's tetralogy, 84, 85
feedback control, 193
feed-forward control, 192
fertilisation, 49, 52–54, 57, 74, 197, 264
fibroblast, 252, 260, 262, 264
flow mediated dilatation, 187–188, 191, 304, 342–343, 348
foetal, 395
foramen of Panizza, 72–73, 287
foramen ovale, 76, 77
force frequency response, 97, 113, 121–122, 129
four-chambered heart, 57, 63–64, 70–75, 78, 87–90, 219
framework convention on tobacco control, 455
Framingham heart study, 235
Frank–Starling, 96, 103, 119–121, 130, 199, 313
frequency-relaxing response, 122

G

gap junctions, 155, 157, 159, 163–164, 189, 191, 194–195, 209, 212, 311
genome size, 417– 419
Germanic, 4–6
Gilgamesh, 20–21, 45–46
gills, 49, 65–66, 68, 281–286, 406
glucose transporter, 1, 309, 356
Greek literature, 3
Gruentzig, Andreas, 86

H

haemerythrins, 383, 385
haemocyanin, 58, 280, 382–385
haemodynamic effects on coronary perfusion, 210
haemoglobin, 374–376, 382–383, 386–390, 393, 395–401, 408–409, 411, 413–417
  allosteric binding, 397
  bisphosphoglyceric acid, 400
  Bohr effect, 399
  class, 390
  embryonic, 392–394
  evolution, 401, 413
  foetal, 392–393, 397
  Haldane effect, 400
  human, 392
  plants, 389
  regulation, 395
  structure, 386
  synthesis, 390–392
hart, 2
healing, 247, 251–252, 258–259, 261–263, 266, 273
hearta, 2
heart attack, 108, 125, 219, 234–239, 241, 243–244, 246, 248, 266, 303–304, 370, 448–449
  blood tests, 238
  electrocardiogram, 237
  troponin T, 238
  diagnosis, 237
  symptoms, 236
heart failure, 18, 27, 113, 129–130, 182, 208, 211–212, 217–218, 230–231, 235, 244, 246–247, 304, 457–458
heart in European languages, 4
heart metaphor, 1–8, 13–15, 19–21, 27, 35–36, 229–230
heart pictographs, 9

heart tube, 53–55, 57, 59, 74, 84, 87, 280
herte, 2
hieroglyphs, 8, 22–23
high blood pressure, 95, 198, 205, 233, 305–306, 319, 438, 458
hirte, 2
hjärta, 2
hjerte, 2
hominids, 429–430
human circulatory system, 290
human dissection, 10, 25
hunter gatherers, 430–431, 434–436, 438, 460
hypertension, 121, 125, 197, 201, 205, 207, 209, 220, 235–236, 247, 303–308, 348, 359, 435–436, 438, 443–444, 457–458
hypertrophic cardiomyopathy, 110, 206
  gene abnormalities, 206
hypertrophic obstructive cardiomyopathy, 207
hypertrophy, 197, 208, 210–212
  coronary circulation, 209
  coronary flow reserve, 209
  coronary morphology, 210
  morphology, 208
  transition to heart failure, 211
hypertrophy as a compensatory mechanism, 201
hypoblast, 53–54, 56

I
inflammation, 250
insect gigantism, 380
intelligent tubes, 324
intercalated disk, 99–100, 102, 155, 212
intimal thickening, 240–241
inward calcium current, 160
isometric and isotonic exercise, 200

J
Japanese, 7, 11, 43, 230–231
Japanese medicine, 11

K
kerd, 2
Krogh, August, 139

L
Laplace, 213
Laplace's law, 187, 200
Late repolarisation, 160
left ventricular function, 232
leukocytes, 369
lifestyle, 437–438, 450, 458–460
linguistic devices, 5
local regulation of muscle blood flow, 185
lonely hearts, 233
long distance running, 200, 432
long-term effects of overload hypertrophy, 207
long-term effects of physiological hypertrophy, 203
lusitropy, 122

M
mammals, 68, 71–74, 90, 105–106, 208, 250, 282–283, 285–291, 296, 299, 312–313, 327, 332, 373–376, 386, 400, 401–409, 413, 415, 418–421, 429
marathon running, 203
Marfan syndrome, 301
mechano-sensing in blood vessels, 347

Mediterranean diet, 440
membrane depolarisation, 144, 149–150, 158, 171, 344, 349–350
mesoderm, 53–56, 60–61, 63, 264, 278–280
mesodermal, 60
Mesopotamian, 8, 13, 15–19, 22, 33, 46
Mesopotamian culture, 19, 45
Mesopotamian medicine, 16
metaphorical meanings, 1, 2, 4, 28, 35
methylation, 51–53, 55
microfibril, 298–301
microRNA, 265
mitral valve, 79–80, 82, 89, 199
molluscs, 64–65, 67, 74, 278, 280–282, 284, 290, 299, 382–385, 389
monocytes, 252, 256, 258–262, 370
muscle pump effects, 185
musician, 31
myocardial infarction, 125, 172–173, 208, 234, 237, 239, 243, 245, 251–252, 254, 262, 266, 276, 369
  electrophysiological dysfunction, 247
  fibroblasts, 261
  healing, 234
  monocytes, 257
  neutrophil infiltration, 257
  new growth, 261
  pro-inflammatory phase, 254
  pump dysfunction, 247
  repair and proliferative phase, 259
myocardial ischaemia, 211, 236, 240, 243, 247–249, 253
myocarial infarction
  complications, 244
myocyte
  regeneration, 218
myocytes, 98–100, 102, 105, 107–109, 113–114, 118, 120–122, 124–126, 135–136, 138, 140, 143–144, 146, 151, 155–156, 197, 208–209, 212, 218–220, 235, 248, 250–251, 253, 256, 261–266, 275, 329, 335
myogenic tone, 304, 326, 339, 343, 348, 351, 355
myosin, 326
  bi-polar configuration, 335
  folded 10S, 358
  side-polar configuration, 335
myosin binding protein-C, 101, 104, 110, 113, 128, 206
myosin light chain, 104–105, 109–110, 116, 129, 339–342, 345–346, 351, 355, 358
myosin polymerisation, 357

N

Na^+/Ca^{2+} exchange pump, 143, 166
nerve conduction, 136
neuromuscular junction, 142
neurovascular unit, 311
neutrophil infiltration, 255, 260
nitric oxide, 88, 188–191, 194, 304, 310–312, 343–345, 347–348, 390
nucleosome, 50–52

O

obesity, 434–436, 438–439, 451, 458
off state, 103–104, 117–118, 122
on state, 103–104, 108, 117–118
open circulatory systems, 280
orderly conduction, 171–173

origin of the heart, 1, 35, 229
osmotic control, 140
osmotic imbalance, 137–138, 140
ostium secundum, 76–77
oxygen consumption, 71, 192–193, 213, 287, 372–374, 383, 432
oxygen transport, 383

P

pacemaker, 95, 112, 118, 120, 123, 151–153, 156–157, 164–171, 209, 248–249
palaeolithic, 430–432, 434
  cardiovascular disease, 433
  diabetes, 433
palermo stone, 27
passive changes in venous capacity, 315
patent ductus arteriosus, 84, 86
pathogen-associated molecule patterns (PAMPs), 253
pathological hypertrophy, 204
perfusion of the heart during exercise, 191
pericardio-peritoneal canal, 68
phaeochromocytoma, 125, 182, 231
phospholamban, 112, 122, 127
photosynthesis, 377, 379, 381, 396
  atmospheric oxygen, 380
  Earth's atmosphere, 377
  evolution of complex, 380
  non-oxygenic, 377
  oxygenic, 377
  prokaryotes, 381
  transmembrane proteins, 381
photosystem II, 379
phylogenetic studies, 140, 147
physical activity guidelines, 202, 221, 431, 436, 438, 446, 448–451, 459
physical inactivity, 449
physiological hypertrophy, 204
pipes, 323–324
  aqueducts, 323
  clay pipes, 323
  lead, 323
platelets, 79, 260, 370
polar gigantism, 380
polyploidy, 208, 220
pregnancy-induced hypertrophy, 197
pressure overloading, 205, 208, 217
primitive atria, 57
progression of LV hypertrophy, 198
prokaryote, 140, 149
proteasomes, 214–216
protein kinase A, 112, 123, 126
protein turnover and muscle growth, 217
pulmonary valve, 72–73, 79, 81–84, 288, 293
pulsatile ejection, 294
pulsatile pumping, 293
Purkinje fibre action potentials, 154, 157, 163

R

re-entry, 172, 174
refractoriness, 162, 170–173, 249
regeneration, 219–220, 250
regeneration of the heart, 213
regional blood flow during exercise, 184
regression of hypertrophy, 202
regulation of calcium ions, 142
regulation of heart size, 196
regulation of the heart, 94, 96, 123, 151
regulatory light chain, 105, 109, 113, 116, 126, 129, 336, 339–340, 345–346, 351, 354, 357–359

religion, 12, 32, 34
repair membrane wounds, 144
repolarisation, 153, 158, 162, 171
reservoir, 64, 67, 183–184, 258, 292, 312–314, 316, 318, 324, 328
resistance arteries, 194–195, 301–306, 308, 311–312
resting membrane potential, 137, 141, 161, 249
retrograde conducted response, 194
romance, 2, 4–6, 229
Rosetta Stone, 22
ryanodine receptors, 127, 212

S

sarcolemmae, 99
sarcolemmal, 212
sarcolemmas, 100, 112, 114, 127, 144, 155, 167
sarcoplasmic reticulum Ca^{2+} ATPase (SERCA), 100, 112–113, 118, 122, 127, 140, 143, 160, 166
secondary hypertension, 306
Semitic, 16
sensing stretch, 348
servicing the heart, 213
shark, 67–68, 295, 375–376, 406–407
shear stress, 187–188, 191, 344, 347–348
sick at heart, 2
signalling pathways, 339
sinoatrial node, 95, 123–124
sinus node, 151, 153, 164–169
sinus venosus, 55–57, 67, 69, 76
Skou, Jens Christian, 139
smooth muscle, 327
  actin–myosin configuration, 336
  actin–myosin contractile elements, 331
  actin–myosin filaments, 331
  actin–myosin interaction, 338
  circular layer, 329
  contractile function, 327
  cytoskeleton, 330
  dense bodies, 331
  intermediate filaments, 330–331
  length adaptability, 327, 356
  longitudinal layer, 329
  structure, 328
smooth muscle, 326
smooth muscle cell
  flexible in structure and function, 352
smooth muscle contraction, 341, 334, 344
  activation of smooth muscle contraction, 339
  synergistic pathways, 344
sodium pump, 138–141, 173
spiral valve, 69–70, 78
splanchnic circulation, 318, 328, 356
splanchnic reservoir, 315–316, 319
  mobilisation, 316
squid, 64, 248, 282, 295, 383–384
stem cell interventions, 263
stretch-induced depolarisation, 350
stretch-mediated vasoconstriction, 349
stroke, 125, 207, 232–233, 240, 303–304, 313–314, 370, 449
sudden cardiac death, 244
Sumerian, 8, 15–16, 19–21, 33, 45–46, 323
sympathetic nerves, 68, 96, 123–124, 192, 315
symptoms, 236

T
takotsubo cardiomyopathy, 230–233, 266
octopus traps, 230
Taussig, Helen, 85
The Edwin Smith Surgical Papyrus, 25
thick filament, 99–105, 107, 109–118, 120, 122, 128–130, 334–339, 353, 357–359
Thomas, Vivian, 85
tinman, 58, 60–63
titin, 100, 103–104, 110–111, 113–114, 117, 120, 129–130
tobacco, 432–433, 438, 451–456, 458
transduction of shear stress, 190
transient receptor potential, 188, 345, 348–349
tricuspid valve, 75, 79–82
tropoelastin, 299–301
tropomyosin, 104, 107–109, 112, 117–118, 123, 128, 206, 213, 238, 336, 339–341, 352–353, 355–356
master actin filament regulator, 355
troponin C, 104, 112
troponin I, 104, 112
troponin T, 104, 112
truncus arteriosus, 57, 78, 88
two-chambered heart, 64, 66–67, 69–70, 74, 282–284, 406

U
ubiquitinated, 216–217
ubiquitin-proteasome system, 214
Uralic, 5–6

V
vascular conductance, 186
vascular regulation, 342
vascular remodelling, 305
vascular resistance, 305
ventricular action potential, 152, 157–158, 160, 162
ventricular fibrillation, 173, 244–245, 248
viscosity of blood, 411
voltage gated ion channels, 146–149, 153, 158, 161, 164, 173
voltage gated potassium channel, 147, 162
volume overload, 199, 205, 208
vulnerable plaques, 242

W
wall stress, 200
weighing of the heart, 30

Z
Z disk, 99–102, 113–114, 130, 215, 335, 337

Lightning Source UK Ltd.
Milton Keynes UK
UKHW020726011122
411442UK00001B/23